D0936848

FRACTAL GEOMETRY

Mathematical Foundations and Applications

FRACTAL GEOMETRY

Mathematical Foundations and Applications

Kenneth Falconer

Reader in Mathematics, University of Bristol

JOHN WILEY & SONS

Chichester • New York • Brisbane • Toronto • Singapore

Reprinted March 1990

Other Wiley Editorial Offices

John Wiley & Sons, Inc., 605 Third Avenue,
New York, NY 10158-0012, USA

Jacaranda Wiley Ltd, G.P.O. Box 859, Brisbane,
Queensland 4001, Australia

John Wiley & Sons (Canada) Ltd, 22 Worcester Road,
Rexdale, Ontario M9W 1L1, Canada

John Wiley & Sons (SEA) Pte Ltd, 37 Jalan Pemimpin 05-04,
Block B, Union Industrial Building, Singapore 2057

Library of Congress Cataloging-in-Publication Data:

Falconer, K. J., 1952–
Fractal geometry : mathematical foundations and applications / Kenneth Falconer.
p. cm.
Includes bibliographical references.
ISBN 0 471 92287 0
1. Fractals. I. Title.
QA614.86.F35 1990
514′.74–dc20 89–37981
CIP

British Library Cataloguing in Publication Data:

Falconer, K. J.
Fractal geometry.
1. Geometrical shapes
I. Title
516′.15

ISBN 0 471 92287 0

Typeset by Thomson Press (India) Ltd
Printed and bound in Great Britain by Courier International, Tiptree Essex

Contents

Preface

I am frequently asked questions such as 'What are fractals?', 'What is fractal dimension?', 'How can one find the dimension of a fractal and what does it tell us anyway?' or 'How can mathematics be applied to fractals?'. This book endeavours to answer some of these questions.

The main aim of the book is to provide a treatment of the mathematics associated with fractals and dimensions at a level which is reasonably accessible to those who encounter fractals in mathematics or science. Although basically a mathematics book, it attempts to provide an intuitive as well as a mathematical insight into the subject.

The book falls naturally into two parts. Part I is concerned with the general theory of fractals and their geometry. Firstly, various notions of dimension and methods for their calculation are introduced. Then geometrical properties of fractals are investigated in much the same way as one might study the geometry of classical figures such as circles or ellipses: locally a circle may be approximated by a line segment, the projection or 'shadow' of a circle is generally an ellipse, a circle typically intersects a straight line segment in two points (if at all), and so on. There are fractal analogues of such properties, usually with dimension playing a key rôle. Thus we consider, for example, the local form of fractals, and projections and intersections of fractals.

Part II of the book contains examples of fractals, to which the theory of the first part may be applied, drawn from a wide variety of areas of mathematics and physics. Topics include self-similar and self-affine sets, graphs of functions, examples from number theory and pure mathematics, dynamical systems, Julia sets, random fractals and some physical applications.

There are many diagrams in the text and frequent illustrative examples. Computer drawings of a variety of fractals are included, and it is hoped that enough information is provided to enable readers with a knowledge of programming to produce further drawings for themselves.

It is hoped that the book will be a useful reference for researchers, providing an accessible development of the mathematics underlying fractals and showing how it may be applied in particular cases. The book covers a wide variety of mathematical ideas that may be related to fractals, and, particularly in Part II, provides a flavour of what is available rather than exploring any one subject in too much detail. The selection of topics is to some extent at the author's whim—there are certainly some important applications that are not included. Some of the material dates back to early in this century whilst some is very recent.

Notes and references are provided at the end of each chapter. The references are by no means exhaustive, indeed complete references on the variety of topics covered would fill a large volume. However, it is hoped that enough information is included to enable those who wish to do so to pursue any topic further.

It would be possible to use the book as a basis for a course on the mathematics of fractals, at postgraduate or, perhaps, final-year undergraduate level, and exercises are included at the end of each chapter to facilitate this. Harder sections and proofs are marked with an asterisk, and may be omitted without interrupting the development.

An effort has been made to keep the mathematics to a level that can be understood by a mathematics or physics graduate, and, for the most part, by a diligent final-year undergraduate. In particular, measure theoretic ideas have been kept to a minimum, and the reader is encouraged to think of measures as 'mass distributions' on sets. Provided that it is accepted that measures with certain (intuitively almost obvious) properties exist, there is little need for technical measure theory in our development.

Results are always stated precisely to avoid the confusion which would otherwise result. Our approach is generally rigorous, but some of the harder or more technical proofs are either just sketched or omitted altogether. (However, a few harder proofs that are not available in that form elsewhere have been included, in particular those on sets with large intersection and on random fractals.) Suitable diagrams can be a help in understanding the proofs, many of which are of a geometric nature. Some diagrams are included in the book; the reader may find it helpful to draw others.

Chapter 1 begins with a rapid survey of some basic mathematical concepts and notation, for example, from the theory of sets and functions, that are used throughout the book. It also includes an introductory section on measure theory and mass distributions which, it is hoped, will be found adequate. The section on probability theory may be helpful for the chapters on random fractals and Brownian motion.

With the wide variety of topics covered it is impossible to be entirely consistent in use of notation and inevitably there sometimes has to be a compromise between consistency within the book and standard usage.

In the last few years fractals have become enormously popular as an art form, with the advent of computer graphics, and as a model of a wide variety of physical phenomena. Whilst it is possible in some ways to appreciate fractals with little or no knowledge of their mathematics, an understanding of the mathematics that can be applied to such a diversity of objects certainly enhances one's appreciation. The phrase 'the beauty of fractals' is often heard—it is the author's belief that much of their beauty is to be found in their mathematics.

It is a pleasure to acknowledge those who have assisted in the preparation of this book. Philip Drazin and Geoffrey Grimmett provided helpful comments on parts of the manuscript. Peter Shiarly gave valuable help with the computer drawings and produced the cover photograph, and Aidan Foss produced some diagrams. I am indebted to Charlotte Farmer, Jackie Cowling and Stuart Gale of John Wiley and Sons for overseeing the production of the book.

Special thanks are due to David Marsh—not only did he make many useful comments on the manuscript and produce many of the computer pictures, but he also typed the entire manuscript in a most expert way.

Finally, I would like to thank my wife Isobel for her support and encouragement, which extended to reading various drafts of the book.

Kenneth J Falconer
Bristol, April 1989

Introduction

In the past, mathematics has been concerned largely with sets and functions to which the methods of classical calculus can be applied. Sets or functions that are not sufficiently smooth or regular have tended to be ignored as 'pathological' and not worthy of study. Certainly, they were regarded as individual curiosities and only rarely were thought of as a class to which a general theory might be applicable.

In recent years this attitude has changed. It has been realized that a great deal can be said, and is worth saying, about the mathematics of non-smooth sets. Moreover, irregular sets provide a much better representation of many natural phenomena than do the figures of classical geometry. Fractal geometry provides a general framework for the study of such irregular sets.

We begin by looking briefly at a number of simple examples of fractals, and note some of their features.

The middle third Cantor set is one of the best known and most easily constructed fractals; nevertheless it displays many typical fractal characteristics. It is constructed from a unit interval by a sequence of deletion operations; see figure 0.1. Let E_0 be the interval $[0,1]$. (Recall that $[a,b]$ denotes the set of real numbers x such that $a \leqslant x \leqslant b$.) Let E_1 be the set obtained by deleting the middle third of E_0, so that E_1 consists of the two intervals $[0,\frac{1}{3}]$ and $[\frac{2}{3},1]$. Deleting the middle thirds of these intervals gives E_2; thus E_2 comprises the four intervals $[0,\frac{1}{9}]$, $[\frac{2}{9},\frac{1}{3}]$, $[\frac{2}{3},\frac{7}{9}]$, $[\frac{8}{9},1]$. We continue in this way, with E_k obtained by deleting the middle third of each interval in E_{k-1}. Thus E_k consists of 2^k intervals each of length 3^{-k}. The *middle third Cantor set* F consists of the numbers that are in E_k for all k; mathematically, F is the intersection $\bigcap_{k=0}^{\infty} E_k$. The Cantor set F may be thought of as the limit of the sequence of sets E_k as k tends to infinity. It is obviously impossible to draw the set F itself, with its infinitesimal detail, so 'pictures of F' tend to be pictures of one of the E_k, which are a good approximation to F when k is reasonably large.

At first glance it might appear that we have removed so much of the interval $[0,1]$ during the construction of F, that nothing remains. In fact, F is an infinite (and indeed uncountable) set, which contains infinitely many numbers in any neighbourhood of each of its points. The middle third Cantor set F consists precisely of those numbers in $[0,1]$ whose base-3 expansion does not contain the digit 1, i.e. all numbers $a_1 3^{-1} + a_2 3^{-2} + a_3 3^{-3} + \cdots$ with $a_i = 0$ or 2 for each i. To see this, note that to get E_1 from E_0 we remove those numbers with $a_1 = 1$, to get E_2 from E_1 we remove those numbers with $a_2 = 1$, and so on.

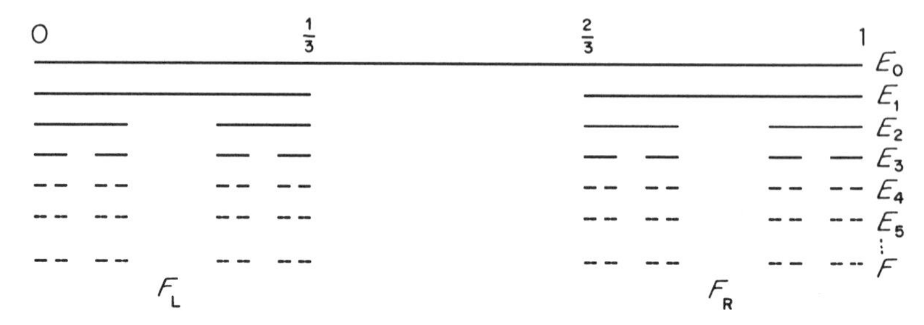

Figure 0.1 Construction of the middle third Cantor set F, by repeated removal of the middle third of intervals. Note that F_L and F_R, the left and right parts of F, are copies of F scaled by a factor $\frac{1}{3}$

We list some of the features of the middle third Cantor set F; as we shall see, similar features are found in many fractals.

(i) F is self-similar. It is clear that the part of F in the interval $[0, \frac{1}{3}]$ and the part of F in $[\frac{2}{3}, 1]$ are geometrically similar to F, scaled by a factor $\frac{1}{3}$. Again, the parts of F in each of the four intervals of E_2 are similar to F but scaled by a factor $\frac{1}{9}$, and so on. The Cantor set contains copies of itself at many different scales.
(ii) The set F has a 'fine structure'; that is, it contains detail at arbitrarily small scales. The more we enlarge the picture of the Cantor set, the more gaps become apparent to the eye.
(iii) Although F has an intricate detailed structure, the actual definition of F is very straightforward.
(iv) F is obtained by a recursive procedure. Our construction consisted of repeatedly removing the middle thirds of intervals. Successive steps give increasingly good approximations E_k to the set F.
(v) The geometry of F is not easily described in classical terms: it is not the locus of the points that satisfy some simple geometric condition, nor is it the set of solutions of any simple equation.
(vi) It is awkward to describe the local geometry of F—near each of its points are a large number of other points, separated by gaps of varying lengths.
(vii) Although F is in some ways quite a large set (it is uncountably infinite), its size is not quantified by the usual measures such as length—by any reasonable definition F has length zero.

Our second example, the von Koch curve, will also be familiar to many readers; see figure 0.2. We let E_0 be a line segment of unit length. The set E_1 consists of the four segments obtained by removing the middle third of E_0 and replacing it by the other two sides of the equilateral triangle based on the removed segment. We construct E_2 by applying the same procedure to each of the segments in E_1, and so on. Thus E_k comes from replacing the middle third of each straight line segment of E_{k-1} by the other two sides of the equilateral triangle. When k is large, the curves E_{k-1} and E_k differ only in fine detail and

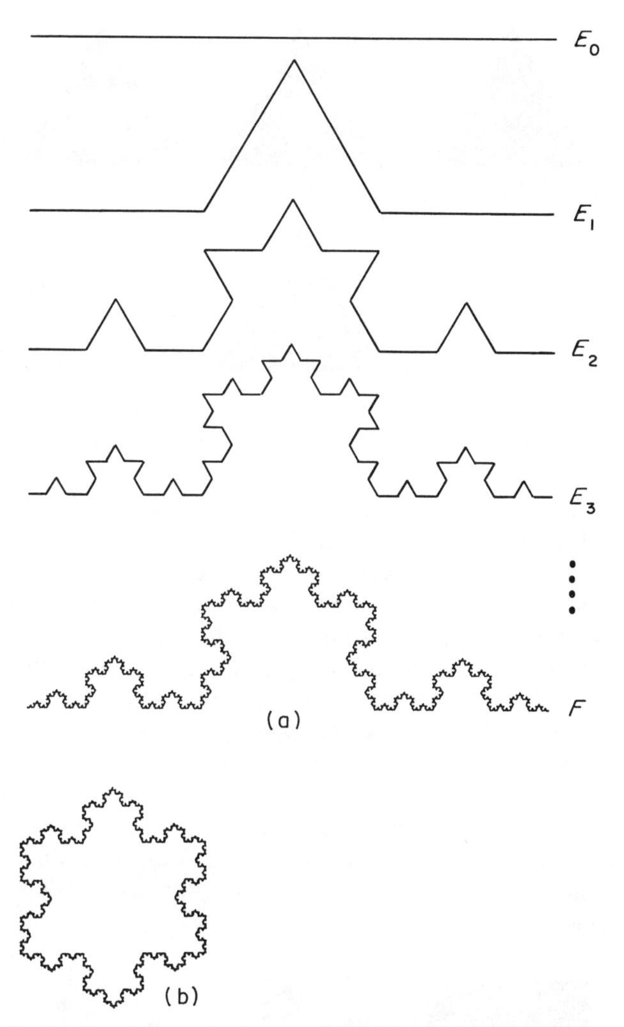

Figure 0.2 (a) Construction of the von Koch curve F. At each stage, the middle third of each interval is replaced by the other two sides of an equilateral triangle. (b) Three von Koch curves fitted together to form a snowflake curve

as k tends to infinity, the sequence of polygonal curves E_k approaches a limiting curve F, called the *von Koch curve*.

The von Koch curve has features in many ways similar to those listed for the middle third Cantor set. It is made up of four 'quarters' each similar to the whole, but scaled by a factor $\frac{1}{3}$. The fine structure is reflected in the irregularities at all scales; nevertheless, this intricate structure stems from a basically simple construction. Whilst it is reasonable to call F a curve, it is much too irregular to have tangents in the classical sense. A simple calculation shows that E_k is of length $(\frac{4}{3})^k$; letting k tend to infinity implies that F has infinite length. On the

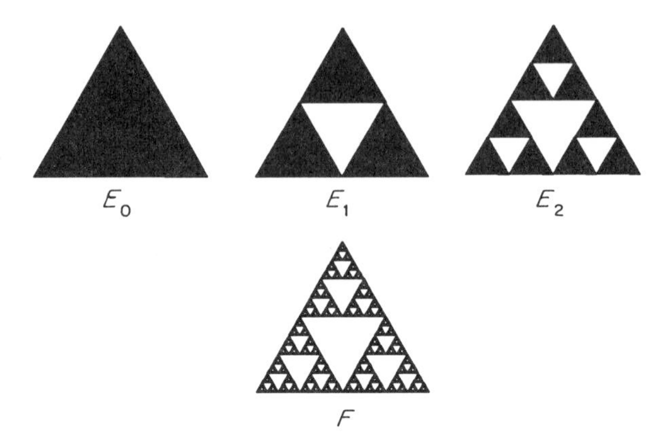

Figure 0.3 Construction of the Sierpiński gasket ($\dim_H F = \dim_B F = \log 3/\log 2$)

other hand, F occupies zero area in the plane, so neither length nor area provides a very useful description of the size of F.

Many other sets may be constructed using such recursive procedures. For example, the Sierpiński gasket is obtained by repeatedly removing (inverted) equilateral triangles from an initial equilateral triangle; see figure 0.3. (For many purposes, it is better to think of this procedure as repeatedly replacing an equilateral triangle by three triangles of half the height.) A plane analogue of the Cantor set, a 'Cantor dust' is illustrated in figure 0.4. At each stage each remaining square is divided into 16 smaller squares of which four are kept and the rest discarded. (Of course, other arrangements or numbers of squares could be used to get different sets.) It should be clear that such examples have properties similar to those mentioned in connection with the Cantor set and the von Koch curve. The example depicted in figure 0.5 is constructed using two different similarity ratios.

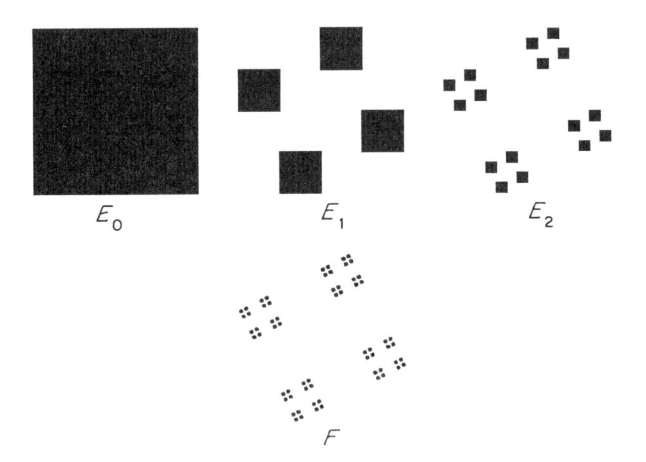

Figure 0.4 Construction of a 'Cantor dust' ($\dim_H F = \dim_B F = 1$)

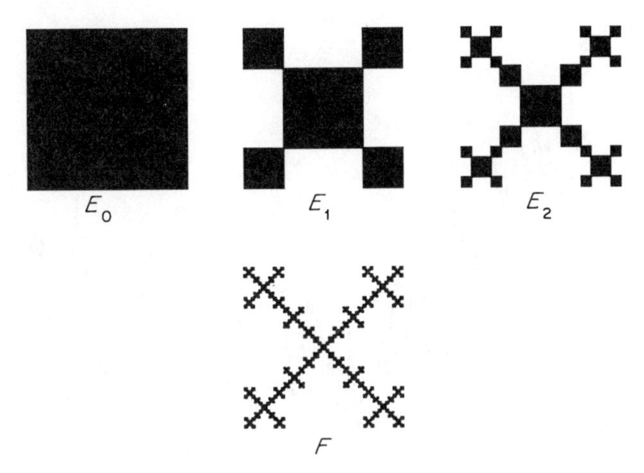

Figure 0.5 Construction of a self-similar fractal with two different similarity ratios

There are many other types of construction, some of which will be discussed in detail later in the book, that also lead to sets with these sorts of properties. The highly intricate structure of the Julia set illustrated in figure 0.6 stems from the single quadratic function $f(z)=z^2+c$ for a suitable constant c. Although the set is not strictly self-similar in the sense that the Cantor set and von Koch curve are, it is 'quasi-self-similar' in that arbitrarily small portions of the set can be magnified and then distorted smoothly to coincide with a large part of the set.

Figure 0.7 shows the graph of the function $f(t)=\sum_{k=0}^{\infty}(\frac{3}{2})^{-k/2}\sin((\frac{3}{2})^k t)$; the infinite summation leads to the graph having a fine structure, rather than being a smooth curve to which classical calculus is applicable.

Some of these constructions may be 'randomized'. Figure 0.8 shows a 'random von Koch curve'—a coin was tossed at each step in the construction to

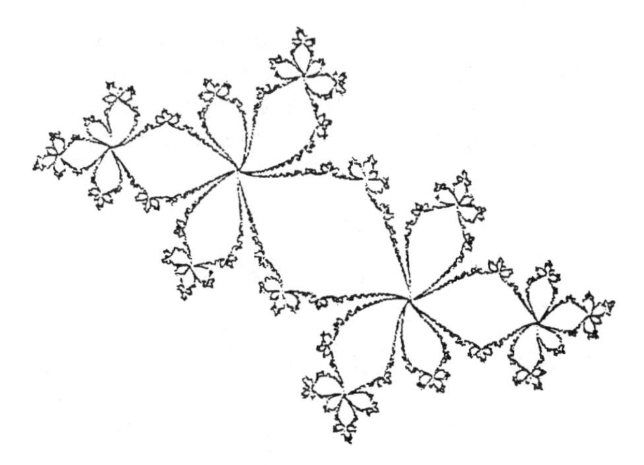

Figure 0.6 A Julia set

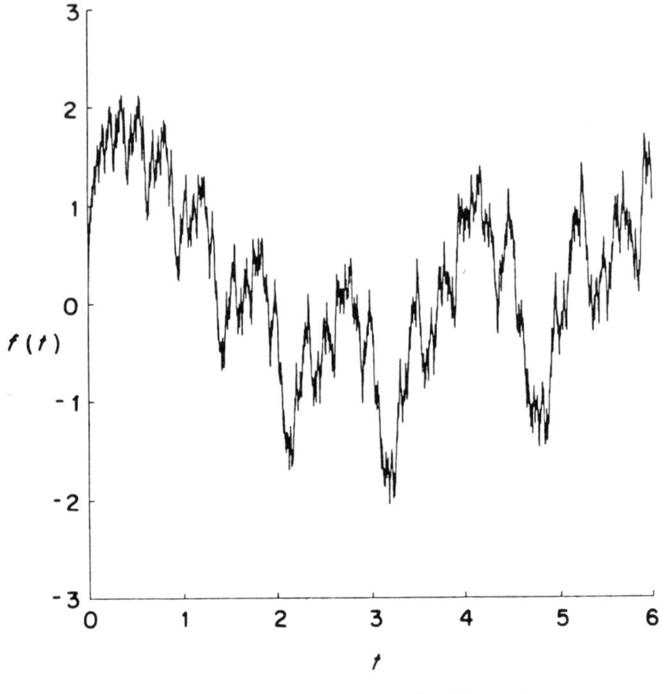

Figure 0.7 Graph of $f(t) = \sum_{k=0}^{\infty} (\frac{3}{2})^{-k/2} \sin((\frac{3}{2})^k t)$

determine on which side of the curve to place the new pair of line segments. This random curve certainly has a fine structure, but the strict self-similarity of the von Koch curve has been replaced by a 'statistical self-similarity'.

These are all examples of sets that are commonly referred to as *fractals*. (The word 'fractal' was coined by Mandelbrot in his fundamental essay from the Latin *fractus*, meaning broken, to describe objects that were too irregular to fit into a traditional geometrical setting.) Properties such as those listed for the Cantor set are characteristic of fractals, and it is sets with such properties that we will have in mind throughout the book. Certainly, any fractal worthy of the name will have a fine structure, i.e. detail at all scales. Many fractals have some

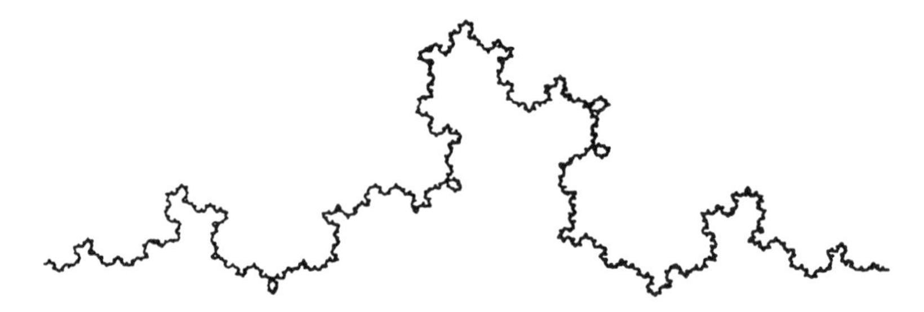

Figure 0.8 A random version of the von Koch curve

degree of self-similarity—they are made up of parts that resemble the whole in some way. Sometimes, the resemblance may be weaker than strict geometrical similarity; for example, the similarity may be approximate or statistical.

Methods of classical geometry and calculus are unsuited to studying fractals and we need alternative techniques. The main tool of fractal geometry is dimension in its many forms. We are familiar enough with the idea that a (smooth) curve is a 1-dimensional object and a surface is 2-dimensional. It is less clear that, for many purposes, the Cantor set should be regarded as having dimension $\log 2/\log 3 = 0.631$ and the von Koch curve as having dimension $\log 4/\log 3 = 1.262$. This latter number is, at least, consistent with the von Koch curve being 'larger than 1-dimensional' (having infinite length) and 'smaller than 2-dimensional' (having zero area).

The following argument gives one (rather crude) interpretation of the meaning of these 'dimensions' indicating how they reflect scaling properties and self-similarity. As figure 0.9 indicates, a line segment is made up of four copies of itself, scaled by a factor $\frac{1}{4}$. The segment has dimension $-\log 4/\log\frac{1}{4} = 1$. A square, however, is made up of four copies of itself scaled by a factor $\frac{1}{2}$ (i.e. with half the side length) and has dimension $-\log 4/\log\frac{1}{2} = 2$. In the same way, the von Koch curve is made up of four copies of itself scaled by a factor $\frac{1}{3}$, and has dimension $-\log 4/\log\frac{1}{3} = \log 4/\log 3$, and the Cantor set may be regarded as comprising four copies of itself scaled by a factor $\frac{1}{9}$ and having dimension $-\log 4/\log\frac{1}{9} = \log 2/\log 3$. In general, a set made up of m copies of itself scaled by a factor r might be thought of as having dimension $-\log m/\log r$. The number obtained in this way is usually referred to as the *similarity dimension* of the set.

Unfortunately, similarity dimension is meaningful only for a small class of strictly self-similar sets. Nevertheless, there are other definitions of dimension that are much more widely applicable. For example, Hausdorff dimension and the box-counting dimensions may be defined for any sets, and, in these four

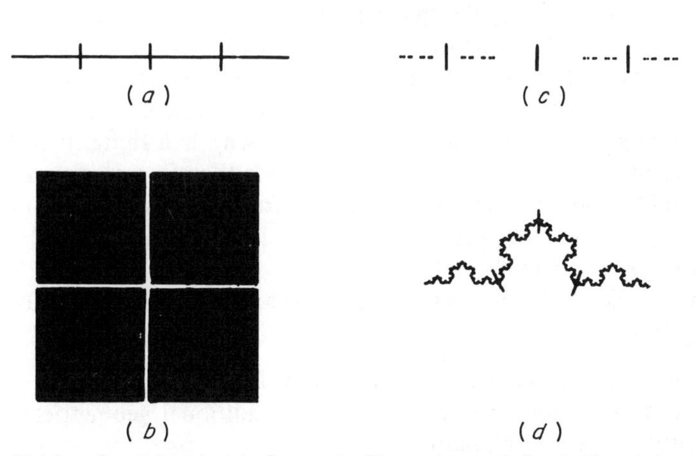

Figure 0.9 Division of certain sets into four parts. The parts are similar to the whole with ratios: $\frac{1}{4}$ for line segment (*a*); $\frac{1}{2}$ for square (*b*); $\frac{1}{9}$ for middle third Cantor set (*c*); $\frac{1}{3}$ for von Koch curve (*d*)

examples, may be shown to equal the similarity dimension. The early chapters of the book are concerned with the definition and properties of Hausdorff and other dimensions, along with methods for their calculation. Very roughly, a dimension provides a description of how much space a set fills. It is a measure of the prominence of the irregularities of a set when viewed at very small scales. A dimension contains much information about the geometrical properties of a set.

A word of warning is appropriate at this point. It is possible to define the 'dimension' of a set in many ways, some satisfactory and others less so. It is important to realize that different definitions may give different values of dimension for the same set, and may also have very different properties. Inconsistent usage has sometimes led to considerable confusion. In particular, warning lights flash in my mind (as in the minds of other mathematicians) whenever the term 'fractal dimension' is seen. Though some authors attach a precise meaning to this, I have known others interpret it inconsistently in a single piece of work. The reader should always be aware of the definition in use in any discussion.

In his original essay, Mandelbrot defined a fractal to be a set with Hausdorff dimension strictly greater than its topological dimension. (The *topological dimension* of a set is always an integer and is 0 if it is totally disconnected, 1 if each point has arbitrarily small neighbourhoods with boundary of dimension 0, and so on.) This definition proved to be unsatisfactory in that it excluded a number of sets that clearly ought to be regarded as fractals. Various other definitions have been proposed, but they all seem to have this same drawback.

My personal feeling is that the definition of a 'fractal' should be regarded in the same way as the biologist regards the definition of 'life'. There is no hard and fast definition, but just a list of properties characteristic of a living thing, such as the ability to reproduce or to move or to exist to some extent independently of the environment. Most living things have most of the characteristics on the list, though there are living objects that are exceptions to each of them. In the same way, it seems best to regard a fractal as a set that has properties such as those listed below, rather than to look for a precise definition which will almost certainly exclude some interesting cases. From the mathematician's point of view, this approach is no bad thing. It is difficult to avoid developing properties of dimension other than in a way that applies to 'fractal' and 'non-fractal' sets alike. For 'non-fractals', however, such properties are of little interest—they are generally almost obvious and could be obtained more easily by other methods.

When we refer to a set F as a fractal, therefore, we will typically have the following in mind.

(i) F has a fine structure, i.e. detail on arbitrarily small scales.
(ii) F is too irregular to be described in traditional geometrical language, both locally and globally.
(iii) Often F has some form of self-similarity, perhaps approximate or statistical.

(iv) Usually, the 'fractal dimension' of F (defined in some way) is greater than its topological dimension.
(v) In most cases of interest F is defined in a very simple way, perhaps recursively.

What can we say about the geometry of as diverse a class of objects as fractals? Classical geometry gives us a clue. In Part I of this book we study certain analogues of familiar geometrical properties in the fractal situation. The orthogonal projection, or 'shadow' of a circle in space onto a plane is, in general, an ellipse. The fractal projection theorems tell us about the 'shadows' of a fractal. For many purposes, a tangent provides a good local approximation to a circle. Though fractals do tend not to have tangents in any sense, it is often possible to say a surprising amount about their local form. Two circles in the plane in 'general position' either intersect in two points or not at all (we regard the case of mutual tangents as 'exceptional'). Using dimension, we can make similar statements about the intersection of fractals. Moving a circle perpendicular to its plane sweeps out a cylinder, with properties that are related to those of the original circle. Similar, and indeed more general, constructions are possible with fractals.

Although classical geometry is of considerable intrinsic interest, it is also called upon widely in other areas of mathematics. For example, circles or parabolae occur as the solution curves of certain differential equations, and a knowledge of the geometrical properties of such curves aids our understanding of the differential equations. In the same way, the general theory of fractal geometry can be applied to the many branches of mathematics in which fractals occur. Various examples of this are given in Part II of the book.

Historically, interest in geometry has been stimulated by its applications to nature. The ellipse assumed importance as the shape of planetary orbits, as did the sphere as the shape of the earth. The geometry of the ellipse and sphere can be applied to these physical situations. Of course, orbits are not quite elliptical, and the earth is not actually spherical, but for many purposes, such as the prediction of planetary motion or the study of the earth's gravitational field, these approximations may be perfectly adequate.

A similar situation pertains with fractals. A glance at the recent physics literature shows the variety of natural objects that are described as fractals—cloud boundaries, topographical surfaces, coastlines, turbulence in fluids, and so on. None of these are actual fractals—their fractal features disappear if they are viewed at sufficiently small scales. Nevertheless, over certain ranges of scale they appear very much like fractals, and at such scales may usefully be regarded as such. The distinction between 'natural fractals' and the mathematical 'fractal sets' that might be used to describe them was emphasized in Mandelbrot's original essay, but this distinction seems to have become somewhat blurred. There are no true fractals in nature. (There are no true straight lines or circles either!)

If the mathematics of fractal geometry is to be really worthwhile, then it should be applicable to physical situations. Progress is being made in this direction and some examples are given towards the end of this book. Although there are natural phenomena that have been explained in terms of fractal mathematics (Brownian motion is a good example), most applications tend to be descriptive rather than predictive. Much of the mathematics used in the study of fractals is not particularly new, though interest in it is. For further progress to be made, development and application of appropriate mathematics deserves a high priority.

Notes and references

Unlike the rest of the book, which consists of fairly solid mathematics, this Introduction contains some of the author's opinions and prejudices which may well not be shared by other workers on fractals. Caveat emptor!

The basic treatise on fractals, which may be appreciated at many levels is the scientific, philosophical and pictorial essay of Mandelbrot (1982) (developed from the original 1975 version), containing a great diversity of natural and mathematical examples of fractals. This essay, in its various versions, has been the inspiration for much of the work that has been done on fractals.

Other books devoted to various aspects of fractals include the mathematical treatment of Falconer (1985a), the beautifully illustrated survey of complex dynamics by Peitgen and Richter (1986), the book by Feder (1988) largely devoted to physical applications, the book edited by Peitgen and Saupe (1988) on computer graphical aspects, and the course book by Barnsley (1988) largely concerned with iterated function schemes. All these contain many further references.

Part I
FOUNDATIONS

Chapter 1 Mathematical background

This chapter reviews some of the basic mathematical ideas and notation that will be used througout the book. Sections 1.1 on set theory and 1.2 on functions are rather concise; readers unfamiliar with this type of material are advised to consult a more detailed text. Measures and mass distributions play an important part in the theory of fractals. A treatment adequate for our needs is given in Section 1.3. By asking the reader to take on trust the existence of certain measures, we can avoid many of the technical difficulties usually associated with measure theory. Some notes on probability theory are given in Section 1.4; an understanding of this is needed in Chapters 15 and 16.

1.1 Basic set theory

In this section we recall some basic notions from set theory and point set topology.

We generally work in *n-dimensional Euclidean space*, $\mathbb{R}^n$, where $\mathbb{R}^1 = \mathbb{R}$ is just the set of real numbers or the 'real line', and $\mathbb{R}^2$ is the (Euclidean) plane. Points in $\mathbb{R}^n$ will generally be denoted by lower case letters x, y, etc, and we will occasionally use the coordinate form $x = (x_1, \ldots, x_n)$, $y = (y_1, \ldots, y_n)$. Addition and scalar multiplication are defined in the usual manner, so that $x + y = (x_1 + y_1, \ldots, x_n + y_n)$ and $\lambda x = (\lambda x_1, \ldots, \lambda x_n)$, where λ is a real scalar. We use the usual *Euclidean distance* or *metric*) on $\mathbb{R}^n$. So if x, y are points of $\mathbb{R}^n$, the distance between them is $|x - y| = (\sum_{i=1}^n |x_i - y_i|^2)^{1/2}$.

Sets, which will generally be subsets of $\mathbb{R}^n$, are denoted by capital letters E, F, U, etc.. In the usual way, $x \in E$ means that the point x belongs to the set E, and $E \subset F$ means that E is a subset of the set F. We write $\{x:\text{condition}\}$ for the set of x for which 'condition' is true. Certain frequently occurring sets have a special notation. The empty set, which contains no elements, is written as $\varnothing$. The integers are denoted by $\mathbb{Z}$ and the rational numbers by $\mathbb{Q}$. We use a superscript $^+$ to denote the positive elements of a set; thus $\mathbb{R}^+$ are the positive real numbers, and $\mathbb{Z}^+$ are the positive integers. Occasionally we refer to the complex numbers $\mathbb{C}$, which for many purposes may be identified with the plane $\mathbb{R}^2$, with $x_1 + ix_2$ corresponding to the point (x_1, x_2).

The *closed ball* of centre x and radius r is defined by $B_r(x) = \{y:|y - x| \leqslant r\}$. Similarly the *open ball* is $B_r^o(x) = \{y:|y - x| < r\}$. Thus the closed ball contains its bounding sphere, but the open ball does not. Of course in $\mathbb{R}^2$ a ball is a disc

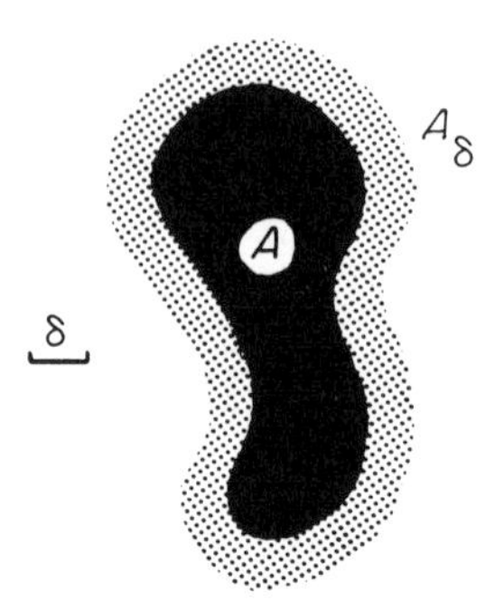

Figure 1.1 A set A and its δ-parallel body A_δ

and in $\mathbb{R}^1$ a ball is just an interval. If $a < b$ we write $[a, b]$ for the *closed interval* $\{x: a \leqslant x \leqslant b\}$ and (a, b) for the *open interval* $\{x: a < x < b\}$. Similarly $[a, b)$ denotes the half-open interval $\{x: a \leqslant x < b\}$, etc.

The *coordinate cube* of side $2r$ and centre $x = (x_1, \ldots, x_n)$ is the set $\{y = (y_1, \ldots, y_n): |y_i - x_i| \leqslant r$ for $i = 1, \ldots, n\}$ (A cube in $\mathbb{R}^2$ is just a square and in $\mathbb{R}^1$ is an interval.)

From time to time we refer to the *δ-parallel body*, A_δ, of a set A, that is the set of points within distance δ of A; thus $A_\delta = \{x: |x - y| \leqslant \delta$ for some y in $A\}$; see figure 1.1.

We write $A \cup B$ for the *union* of the sets A and B, i.e. the set of points belonging to either A or B. Similarly, we write $A \cap B$ for their *intersection*, the points in both A and B. More generally, $\bigcup_\alpha A_\alpha$ denotes the union of an arbitrary collection of sets $\{A_\alpha\}$, i.e. those points in at least one of the sets A_α, and $\bigcap_\alpha A_\alpha$ denotes their intersection, consisting of the set of points common to all of the A_α. A collection of sets is *disjoint* if the intersection of any pair is the empty set. The *difference* $A \backslash B$ of A and B consists of the points in A but not B. The set $\mathbb{R}^n \backslash A$ is termed the *complement* of A.

The set of all ordered pairs $\{(a, b): a \in A$ and $b \in B\}$ is called the (*Cartesian*) *product* of A and B and is denoted by $A \times B$. If $A \subset \mathbb{R}^n$ and $B \subset \mathbb{R}^m$ then $A \times B \subset \mathbb{R}^{n+m}$.

If A and B are subsets of $\mathbb{R}^n$ and λ is a real number, we define the *vector sum* of the sets as $A + B = \{x + y: x \in A$ and $y \in B\}$ and we define the *scalar multiple* as $\lambda A = \{\lambda x: x \in A\}$.

An infinite set A is *countable* if its elements can be listed in the form $x_1, x_2, \ldots$ with every element of A appearing at a specific place in the list; otherwise the set is *uncountable*. The sets $\mathbb{Z}$ and $\mathbb{Q}$ are countable but $\mathbb{R}$ is uncountable.

If A is any set of real numbers then the *supremum* $\sup A$ is the least number m such that $x \leqslant m$ for every x in A, or is ∞ if no such number exists. Similarly, the *infimum* $\inf A$ is the greatest number m such that $m \leqslant x$ for all x in A, or is $-\infty$. Intuitively the supremum and infimum are thought of as the maximum and minimum of the set, though it is important to realize that $\sup A$ and $\inf A$ need not be members of the set A itself. We write $\sup_{x \in B}(\ )$ for the supremum of the quantity in brackets, which may depend on x, as x ranges over the set B.

We define the *diameter* $|A|$ of a (non-empty) subset of $\mathbb{R}^n$ as the greatest distance apart of pairs of points in A. Thus $|A| = \sup\{|x-y| : x, y \in A\}$. A set A is *bounded* if it has finite diameter, or, equivalently, if A is contained in some (sufficiently large) ball.

Convergence of sequences is defined in the usual way. A sequence $\{x_k\}$ in $\mathbb{R}^n$ *converges* to a point x of $\mathbb{R}^n$ as $k \to \infty$ if, given $\varepsilon > 0$, there exists a number K such that $|x_k - x| < \varepsilon$ whenever $k > K$, that is if $|x_k - x|$ converges to 0. The number x is called the *limit* of the sequence, and we write $x_k \to x$ or $\lim_{k\to\infty} x_k = x$.

The ideas of 'open' and 'closed' that have been mentioned in connection with balls apply to much more general sets. Intuitively, a set is closed if it contains its boundary and open if it contains none of its boundary points. More precisely, a subset A of $\mathbb{R}^n$ is *open* if, for all points x in A there is some ball $B_r(x)$, centred at x and of positive radius, that is contained in A. A set is *closed* if, whenever $\{x_k\}$ is a sequence of points of A converging to a point x of $\mathbb{R}^n$, then x is in A; see figure 1.2. The empty set $\varnothing$ and $\mathbb{R}^n$ are regarded as both open and closed.

It may be shown that a set is open if and only if its complement is closed. The union of any collection of open sets is open, as is the intersection of any finite number of open sets. The intersection of any collection of closed sets is closed, as is the union of any finite number of closed sets.

A set A is called a *neighbourhood* of a point x if there is some (small) ball $B_r(x)$ centred at x and contained in A.

The intersection of all the closed sets containing a set A is called the *closure* of A, written $\bar{A}$. The union of all the open sets contained in A is the *interior* $\operatorname{int}(A)$ of A. The closure of A is thought of as the smallest closed set containing A, and the interior as the largest open set contained in A. The *boundary* ∂A of A is given by $\partial A = \bar{A} \backslash \operatorname{int}(A)$.

A set B is a *dense* subset of A if $B \subset A \subset \bar{B}$, i.e. if there are points of B arbitrarily close to each point of A.

A set A is *compact* if any collection of open sets which covers A (i.e. with union containing A) has a finite subcollection which also covers A. Technically, compactness is an extremely useful property that enables infinite sets of conditions to be reduced to finitely many. However, as far as most of this book

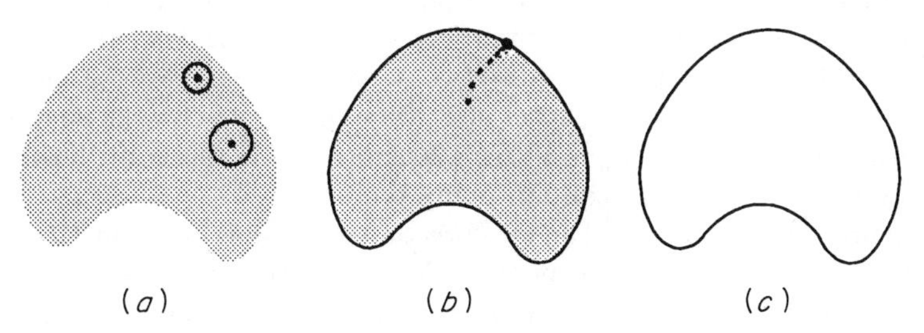

Figure 1.2 (*a*) An open set—there is a ball contained in the set centred at each point of the set. (*b*) A closed set—the limit of any convergent sequence of points from the set lies in the set. (*c*) The boundary of the set in (*a*) or (*b*)

is concerned, it is enough to think of a compact subset of $\mathbb{R}^n$ as one that is both closed and bounded.

The intersection of any collection of compact sets is compact. It may be shown that if $A_1 \supset A_2 \supset \cdots$ is a decreasing sequence of compact sets then the intersection $\bigcap_{i=1}^{\infty} A_i$ is non-empty. Moreover, if $\bigcap_{i=1}^{\infty} A_i$ is contained in V for some open set V, then then finite intersection $\bigcap_{i=1}^{k} A_i$ is contained in V for some k.

A subset A of $\mathbb{R}^n$ is *connected* if there do not exist open sets U and V such that $U \cup V$ contains A with $A \cap U$ and $A \cap V$ disjoint and non-empty. Intuitively, we think of a set A as connected if it consists of just one 'piece'. The largest connected subset of A containing a point x is called the *connected component* of x. The set A is *totally disconnected* if the connected component of each point consists of just that point. This will certainly be so if for any pair of points x and y in A we can find disjoint open sets U and V such that $x \in U$, $y \in V$ and $A \subset U \cap V$.

There is one further class of set that must be mentioned though its precise definition is indirect and should not concern the reader unduly. The class of *Borel sets* is the smallest collection of subsets of $\mathbb{R}^n$ with the following properties:

(a) every open set and every closed set is a Borel set;
(b) the union of every finite or countable collection of Borel sets is a Borel set, and the intersection of every finite or countable collection of Borel sets is a Borel set.

Throughout this book, virtually all of the subsets of $\mathbb{R}^n$ that will be of any interest to us will be Borel sets. Any set that can be constructed using a sequence of countable unions or intersections starting with the open sets or closed sets will certainly be Borel. The reader will not go far wrong in work of the sort described in this book by assuming that all the sets encountered are Borel sets.

1.2 Functions and limits

Let X and Y be any sets. A *mapping*, *function* or *transformation* f from X to Y is a rule or formula that associates a point $f(x)$ of Y with each point x of X. We write $f: X \to Y$ to denote this situation; X is called the *domain* of f and Y is called the *codomain*. If A is any subset of X we write $f(A)$ for the *image* of A, given by $\{f(x): x \in A\}$. If B is a subset of Y, we write $f^{-1}(B)$ for the *inverse image* or *pre-image* of B, i.e. the set $\{x \in X: f(x) \in B\}$; note that in this context the inverse image of a single point can contain many points.

A function $f: X \to Y$ is called an *injection* or a *one-to-one* function if $f(x) \neq f(y)$ whenever $x \neq y$, i.e. different elements of X are mapped to different elements of Y. The function is called a *surjection* or an *onto* function if, for every y in Y, there is an element x in X with $f(x) = y$, i.e. every element of Y is the image of some point in X. A function that is both an injection and a surjection is called a *bijection* or *one-to-one correspondence* between X and Y. If $f: X \to Y$

is a bijection then we may define the *inverse function* $f^{-1}:Y \to X$ by taking $f^{-1}(y)$ as the unique element of X such that $f(x) = y$. In this situation, $f^{-1}(f(x)) = x$ for x in X and $f(f^{-1}(y)) = y$ for y in Y.

The *composition* of the functions $f:X \to Y$ and $g:Y \to Z$ is the function $g \circ f:X \to Z$ given by $(g \circ f)(x) = g(f(x))$. This definition extends to the composition of any finite number of functions in the obvious way.

Certain functions from $\mathbb{R}^n$ to $\mathbb{R}^n$ have a particular geometric significance; often in this context they are referred to as transformations and are denoted by capital letters. Their effects are shown in figure 1.3. The transformation $S:\mathbb{R}^n \to \mathbb{R}^n$ is called a *congruence* or *isometry* if it preserves distances, i.e. if $|S(x) - S(y)| = |x - y|$ for x, y in $\mathbb{R}^n$. Congruences also preserve angles, and transform sets into geometrically congruent ones. Special cases include *translations*, which are of the form $S(x) = x + a$ and have the effect of shifting points parallel to the vector a, *rotations* which have a centre a such that $|S(x) - a| = |x - a|$ for all x (for convenience we also regard the identity transformation given by $I(x) = x$ as a rotation) and *reflections* which map points to their mirror images in some $(n - 1)$-dimensional plane. A congruence that may be achieved by a combination of a rotation and a translation, i.e. does not involve reflection, is called a *rigid motion* or *direct congruence*. A transformation $S:\mathbb{R}^n \to \mathbb{R}^n$ is a *similarity* if there is a constant c such that $|S(x) - S(y)| = c|x - y|$ for all x, y in $\mathbb{R}^n$. A similarity transforms sets into geometrically similar ones.

A transformation $T:\mathbb{R}^n \to \mathbb{R}^n$ is *linear* if $T(x + y) = T(x) + T(y)$ and $T(\lambda x) = \lambda T(x)$ for all $x, y \in \mathbb{R}^n$ and $\lambda \in \mathbb{R}$; linear transformations may be represented by matrices in the usual way. Such a linear transformation is

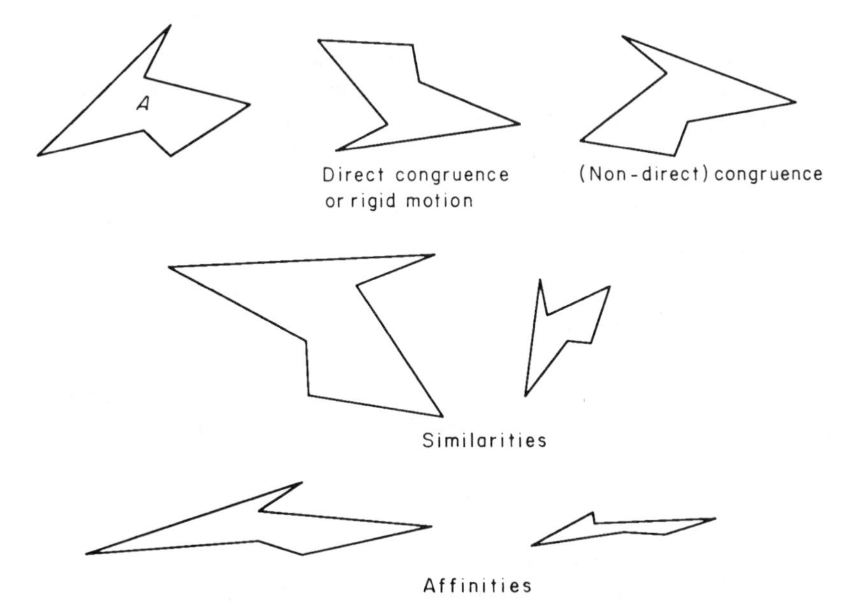

Figure 1.3 The effect of various transformations on a set A

non-singular if $T(x)=0$ if and only if $x=0$. If $S:\mathbb{R}^n \to \mathbb{R}^n$ is of the form $S(x)=T(x)+a$, where T is a non-singular linear transformation and a is a point in $\mathbb{R}^n$, then S is called an *affine transformation* or an *affinity*. An affinity may be thought of as a shearing transformation; its contracting or expanding effect need not be the same in every direction.

It is worth pointing out that such classes of transformation form groups under composition of mappings. For example, the composition of two translations is a translation, the identity transformation is trivially a translation, and the inverse of a translation is a translation. Finally, the associative law $S\circ(T\circ U)=(S\circ T)\circ U$ holds for all translations S, T, U. Similar group properties hold for the congruences, the rigid motions, the similarities and the affinities.

A function $f:X\to Y$ is called a *Hölder function of exponent* α if

$$|f(x)-f(y)| \leqslant c|x-y|^{\alpha} \qquad (x, y \in X)$$

for some constant c. The function f is called a *Lipschitz function* if α may be taken to be equal to 1, and a *bi-Lipschitz function* if

$$c_1|x-y| \leqslant |f(x)-f(y)| \leqslant c_2|x-y| \qquad (x, y \in X)$$

for $0<c_1\leqslant c_2<\infty$.

We next remind readers of the basic ideas of limits and continuity of functions. Let X and Y be subsets of $\mathbb{R}^n$ and $\mathbb{R}^m$ respectively, let $f:X\to Y$ be a function, and let a be a point of $\bar{X}$. We say that $f(x)$ has *limit* y (or *tends* to y, or *converges* to y) as x tends to a, if, given $\varepsilon>0$, there exists $\delta>0$ such that $|f(x)-y|<\varepsilon$ for all $x\in X$ with $|x-a|<\delta$. We denote this by writing $f(x)\to y$ as $x\to a$ or by $\lim_{x\to a} f(x)=y$. For a function $f:X\to\mathbb{R}$ we say that $f(x)$ *tends to infinity* (written $f(x)\to\infty$) as $x\to a$ if, given M, there exists $\delta>0$ such that $f(x)>M$ whenever $|x-a|<\delta$. The definition of $f(x)\to-\infty$ is similar.

Suppose, now, that $f:\mathbb{R}^+\to\mathbb{R}$. We shall frequently be interested in the values of such functions for small positive values of x. Note that if $f(x)$ is increasing as x decreases, then $\lim_{x\to 0} f(x)$ exists either as a finite limit or as ∞, and if $f(x)$ is decreasing as x decreases then $\lim_{x\to 0} f(x)$ exists and is finite or $-\infty$. Of course, $f(x)$ can fluctuate wildly for small x and $\lim_{x\to 0} f(x)$ need not exist at all. We use lower and upper limits to describe such fluctuations. We define the *lower limit* as

$$\varliminf_{x\to 0} f(x) \equiv \lim_{x\to 0}(\inf\{f(x):0<x<r\}).$$

Since $\inf\{f(x):0<x<r\}$ is either $-\infty$ for all positive r or else increases as r decreases, $\varliminf_{x\to 0} f(x)$ always exists. Similarly, the *upper limit* is defined as

$$\varlimsup_{x\to 0} f(x) \equiv \lim_{x\to 0}(\sup\{f(x):0<x<r\}).$$

The lower and upper limits exist (as real numbers or $-\infty$ or ∞) for any function f, and are indicative of the variation in values of f for x close to 0; see figure 1.4. If $\varliminf_{x\to 0} f(x)=\varlimsup_{x\to 0} f(x)$ then $\lim_{x\to 0} f(x)$ exists and equals this common

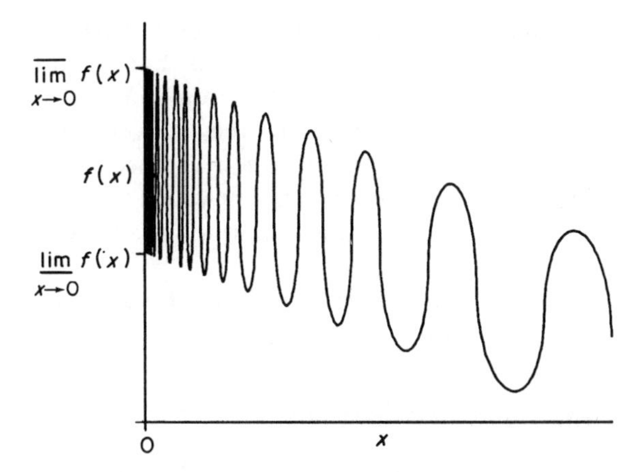

Figure 1.4 The upper and lower limits of a function

value. In the same way, it is possible to define lower and upper limits as $x \to a$ for functions $f: X \to \mathbb{R}$ where X is a subset of $\mathbb{R}^n$ with a in $\bar{X}$.

We often need to compare two functions $f, g: \mathbb{R}^+ \to \mathbb{R}$ for small values. We write $f(x) \sim g(x)$ to mean that $f(x)/g(x) \to 1$ as $x \to 0$. We will often have that $f(x) \sim x^s$; in other words that f obeys an approximate power law of exponent s when x is small. We use the notation $f(x) \simeq g(x)$ more loosely, to mean that $f(x)$ and $g(x)$ are approximately equal in some sense, to be specified in the particular circumstances.

Recall that the function $f: X \to Y$ is *continuous* at a point a of X if $f(x) \to f(a)$ as $x \to a$, and is *continuous on* X if it is continuous at all points of X. If $f: X \to Y$ is a continuous bijection with continuous inverse $f^{-1}: Y \to X$ then f is called a *homeomorphism*, and X and Y are termed *homeomorphic* sets.

The function $f: \mathbb{R} \to \mathbb{R}$ is *differentiable* at x with the number $f'(x)$ as *derivative* if

$$\lim_{h\to 0} \frac{f(x+h) - f(x)}{h} = f'(x).$$

In particular, the mean-value theorem applies: given $a < b$ and f differentiable on $[a, b]$ there exists c with $a < c < b$ such that

$$\frac{f(b) - f(a)}{b - a} = f'(c)$$

(intuitively, any chord of the graph of f is parallel to the slope of f at some intermediate point). A function f is *continuously differentiable* if $f'(x)$ is continuous in x.

More generally, if $f: \mathbb{R}^n \to \mathbb{R}^n$, we say that f is *differentiable* at x with *derivative* the linear mapping $f'(x): \mathbb{R}^n \to \mathbb{R}^n$ if

$$\lim_{|h|\to 0} \frac{|f(x+h) - f(x) - f'(x)h|}{|h|} = 0.$$

Occasionally, we shall be interested in the convergence of a sequence of functions $f_k:X \to Y$ where X and Y are subsets of Euclidean spaces. We say that functions f_k converge *pointwise* to a function $f:X \to Y$ if $f_k(x) \to f(x)$ as $k \to \infty$ for each x in X. We say that the convergence is *uniform* if $\sup_{x \in X} |f_k(x) - f(x)| \to 0$ as $k \to \infty$. Uniform convergence is a rather stronger property than pointwise convergence; the rate at which the limit is approached is uniform across X. If the functions f_k are continuous and converge uniformly to f, then f is continuous.

Finally, we remark that logarithms will always be to base e. The identity $a^b = c^{b \log a / \log c}$ will often be used.

1.3 Measures and mass distributions

Anyone studying the mathematics of fractals will not get far before encountering measures in some form or other. Many people are put off by the seemingly technical nature of measure theory—often unnecessarily so, since for most fractal applications only a few basic ideas are needed. Moreover, these ideas are often already familiar in the guise of the mass or charge distributions encountered in basic physics.

We need only be concerned with measures on subsets of $\mathbb{R}^n$. Basically a measure is just a way of ascribing a numerical 'size' to sets, such that if a set is decomposed into a finite or countable number of pieces in a reasonable way, then the size of the whole is the sum of the sizes of the pieces.

We call μ a *measure* on $\mathbb{R}^n$ if μ assigns a non-negative number, possibly ∞, to each subset of $\mathbb{R}^n$ such that:

(*a*) $\mu(\varnothing) = 0$; (1.1)

(*b*) $\mu(A) \leqslant \mu(B)$ if $A \subset B$; (1.2)

(*c*) If $A_1, A_2, \ldots$ is countable (or finite) sequence of sets then

$$\mu\left(\bigcup_{i=1}^{\infty} A_i\right) \leqslant \sum_{i=1}^{\infty} \mu(A_i) \tag{1.3}$$

with equality in (1.3), i.e.

$$\mu\left(\bigcup_{i=1}^{\infty} A_i\right) = \sum_{i=1}^{\infty} \mu(A_i) \tag{1.4}$$

if the A_i are disjoint Borel sets.

We call $\mu(A)$ the *measure* of the set A, and think of $\mu(A)$ as the size of A measured in some way. Condition (*a*) says that the empty set has zero measure, condition (*b*) says 'the larger the set, the larger the measure' and (*c*) says that if a set is a union of a countable number of pieces (which may overlap) then the sum of the measure of the pieces is at least equal to the measure of the whole. If a set is decomposed into a countable number of disjoint Borel sets then the total measure of the pieces equals the measure of the whole.

Technical note. For the measures that we shall encounter, (1.4) generally holds for a much wider class of sets than just the Borel sets, in particular for all images of Borel sets under continuous functions. However, for reasons that need not concern us here, we cannot in general require that (1.4) holds for every countable collection of disjoint sets A_i. The reader who is familiar with measure theory will realize that our definition of a measure on $\mathbb{R}^n$ is the definition of what would normally be termed 'an outer measure on $\mathbb{R}^n$ for which the Borel sets are measurable'. However, to save frequent referral to 'measurable sets' it is convenient to have $\mu(A)$ defined for every set A, and, since we are usually interested in measures of Borel sets, it is enough to have (1.4) holding for Borel sets rather than for a larger class. If μ is defined and satisfies (1.1)–(1.4) for the Borel sets, the definition of μ may be extended to an outer measure on all sets in such a way that (1.1)–(1.3) hold, so our definition is consistent with the usual one.

If $A \subset B$ then A may be expressed as a disjoint union $A = B \cup (A \backslash B)$, so it is immediate from (1.4) that, if A and B are Borel sets,

$$\mu(A \backslash B) = \mu(A) - \mu(B). \tag{1.5}$$

Similarly, if $A_1 \subset A_2 \subset \cdots$ is an increasing sequence of Borel sets then

$$\lim_{i \to \infty} \mu(A_i) = \mu\left(\bigcup_{i=1}^{\infty} A_i \right). \tag{1.6}$$

To see this, note that $\bigcup_{i=1}^{\infty} A_i = A_1 \cup (A_2 \backslash A_1) \cup (A_3 \backslash A_2) \cup \ldots$, with this union disjoint, so that

$$\begin{aligned} \mu\left(\bigcup_{i=1}^{\infty} A_i \right) &= \mu(A_1) + \sum_{i=1}^{\infty} (\mu(A_{i+1}) - \mu(A_i)) \\ &= \mu(A_1) + \lim_{k \to \infty} \sum_{i=1}^{k} (\mu(A_{i+1}) - \mu(A_i)) \\ &= \lim_{k \to \infty} \mu(A_k). \end{aligned}$$

More generally, it follows that if, for $\delta > 0$, A_δ are Borel sets that are increasing as δ decreases, i.e. $A_{\delta'} \subset A_\delta$ for $0 < \delta < \delta'$, then

$$\lim_{\delta \to 0} \mu(A_\delta) = \mu\left(\bigcup_{\delta > 0} A_\delta \right). \tag{1.7}$$

We think of the support of a measure as the set on which the measure is concentrated. Formally, the *support* of μ is the smallest closed set X such that $\mu(\mathbb{R}^n \backslash X) = 0$. The support of a measure is always closed and x is in the support if and only if $\mu(B_r(x)) > 0$ for all positive radii r. We say that μ is a measure *on* a set A if A contains the support of μ.

A measure on a bounded subset of $\mathbb{R}^n$ for which $0 < \mu(\mathbb{R}^n) < \infty$ will be called a *mass distribution*, and we think of $\mu(A)$ as the mass of the set A. We often think of this intuitively: we take a finite mass and spread it in some way across a set X to get a mass distribution on X; the conditions for a measure will then be satisfied.

We give some examples of measures and mass distributions. In general, we omit the proofs that measures with the stated properties exist. Much of technical measure theory concerns the existence of such measures, but, as far as applications go, their existence is intuitively reasonable, and can be taken on trust.

Example 1.1. The counting measure

For each subset A of $\mathbb{R}^n$ let $\mu(A)$ be the number of points in A if A is finite, and ∞ otherwise. Then μ is a measure on $\mathbb{R}^n$.

Example 1.2. Point mass

Let a be a point in $\mathbb{R}^n$ and define $\mu(A)$ to be 1 if A contains a, and 0 otherwise. Then μ is a mass distribution, thought of as a point mass concentrated at a.

Example 1.3. Lebesgue measure on $\mathbb{R}$

Lebesgue measure $\mathscr{L}^1$ extends the idea of 'length' to a large collection of subsets of $\mathbb{R}$ that includes the Borel sets. For open and closed intervals, we take $\mathscr{L}^1(a,b) = \mathscr{L}^1[a,b] = b-a$. If $A = \bigcup_i [a_i, b_i]$ is a finite or countable union of disjoint intervals, we let $\mathscr{L}^1(A) = \sum(b_i - a_i)$ be the length of A thought of as the sum of the length of the intervals. This leads us to the definition of the *Lebesgue measure* $\mathscr{L}^1(A)$ of an arbitrary set A. We define

$$\mathscr{L}^1(A) = \inf\left\{\sum_{i=1}^{\infty}(b_i - a_i) : A \subset \bigcup_{i=1}^{\infty}[a_i, b]\right\}$$

that is, we look at all coverings of A by countable collections of intervals, and take the smallest total interval length possible. It is not hard to see that (1.1)–(1.3) hold; it is rather harder to show that (1.4) holds for disjoint Borel sets A_i, and we avoid this question here. (In fact, (1.4) holds for a much larger class of sets than the Borel sets, 'the Lebesgue measurable sets', but not for all subsets of $\mathbb{R}$.) Lebesgue measure on $\mathbb{R}$ is generally though of as 'length', and we often write length(A) for $\mathscr{L}^1(A)$ when we wish to emphasize this intuitive meaning.

Example 1.4. Lebesgue measure on $\mathbb{R}^n$

If $A = \{(x_1, \ldots, x_n) \in \mathbb{R}^n : a_i \leqslant x_i \leqslant b_i\}$ is a 'coordinate parallelepiped' in $\mathbb{R}^n$, the n-dimensional volume of A is given by

$$\text{vol}^n(A) = (b_1 - a_1)(b_2 - a_2)\cdots(b_n - a_n).$$

(Of course, vol^1 is length, as in Example 1.3, vol^2 is area and vol^3 is the usual 3-dimensional volume.) Then *n-dimensional Lebesgue measure* $\mathscr{L}^n$ may be thought of as the extension of n-dimensional volume to a large class of sets. Just as in Example 1.3, we obtain a measure on $\mathbb{R}^n$ by defining

$$\mathscr{L}^n(A) = \inf\left\{ \sum_{i=1}^{\infty} \text{vol}^n(A_i) : A \subset \bigcup_{i=1}^{\infty} A_i \right\}$$

where the infimum is taken over all coverings of A by coordinate parallelepipeds A_i. We get that $\mathscr{L}^n(A) = \text{vol}^n(A)$ if A is a coordinate parallelepiped or, indeed, any set for which the volume can be determined by the usual rules of mensuration. Again, to aid intuition, we sometimes write area(A) in place of $\mathscr{L}^2(A)$, vol(A) for $\mathscr{L}^3(A)$ and $\text{vol}^n(A)$ for $\mathscr{L}^n(A)$.

Sometimes, we need to define 'k-dimensional' volume on a k-dimensional plane X in $\mathbb{R}^n$; this may be done by identifying X with $\mathbb{R}^k$ and using $\mathscr{L}^k$ on subsets of X in the obvious way.

Example 1.5. Uniform mass distribution on a line segment

Let L be a line segment of unit length in the plane. Define $\mu(A) = \mathscr{L}^1(L \cap A)$ i.e. the 'length' of intersection of A with L. Then μ is a mass distribution with support L, since $\mu(A) = 0$ if $A \cap L = \varnothing$. We may think of μ as unit mass spread evenly along the line segment L.

Example 1.6. Restriction of a measure

Let μ be a measure on $\mathbb{R}^n$ and E a Borel subset of $\mathbb{R}^n$. We may define a measure ν on $\mathbb{R}^n$, called the *restriction of μ to E*, by $\nu(A) = \mu(E \cap A)$ for any set A. Then ν is a measure on $\mathbb{R}^n$ with support contained in $\bar{E}$.

As far as this book is concerned, the most important measures we shall meet are the s-dimensional Hausdorff measures $\mathscr{H}^s$ on subsets of $\mathbb{R}^n$, where $0 \leqslant s \leqslant n$. These measures, which are introduced in Section 2.1, are a generalization of Lebesgue measures to dimensions that are not necessarily integral.

The following method is often used to construct a mass distribution on a subset of $\mathbb{R}^n$. It involves repeated subdivision of a mass between parts of a bounded Borel set E. Let $\mathscr{E}_0$ consist of the single set E. For $k = 1, 2, \ldots$ we let $\mathscr{E}_k$ be a collection of disjoint Borel subsets of E such that each set U in $\mathscr{E}_k$ is contained in one of the sets of $\mathscr{E}_{k-1}$ and contains a finite number of the sets in $\mathscr{E}_{k+1}$. We assume that the maximum diameter of the sets in $\mathscr{E}_k$ tends to 0 as $k \to \infty$. We define a mass distribution on E by repeated subdivision; see figure 1.5. We let $\mu(E)$ satisfy $0 < \mu(E) < \infty$, and we split this mass between the sets $U_1, \ldots, U_m$ in $\mathscr{E}_1$ by defining $\mu(U_i)$ in such a way that $\sum_{i=1}^m \mu(U_i) = \mu(E)$. Similarly, we assign masses to the sets of $\mathscr{E}_2$ so that if $U_1, \ldots, U_m$ are the sets of $\mathscr{E}_2$ contained in a set U of $\mathscr{E}_1$, then $\sum_{i=1}^m \mu(U_i) = \mu(U)$. In general, we assign masses so that

$$\sum_i \mu(U_i) = \mu(U) \tag{1.8}$$

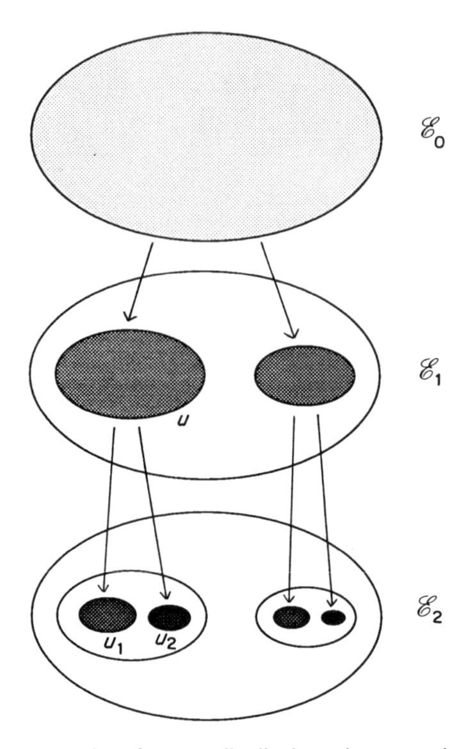

Figure 1.5 Steps in the construction of a mass distribution μ by repeated subdivision. The mass on the sets of $\mathscr{E}_k$ is divided between the sets of $\mathscr{E}_{k+1}$; so, for example, $\mu(U) = \mu(U_1) + \mu(U_2)$

for each set U of $\mathscr{E}_k$, where the $\{U_i\}$ are the disjoint sets in $\mathscr{E}_{k+1}$ contained in U. For each k, we let E_k be the union of the sets in $\mathscr{E}_k$, and we let $\mu(\mathbb{R}^n \backslash E_k) = 0$.

Let $\mathscr{E}$ denote the collection of sets that belong to $\mathscr{E}_k$ for some k together with the sets $\mathbb{R}^n \backslash E_k$. The above procedure defines the mass $\mu(A)$ of every set A in $\mathscr{E}$, and it should seem reasonable that, by building up sets from the sets in $\mathscr{E}$, it specifies enough about the distribution of the mass μ across E to determine $\mu(A)$ for any (Borel) set A. This is indeed the case, as the following proposition states.

Proposition 1.7

Let μ be defined on a collection of sets $\mathscr{E}$ as above. Then the definition of μ may be extended to all subsets of $\mathbb{R}^n$ so that μ becomes a measure. The value of $\mu(A)$ is uniquely determined if A is a Borel set. The support of μ is contained in $\bigcap_{k=1}^{\infty} \bar{E}_k$.

Note on Proof. If A is any subset of $\mathbb{R}^n$, let

$$\mu(A) = \inf\left\{\sum_i \mu(U_i) : A \subset \bigcup_i U_i \text{ and } U_i \in \mathscr{E}\right\}. \tag{1.9}$$

(Thus we take the smallest value we can of $\sum_{i=1}^{\infty} \mu(U_i)$ where the sets U_i are in

$\mathscr{E}$ and cover A; we have already defined $\mu(U_i)$ for such U_i.) It is not difficult to see that if A is one of the sets in $\mathscr{E}$, then (1.9) reduces to the mass $\mu(A)$ specified in the construction. The complete proof that μ satisfies all the conditions of a measure and that its values on the sets of $\mathscr{E}$ determine its values on the Borel sets is quite involved, and need not concern us here. Since $\mu(\mathbb{R}^n \backslash E_k) = 0$, we have $\mu(A) = 0$ if A is an open set that does not intersect E_k for some k, so the support of μ is in $\bar{E}_k$ for all k. □

Example 1.8

Let $\mathscr{E}_k$ denote the collection of 'binary intervals' of length 2^{-k} of the form $[r2^{-k}, (r+1)2^{-k})$ where $0 \leqslant r \leqslant 2^k - 1$. If we take $\mu[r2^{-k}, (r+1)2^{-k}) = 2^{-k}$ in the above construction, we get that μ is Lebesgue measure on $[0, 1]$.

Note on calculation. Clearly, if I is an interval in $\mathscr{E}_k$ of length 2^{-k} and I_1, I_2 are the two subintervals of I in $\mathscr{E}_{k+1}$ of length 2^{-k-1}, we have $\mu(I) = \mu(I_1) + \mu(I_2)$ which is (1.8). By Proposition 1.7 μ extends to a mass distribution on $[0, 1]$. We have $\mu(I) = \text{length}(I)$ for I in $\mathscr{E}$, and it may be shown that this implies that μ coincides with Lebesgue measure on any set. □

We say that a property holds for *almost all* x, or *almost everywhere*, (with respect to a measure μ) if the set for which the property fails has μ-measure zero. For example, we might say that almost all real numbers are irrational with respect to Lebesgue measure. The rational numbers $\mathbb{Q}$ are countable; they may be listed as $x_1, x_2, \ldots$, say, so that $\mu(\mathbb{Q}) = \sum_{i=1}^{\infty} \mu\{x_i\} = 0$.

Although we shall usually be interested in measures in their own right, we shall sometimes need to integrate functions with respect to measures. There are technical difficulties concerning which functions can be integrated. We may get around these difficulties by assuming that, for $f: D \to \mathbb{R}$ a function defined on a Borel subset D of $\mathbb{R}^n$, the set $f^{-1}(-\infty, a] = \{x \in D: f(x) \leqslant a\}$ is a Borel set for all real numbers a. A very large class of functions satisfies this condition, including all continuous functions (for which $f^{-1}(-\infty, a]$ is closed and therefore a Borel set). We make the assumption throughout this book that all functions to be integrated satisfy this condition; certainly this is true of functions that are likely to be encountered in practice.

To define integration we first suppose that $f: D \to \mathbb{R}$ is a *simple function*, i.e. one that takes only finitely many values $a_1, \ldots, a_k$. We define the *integral with respect to the measure* μ of a non-negative simple function f as

$$\int f \, d\mu = \sum_{i=1}^{k} a_i \mu\{x: f(x) = a_i\}.$$

The integral of more general functions is defined using approximation by simple functions. If $f: D \to \mathbb{R}$ is a non-negative function, we define its integral as

$$\int f \, d\mu = \sup \left\{ \int g \, d\mu : g \text{ is simple, } 0 \leqslant g \leqslant f \right\}.$$

To complete the definition, if f takes both positive and negative values, we let $f_+(x) = \max\{f(x), 0\}$ and $f_-(x) = \max\{-f(x), 0\}$, so that $f = f_+ - f_-$, and define

$$\int f \,\mathrm{d}\mu = \int f_+ \,\mathrm{d}\mu - \int f_- \,\mathrm{d}\mu$$

provided that $\int f_+ \,\mathrm{d}\mu$ and $\int f_- \,\mathrm{d}\mu$ are both finite.

All the usual properties hold for integrals, for example,

$$\int (f+g)\,\mathrm{d}\mu = \int f \,\mathrm{d}\mu + \int g \,\mathrm{d}\mu$$

and

$$\int \lambda f \,\mathrm{d}\mu = \lambda \int f \,\mathrm{d}\mu$$

if λ is a scalar. We also have the monotone convergence theorem, that if $f_k: D \to \mathbb{R}$ is an increasing sequence of non-negative functions converging (pointwise) to f, then

$$\lim_{k\to\infty} \int f_k \,\mathrm{d}\mu = \int f \,\mathrm{d}\mu.$$

If A is a Borel subset of D, we define integration over the set A by

$$\int_A f \,\mathrm{d}\mu = \int f \chi_A \,\mathrm{d}\mu$$

where $\chi_A: \mathbb{R}^n \to \mathbb{R}$ is the 'indicator function' such that $\chi_A(x) = 1$ if x is in A and $\chi_A(x) = 0$ otherwise.

Note that, if $f(x) \geqslant 0$ and $\int f \,\mathrm{d}\mu = 0$, then $f(x) = 0$ for μ-almost all x.

As usual, integration is denoted in various ways, such as $\int f \,\mathrm{d}\mu$, $\int f$ or $\int f(x)\,\mathrm{d}\mu(x)$, depending on the emphasis required. When μ is n-dimensional Lebesgue measure $\mathscr{L}^n$, we usually write $\int f \,\mathrm{d}x$ or $\int f(x)\,\mathrm{d}x$ in place of $\int f \,\mathrm{d}\mathscr{L}^n$.

On a couple of occasions we shall need to use Egoroff's theorem. Let D be a Borel subset of $\mathbb{R}^n$ and μ a measure with $\mu(D) < \infty$. Let $f_1, f_2, \ldots$ and f be functions from D to $\mathbb{R}$ such that $f_k(x) \to f(x)$ for each x in D. Egoroff's theorem states that for any $\delta > 0$, there is a Borel subset E of D such that $\mu(D \backslash E) < \delta$ and such that the sequence $\{f_k\}$ converges uniformly to f on E, i.e. with $\sup_{x\in E} |f_k(x) - f(x)| \to 0$ as $k \to \infty$. For the measures that we shall be concerned with, it may be shown that we can always take the set E to be compact.

1.4 Notes on probability theory

For an understanding of some of the later chapters of the book, a basic knowledge of probability theory is necessary. We give a brief survey of the concepts needed.

Probability theory starts with the idea of an *experiment* or *trial*; that is, an action whose outcome is, for all practical purposes, not predetermined. Mathematically, such an experiment is described by a probability space, which has three components: the set of all possible outcomes of the experiment, the list of all the events that may occur as consequences of the experiment, and an assessment of likelihood of these events. For example, if a die is thrown, the possible outcomes are $\{1,2,3,4,5,6\}$, the list of events includes 'a 3 is thrown', 'an even number is thrown' and 'at least a 4 is thrown'. For a 'fair die' it may be reasonable to assess the six possible outcomes as equally likely.

The set of all possible outcomes of an experiment is called the *sample space*, denoted by Ω. Questions of interest concerning the outcome of an experiment can always be phrased in terms of subsets of Ω; in the above example 'is an odd number thrown?' asks 'is the outcome in the subset $\{1,3,5\}$?'. Associating events dependent on the outcome of the experiment with subsets of Ω in this way, it is natural to think of the union $A \cup B$ as 'either A or B occurs', the intersection $A \cap B$ as 'both A and B occur', and the complement $\Omega \backslash A$ as the event 'A does not occur', for any events A and B. In general, there is a collection $\mathscr{F}$ of subsets of Ω that particularly interest us, which we call *events*. In the example of the die, $\mathscr{F}$ would normally be the collection of all subsets of Ω, but in more complicated situations a relatively small collection of subsets might be relevant. Usually, $\mathscr{F}$ satisfies certain conditions; for example, if the occurrence of an event interests us, then so does its non-occurrence, so if A is in $\mathscr{F}$, we would expect the complement $\Omega \backslash A$ also to be in $\mathscr{F}$. We call a (non-empty) collection $\mathscr{F}$ of subsets of the sample space Ω an *event space* if

$$\Omega \backslash A \in \mathscr{F} \quad \text{whenever } A \in \mathscr{F} \tag{1.10}$$

and

$$\bigcup_{i=1}^{\infty} A_i \in \mathscr{F} \quad \text{whenever } A_i \in \mathscr{F} \ (1 \leqslant i < \infty). \tag{1.11}$$

It follows from these conditions that $\varnothing$ and Ω are in $\mathscr{F}$, and that $A \backslash B$ and $\bigcap_{i=1}^{\infty} A_i$ are in $\mathscr{F}$ whenever A, B and A_i are in $\mathscr{F}$. As far as our applications are concerned, we do not, in general, specify $\mathscr{F}$ precisely—this avoids technical difficulties connected with the existence of suitable event spaces.

Next, we associate probabilities with the events of $\mathscr{F}$, with $\mathsf{P}(A)$ thought of as the probability, or likelihood, that the event A occurs. We call P a *probability* or *probability measure* if P assigns a number $\mathsf{P}(A)$ to each A in $\mathscr{F}$, such that the following conditions hold:

$$0 \leqslant \mathsf{P}(A) \leqslant 1 \text{ for all } A \in \mathscr{F} \tag{1.12}$$

$$\mathsf{P}(\varnothing) = 0 \text{ and } \mathsf{P}(\Omega) = 1 \tag{1.13}$$

and, if $A_1, A_2, \ldots$ are disjoint events in $\mathscr{F}$,

$$\mathsf{P}\left(\bigcup_{i=1}^{\infty} A_i\right) = \sum_{i=1}^{\infty} \mathsf{P}(A_i). \tag{1.14}$$

It should seem natural for any definition of probability to satisfy these conditions.

We call a triple $(\Omega, \mathscr{F}, \mathsf{P})$ a *probability space* if $\mathscr{F}$ is an event space of subsets of Ω and P is a probability measure defined on the sets of $\mathscr{F}$.

For the die-throwing experiment we might have $\Omega = \{1, 2, 3, 4, 5, 6\}$ with the event space consisting of all subsets of Ω, and with $\mathsf{P}(A) = \frac{1}{6} \times$ number of elements in A. This describes the 'fair die' situation with each outcome equally likely.

The resemblance of the definition of probability to the definition of a measure in (1.1), (1.2) and (1.4) and the use of the term probability measure is no coincidence. Probabilities and measures may be put into the same context, with Ω corresponding to $\mathbb{R}^n$ and with the event space corresponding to the Borel sets.

In our applications later on in the book, we shall be particularly interested in events (on rather large sample spaces) that are virtually certain to occur. We say that an event A occurs *with probability* 1, or *almost surely* if $\mathsf{P}(A) = 1$.

Sometimes, we may possess partial information about the outcome of an experiment; for example, we might be told that the number showing on the die is even. This leads us to reassess the probabilities of the various events. If A and B are in $\mathscr{F}$ with $\mathsf{P}(B) > 0$, the (*conditional*) *probability of A given B*, denoted by $\mathsf{P}(A \mid B)$, is defined by

$$\mathsf{P}(A \mid B) = \frac{\mathsf{P}(A \cap B)}{\mathsf{P}(B)}. \tag{1.15}$$

This is thought of as the probability of A given that the event B is known to occur; as would be expected $\mathsf{P}(B \mid B) = 1$. It is easy to show that $(\Omega, \mathscr{F}, \mathsf{P}')$ is a probability space, where $\mathsf{P}'(A) = \mathsf{P}(A \mid B)$. We also have the partition formula: if $B_1, B_2, \ldots$ are disjoint events with $\bigcup_i B_i = \Omega$ and $\mathsf{P}(B_i) > 0$ for all i, then, for an event A,

$$\mathsf{P}(A) = \sum_i \mathsf{P}(A \mid B_i)\mathsf{P}(B_i). \tag{1.16}$$

In the case of the 'fair die' experiment, if B_1 is the event 'an even number is thrown' B_2 is 'an odd number is thrown' and A is 'at least 4 is thrown', then

$$\mathsf{P}(A \mid B_1) = \mathsf{P}(4 \text{ or } 6 \text{ is thrown})/\mathsf{P}(2, 4 \text{ or } 6 \text{ is thrown}) = \tfrac{2}{6}/\tfrac{3}{6} = \tfrac{2}{3}.$$
$$\mathsf{P}(A \mid B_2) = \mathsf{P}(5 \text{ is thrown})/\mathsf{P}(1, 3 \text{ or } 5 \text{ is thrown}) = \tfrac{1}{6}/\tfrac{3}{6} = \tfrac{1}{3}$$

from which (1.16) is easily verified.

We think of two events as independent if the occurrence of one does not affect the probability that the other occurs, i.e. if $\mathsf{P}(A \mid B) = \mathsf{P}(A)$ and $\mathsf{P}(B \mid A) = \mathsf{P}(B)$. Using (1.15), we are led to make the definition that two events A and B in a probability space are *independent* if

$$\mathsf{P}(A \cap B) = \mathsf{P}(A)\mathsf{P}(B). \tag{1.17}$$

More generally, an arbitrary collection of events is independent if for every finite subcollection $\{A_k : k \in J\}$ we have

$$\mathsf{P}\left(\bigcap_{k\in J} A_k\right) = \prod_{k\in J} \mathsf{P}(A_i). \tag{1.18}$$

In the die example, it is easy to see that 'a throw of at least 5' and 'an even number is thrown' are independent events, but 'a throw of at least 4' and 'an even number is thrown' are not.

The idea of a random variable and its expectation (or average or mean) is fundamental to probability theory. Essentially, a random variable X is a real-valued function on a sample space. In the die example, X might represent the score on the die. Alternatively it might represent the reward for throwing a particular number, for example $X(\omega) = 0$ if $\omega = 1, 2, 3$, or 4, $X(5) = 1$ and $X(6) = 2$. The outcome of an experiment determines a value of the random variable. The expectation of the random variable is the average of these values weighted according to the likelihood of each outcome.

The precise definition of a random variable requires a little care. We say that X is a *random variable* on a probability space $(\Omega, \mathscr{F}, \mathsf{P})$ if $X : \Omega \to \mathbb{R}$ is a function such that $X^{-1}((-\infty, a])$ is an event in $\mathscr{F}$ for each real number a; in other words, the set of ω in Ω with $X(\omega) \leqslant a$ is in the event space. This condition is equivalent to saying that $X^{-1}(E)$ is in $\mathscr{F}$ for any Borel set E. In particular, for any such E the probability that the random variable X takes a value in E, i.e. $\mathsf{P}(\{\omega : X(\omega) \in E\})$, is defined. It may be shown that $\mathsf{P}(\{\omega : X(\omega) \in E\})$ is determined for all Borel sets E from a knowledge of $\mathsf{P}(\{\omega : X(\omega) \leqslant a\})$ for each real number a. Note that it is usual to abbreviate expressions such as $\mathsf{P}(\{\omega : X(\omega) \in E\})$ to $\mathsf{P}(X \in E)$.

It is not difficult to show that if X and Y are random variables on $(\Omega, \mathscr{F}, \mathsf{P})$ and λ is a real number, then $X + Y, X - Y, XY$ and λX are all random variables (these are defind in the obvious way, for example $(X + Y)(\omega) = X(\omega) + Y(\omega)$ for each $\omega \in \Omega$). Moreover, if $X_1, X_2, \ldots$ is a sequence of random variables with $X_k(\omega)$ increasing and bounded for each ω, then $\lim_{k\to\infty} X_k$ is a random variable.

A collection of random variables $\{X_k\}$ is *independent* if, for any Borel sets E_k, the events $\{(X \in E_k)\}$ are independent in the sense of (1.18); that is if, for every finite set of indices J,

$$\mathsf{P}(X_k \in E_k \text{ for all } k \in J) = \prod_{k\in J} \mathsf{P}(X_k \in E_k).$$

Intuitively, X and Y are independent if the probability of Y taking any particular value is unaffected by a knowledge of the value of X. Consider the probability space representing two successive throws of a die, with sample space $\{(x, y) : x, y = 1, 2, \ldots, 6\}$ and probability measure P defined by $\mathsf{P}\{(x, y)\} = \frac{1}{36}$ for each pair (x, y). If X and Y are the random variables given by the scores on successive throws, then X and Y are independent, modelling the assumption that one throw does not affect the other. However, X and $X + Y$ are not independent—this reflects that the bigger the score for the first throw, the greater the chance of a high total score.

The formal definition of the expectation of a random variable is analogous to the definition of the integral of a function; indeed, expectation is really the integral of the random variable with respect to the probability measure. Let X be a random variable on a probability space $(\Omega, \mathscr{F}, \mathsf{P})$. First suppose that $X(\omega) \geqslant 0$ for all ω in Ω and that X takes only finitely many values $x_1, \ldots, x_k$; we call such a random variable *simple*. We define the *expectation*, *mean* or *average* $\mathsf{E}(X)$ of X as

$$\mathsf{E}(X) = \sum_{i=1}^{k} x_i \mathsf{P}(X = x_i). \tag{1.19}$$

The expectation of an arbitrary random variable is defined using approximation by simple random variables. Thus for a non-negative random variable X

$$\mathsf{E}(X) = \sup\{\mathsf{E}(Y): Y \text{ is a simple random variable with } 0 \leqslant Y(\omega) \leqslant X(\omega) \text{ for all } \omega \in \Omega\}.$$

Lastly, if X takes both positive and negative values, we let $X_+ = \max\{X, 0\}$ and $X_- = \max\{-X, 0\}$, so that $X = X_+ - X_-$, and define

$$\mathsf{E}(X) = \mathsf{E}(X_+) - \mathsf{E}(X_-)$$

provided that both $\mathsf{E}(X_+) < \infty$ and $\mathsf{E}(X_-) < \infty$.

The random variable X representing the score of a fair die is a simple random variable, since $X(\omega)$ takes just the values $1, \ldots, 6$. Thus

$$\mathsf{E}(X) = \sum_{i=1}^{6} i \times \tfrac{1}{6} = 3\tfrac{1}{2}.$$

Expectation satisfies certain basic properties, analogous to those for the integral. If $X_1, X_2, \ldots$ are random variables then

$$\mathsf{E}(X_1 + X_2) = \mathsf{E}(X_1) + \mathsf{E}(X_2)$$

and, more generally,

$$\mathsf{E}\left(\sum_{i=1}^{k} X_i\right) = \sum_{i=1}^{k} \mathsf{E}(X_i).$$

If λ is a constant

$$\mathsf{E}(\lambda X) = \lambda \mathsf{E}(X)$$

and if the sequence of non-negative random variables $X_1, X_2, \ldots$ is increasing with $X = \lim_{k \to \infty} X_k$ a (finite) random variable, then

$$\lim_{k \to \infty} \mathsf{E}(X_k) = \mathsf{E}(X).$$

Provided that X_1 and X_2 are independent, we also have

$$\mathsf{E}(X_1 X_2) = \mathsf{E}(X_1)\mathsf{E}(X_2).$$

Thus if X_i represents that kth throw of a fair die in a sequence of throws, the expectation of the sum of the first k throws is $\mathsf{E}(X_1 + \cdots + X_k) = \mathsf{E}(X_1) + \cdots + \mathsf{E}(X_k) = 3\tfrac{1}{2} \times k$.

We define the *conditional expectation* $\mathsf{E}(X|B)$ of X given an event B with $\mathsf{P}(B)>0$ in a similar way, but starting with

$$\mathsf{E}(X|B)=\sum_{i=1}^{k} x_i \mathsf{P}(X=x_i|B) \tag{1.20}$$

in place of (1.19). We get a partition formula resembling (1.16)

$$\mathsf{E}(X)=\sum_{i} \mathsf{E}(X|B_i)\mathsf{P}(B_i) \tag{1.21}$$

where $B_1, B_2, \ldots$ are disjoint events with $\bigcup_i B_i=\Omega$ and $\mathsf{P}(B_i)>0$.

It is often useful to have an indication of the fluctuation of a random variable across a sample space. Thus we introduce the *variance* of the random variable X as

$$\begin{aligned}\operatorname{var}(X)&=\mathsf{E}((X-\mathsf{E}(X))^2)\\&=\mathsf{E}(X^2)-\mathsf{E}(X)^2\end{aligned}$$

by a simple calculation. Using the properties of expectation, we get

$$\operatorname{var}(\lambda X)=\lambda^2 \operatorname{var} X$$

for any real number λ, and

$$\operatorname{var}(X+Y)=\operatorname{var}(X)+\operatorname{var}(Y)$$

provided that X and Y are independent.

If the probability distribution of a random variable is given by an integral, i.e.

$$\mathsf{P}(X\leqslant x)=\int_{-\infty}^{x} f(u)\,\mathrm{d}u \tag{1.22}$$

the function f is called the *probability density function* for X. It may be shown from the definition of expectation that

$$\mathsf{E}(X)=\int_{-\infty}^{\infty} u f(u)\,\mathrm{d}u$$

and

$$\mathsf{E}(X^2)=\int_{-\infty}^{\infty} u^2 f(u)\,\mathrm{d}u$$

which allows $\operatorname{var}(X)=\mathsf{E}(X^2)-\mathsf{E}(X)^2$ to be calculated.

Note that the density function tells us about the distribution of the random variable X without reference to the underlying probability space, which, for many purposes, is irrelevant. We may express the probability that X belongs to any Borel set E in terms of the density function as

$$\mathsf{P}(X\in E)=\int_{E} f(u)\,\mathrm{d}u.$$

We say that a random variable X has *uniform distribution* on the interval (a, b) if

$$\mathsf{P}(X \leqslant x) = \frac{1}{b-a}\int_a^x \mathrm{d}u \qquad (a < x < b). \tag{1.23}$$

Thus the probability of X lying in an subinterval of (a, b) is proportional to the length of the interval. In this case, we get that $\mathsf{E}(X) = \frac{1}{2}(a+b)$ and $\operatorname{var}(X) = \frac{1}{12}(b-a)^2$.

A random variable X has *normal* or *Gaussian distribution* of mean m and variance σ^2 if

$$\mathsf{P}(X \leqslant x) = (2\pi)^{-1/2}\sigma^{-1}\int_{-\infty}^x \exp(-(u-m)^2/2\sigma^2)\,\mathrm{d}u. \tag{1.24}$$

It may be verified by integration that $\mathsf{E}(X) = m$ and $\operatorname{var}(X) = \sigma^2$. If X_1 and X_2 are independent normally distributed random variables of means m_1 and m_2 and variances σ_1^2 and σ_2^2 respectively, then $X_1 + X_2$ is normal with mean $m_1 + m_2$ and variance $\sigma_1^2 + \sigma_2^2$, and λX_1 is normal with mean λm_1 and variance $\lambda^2\sigma_1^2$, for any real number λ. The property that sums and scalar multiples of normal random variables are normal characterizes the normal distribution.

If we throw a fair die a large number of times, we might expect the average score thrown to be very close to $3\frac{1}{2}$, the expectation or mean outcome of each throw. Moreover, the larger the number of throws, the closer the average should be to the mean. This 'law of averages' is made precise as the strong law of large numbers.

Let $(\Omega, \mathscr{F}, \mathsf{P})$ be a probability space. Let $X_1, X_2, \ldots$ be random variables that are independent and that have identical distribution (i.e. for any set E, $\mathsf{P}(X_i \in E)$ is the same for all i), with expectation m and variance σ^2, both assumed finite. For each k we may form the random variable $S_k = X_1 + \cdots + X_k$, so that the random variable $(1/k)S_k$ is the average of the first k trials. The *strong law of large numbers* states that, with probability 1,

$$\lim_{k\to\infty} \frac{1}{k} S_k = m. \tag{1.25}$$

We can also say a surprising amount about the distribution of the random variable S_k when k is large. It may be shown that S_k has approximately the normal distribution with mean km and variance $k\sigma^2$. This is the content of the *central limit theorem*, which states that, for any real number x,

$$\mathsf{P}\left(\frac{S_k - km}{\sigma\sqrt{k}} \leqslant x\right) \to \int_{-\infty}^x (2\pi)^{-1/2}\exp(-\tfrac{1}{2}u^2)\,\mathrm{d}u \qquad \text{as } k \to \infty. \tag{1.26}$$

An important aspect of the normal distribution now becomes clear—it is the form of distribution approached by sums of a large number of independent identically distributed random variables.

We may apply these results to the experiment consisting of an infinite sequence of die throws. Let Ω be the set of all infinite sequences $\{\omega = (\omega_1, \omega_2, \ldots): \omega_i = 1, 2, \ldots, 6\}$ (we think of ω_i as the outcome of the kth throw). It is possible to define an event space $\mathscr{F}$ and probability measure P in such a way that for any given k and sequence $\omega_1, \ldots, \omega_k$ ($\omega_i = 1, 2, \ldots, 6$), the event 'the first k throws are $\omega_1, \ldots, \omega_k$' is in $\mathscr{F}$ and has probability $(\frac{1}{6})^{-k}$. Let X_k be the random variable given by the outcome of the kth throw, so that $X_k(\omega) = \omega_k$. It is easy to see that the X_k are independent and identically distributed, with mean $m = 3\frac{1}{2}$ and variance $2\frac{11}{12}$. The strong law of large numbers tells us that, with probability 1, the average of the first k throws, $(1/k)S_k$, converges to $3\frac{1}{2}$ as k tends to infinity, and the central limit theorem tells us that, when k is large, the sum S_k is approximately normally distributed, with mean $3\frac{1}{2} \times k$ and variance $2\frac{11}{12} \times k$. Thus if we repeat the experiment of throwing k dice a large number of times, the sum of the k throws will have a distribution close to the normal distribution, in the sense of (1.26).

1.5 Notes and references

The material outlined in this chapter is covered at various levels of sophistication in numerous undergraduate mathematical texts. Almost any book on mathematical analysis, for example Apostol (1974), contains the basic theory of sets and functions. A thorough treatment of measure and probability theory may be found in Kingman and Taylor (1966) and in Billingsley (1979). For probability theory, the book by Grimmett and Stirzaker (1982) may be found helpful.

Exercises

The following exercises do no more than emphasize some of the many facts that have been mentioned in this chapter.

1.1 Show that the union of any collection of open subsets of $\mathbb{R}^n$ is open and that the intersection of any finite collection of open sets is open. Show that a subset of $\mathbb{R}^n$ is closed if and only if its complement is open and hence deduce the corresponding result for unions and intersection of closed sets.

1.2 Show that if $A_1 \supset A_2 \supset \cdots$ is a decreasing sequence of non-empty compact subsets of $\mathbb{R}^n$ then $\bigcap_{k=1}^{\infty} A_k$ is a non-empty compact set.

1.3 Show that the half-open interval $\{x \in \mathbb{R}: 0 \leqslant x < 1\}$ is a Borel subset of $\mathbb{R}$.

1.4 Let F be the set of numbers in $[0, 1]$ whose decimal expansions contain the digit 5 infinitely many times. Show that F is a Borel set.

1.5 Show that the composition of two rotations in the plane is either a rotation or a translation.

1.6 Find $\underline{\lim}_{x\to 0} f(x)$ and $\overline{\lim}_{x\to 0} f(x)$ where $f(x)$ is given by: (i) $\sin(x)$; (ii) $\sin(1/x)$; (iii) $x^2 + (3 + x)\sin(1/x)$.

1.7 Let $f, g:[0,1]\to\mathbb{R}$ be Lipschitz functions. Show that the functions defined on $[0,1]$ by $f(x)+g(x)$ and $f(x)g(x)$ are also Lipschitz.

1.8 Let $f:\mathbb{R}\to\mathbb{R}$ be differentiable with $|f'(x)|\leqslant c$ for all x. Show that f is a Lipschitz function.

1.9 Let $A_1, A_2, \ldots,$ be a decreasing sequence of Borel subsets of $\mathbb{R}^n$ and let $A=\bigcap_{k=1}^{\infty} A_k$. If μ is a measure on $\mathbb{R}^n$ with $\mu(A_1)<\infty$, show that $\mu(A_k)\to\mu(A)$ as $k\to\infty$.

1.10 Let $f:[0,1]\to\mathbb{R}$ be a continuous function. For A a subset of $\mathbb{R}^2$ define $\mu(A)=\mathscr{L}\{x:(x,f(x))\in A\}$, where $\mathscr{L}$ is Lebesgue measure. Show that μ is a mass distribution on $\mathbb{R}^2$ supported by the graph of f.

1.11 Let D be a Borel subset of $\mathbb{R}^n$ and let μ be a measure on D with $\mu(D)<\infty$. Let $f_k:D\to\mathbb{R}$ be a sequence of functions such that $f_k(x)\to f(x)$ for all x in D. Prove Egoroff's theorem: that given $\varepsilon>0$ there exists a Borel subset A of D with $\mu(D\backslash A)<\varepsilon$ such that $f_k(x)$ converges to $f(x)$ uniformly for x in A.

1.12 Prove that if μ is a measure on D and $f:D\to\mathbb{R}$ satisfies $f(x)\geqslant 0$ for all x in D and $\int_D f\,\mathrm{d}\mu=0$ then $f(x)=0$ for μ-almost all x.

1.13 If X is a random variable show that $\mathsf{E}((X-\mathsf{E}(X))^2)=\mathsf{E}(X^2)-\mathsf{E}(X)^2$ (these numbers equalling the variance of X).

1.14 Verify that if X has the uniform distribution (see (1.23)) then $\mathsf{E}(X)=\frac{1}{2}(a+b)$ and $\mathrm{var}(X)=(b-a)^2/12$.

1.15 Let $A_1, A_2, \ldots$ be a sequence of independent events in some probability space such that $\mathsf{P}(A_k)=p$ for all k, where $0<p<1$. Let N_k be the random variable defined by taking $N_k(\omega)$ to equal the number of i with $1\leqslant i\leqslant k$ for which $\omega\in A_i$. Use the strong law of large numbers to show that, with probability 1, $N_k/k\to p$ as $k\to\infty$. Deduce that the proportion of occurrences of an event in a sequence of independent trials converges to the probability of the event.

1.16 A fair die is thrown 6000 times. Use the central limit theorem to estimate the probability that at least 1050 sixes are thrown. (A numerical method will be needed if the integral obtained is to be evaluated).

Chapter 2 Hausdorff measure and dimension

Of the wide variety of 'fractal dimensions' in use, the definition of Hausdorff, based on a construction of Carathéodory, is the oldest and probably the most important. Hausdorff dimension has the advantage of being defined for any set, and is mathematically convenient, as it is based on measures, which are relatively easy to manipulate. A major disadvantage is that in many cases it is hard to calculate or to estimate by computational methods. However, for an understanding of the mathematics of fractals, familiarity with Hausdorff measure and dimension is essential.

2.1 Hausdorff measure

Recall that if U is any non-empty subset of n-dimensional Euclidean space, $\mathbb{R}^n$, the *diameter* of U is defined as $|U| = \sup\{|x-y| : x, y \in U\}$, i.e. the greatest distance apart of any pair of points in U. If $\{U_i\}$ is a countable (or finite) collection of sets of diameter at most δ that cover F, i.e. $F \subset \bigcup_{i=1}^{\infty} U_i$ with $0 < |U_i| \leqslant \delta$ for each I, we say that $\{U_i\}$ is a *δ-cover* of F.

Suppose that F is a subset of $\mathbb{R}^n$ and s is a non-negative number. For any $\delta > 0$ we define

$$\mathscr{H}^s_\delta(F) = \inf\left\{ \sum_{i=1}^{\infty} |U_i|^s : \{U_i\} \text{ is a } \delta\text{-cover of } F \right\}. \tag{2.1}$$

Thus we look at all covers of F by sets of diameter at most δ and seek to minimize the sum of the sth powers of the diameters (figure 2.1). As δ decreases, the class of permissible covers of F in (2.1) is reduced. Therefore, the infimum $\mathscr{H}^s_\delta(F)$ increases, and so approaches a limit as $\delta \to 0$. We write

$$\mathscr{H}^s(F) = \lim_{\delta \to 0} \mathscr{H}^s_\delta(F). \tag{2.2}$$

This limit exists for any subset F of $\mathbb{R}^n$, though the limiting value can be (and usually is) 0 or ∞. We call $\mathscr{H}^s(F)$ the *s-dimensional Hausdorff measure* of F.

With a certain amount of effort, $\mathscr{H}^s$ may be shown to be a measure; see section 1.3. In particular, $\mathscr{H}^s(\varnothing) = 0$, if E is contained in F then $\mathscr{H}^s(E) \leqslant \mathscr{H}^s(F)$,

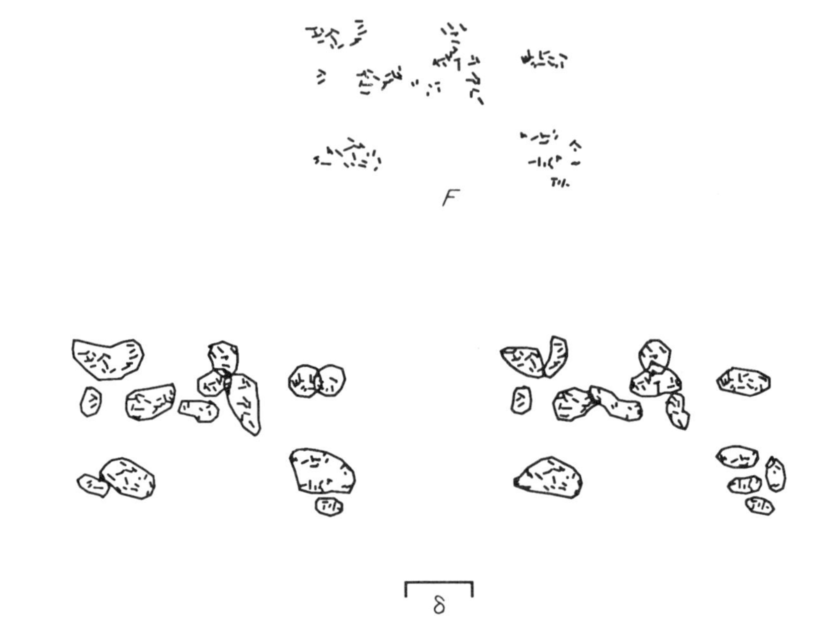

Figure 2.1 A set F and two possible δ-covers for F. The infimum of $\Sigma|U_i|^s$ over all such δ-covers $\{U_i\}$ gives $\mathscr{H}^s_\delta(F)$

and if $\{F_i\}$ is any countable collection of disjoint Borel sets, then

$$\mathscr{H}^s\left(\bigcup_{i=1}^{\infty} F_i\right) = \sum_{i=1}^{\infty} \mathscr{H}^s(F_i). \tag{2.3}$$

Hausdorff measures generalize the familiar ideas of length, area, volume, etc. It may be shown that, for subsets of $\mathbb{R}^n$, n-dimensional Hausdorff measure is, to within a constant multiple, just n-dimensional Lebesgue measure, i.e. the usual n-dimensional volume. More precisely, if F is a Borel subset of $\mathbb{R}^n$, then

$$\mathscr{H}^n(F) = c_n \operatorname{vol}^n(F) \tag{2.4}$$

where the constant $c_n = \pi^{\frac{1}{2}n}/2^n(\frac{1}{2}n)!$ is the volume of an n-dimensional ball of diameter 1. Similarly, for 'nice' lower-dimensional subsets of $\mathbb{R}^n$, we have that $\mathscr{H}^0(F)$ is the number of points in F; $\mathscr{H}^1(F)$ gives the length of a smooth curve F; $\mathscr{H}^2(F) = \frac{1}{4}\pi \times \text{area}(F)$ if F is a smooth surface; $\mathscr{H}^3(F) = \frac{4}{3}\pi \times \text{vol}(F)$; and $\mathscr{H}^m(F) = c_m \times \text{vol}^m(F)$ if F is a smooth m-dimensional submanifold of $\mathbb{R}^n$ (i.e. an m-dimensional surface in the classical sense).

The scaling properties of length, area and volume are well known. On magnification by a factor λ, the length of a curve is multiplied by λ, the area of a plane region is multiplied by λ^2 and the volume of a 3-dimensional object is multiplied by λ^3. As might be anticipated, s-dimensional Hausdorff measure scales with a factor λ^s (figure 2.2). Such scaling properties are fundamental to the theory of fractals.

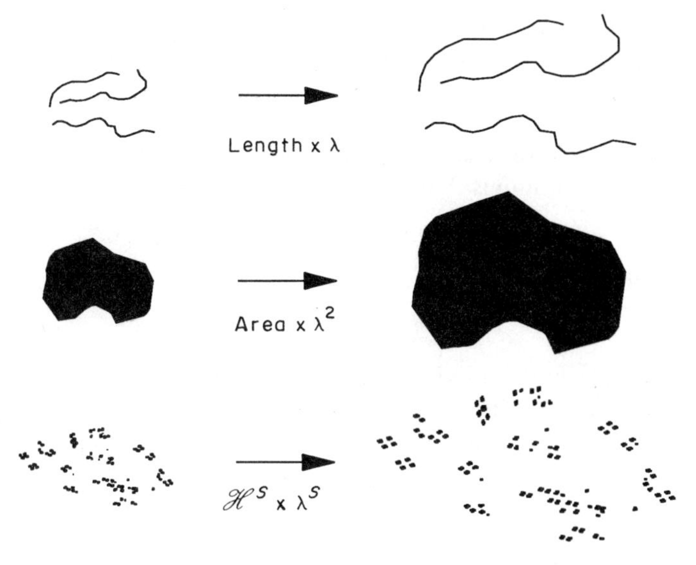

Figure 2.2 Scaling sets by a factor λ increases length by a factor λ, area by a factor λ^2, and s-dimensional Hausdorff measure by a factor λ^s

Scaling property 2.1

If $F \subset \mathbb{R}^n$ and $\lambda > 0$ then

$$\mathcal{H}^s(\lambda F) = \lambda^s \mathcal{H}^s(F) \tag{2.5}$$

where $\lambda F = \{\lambda x : x \in F\}$, i.e. the set F scaled by a factor λ.

Proof. If $\{U_i\}$ is a δ-cover of F then $\{\lambda U_i\}$ is a $\lambda\delta$-cover of λF. Hence

$$\begin{aligned}\mathcal{H}^s_{\lambda\delta}(\lambda F) \leqslant \Sigma |\lambda U_i|^s &= \lambda^s \Sigma |U_i|^s \\ &\leqslant \lambda^s \mathcal{H}^s_\delta(F)\end{aligned}$$

since this holds for any δ-cover $\{U_i\}$. Letting $\delta \to 0$ gives that $\mathcal{H}^s(\lambda F) \leqslant \lambda^s \mathcal{H}^s(F)$. Replacing λ by $1/\lambda$ and F by λF gives the opposite inequality required. □

A similar argument gives the following basic estimate of the effect of more general transformations on the Hausdorff measures of sets.

Proposition 2.2

Let $F \subset \mathbb{R}^n$ and $f: F \to \mathbb{R}^m$ be a mapping such that

$$|f(x) - f(y)| \leqslant c|x - y|^\alpha \qquad (x, y \in F) \tag{2.6}$$

for constants $c > 0$ and $\alpha > 0$. Then for each s

$$\mathcal{H}^{s/\alpha}(f(F)) \leqslant c^{s/\alpha} \mathcal{H}^s(F). \tag{2.7}$$

Proof. If $\{U_i\}$ is a δ-cover of F, then, since $|f(F\cap U_i)|\leqslant c|U_i|^\alpha$, it follows that $\{f(F\cap U_i)\}$ is an ε-cover of $f(F)$, where $\varepsilon = c\delta^\alpha$. Thus $\sum_i |f(F\cap U_i)|^{s/\alpha}\leqslant c^{s/\alpha}\sum_i |U_i|^s$, so that $\mathscr{H}^{s/\alpha}_\varepsilon(f(F))\leqslant c^{s/\alpha}\mathscr{H}^s_\delta(F)$. As $\delta\to 0$, so $\varepsilon\to 0$, giving (2.7). □

Condition (2.6) is known as a *Hölder condition of exponent* α; such a condition implies that f is continuous. Particularly important is the case $\alpha = 1$, i.e.

$$|f(x)-f(y)|\leqslant c|x-y| \qquad (x, y\in F) \tag{2.8}$$

when f is called a *Lipschitz mapping*, and

$$\mathscr{H}^s(f(E))\leqslant c^s\mathscr{H}^s(F). \tag{2.9}$$

Any differentiable function with bounded derivative is necessarily Lipschitz as a consequence of the mean-value theorem. If f is an isometry, i.e. $|f(x)-f(y)| = |x-y|$, then $\mathscr{H}^s(f(F)) = \mathscr{H}^s(F)$. In particular, Hausdorff measures are translation invariant (i.e. $\mathscr{H}^s(F+z) = \mathscr{H}^s(F)$, where $F+z = \{x+z: x\in F\}$), and rotation invariant, as would certainly be expected.

2.2 Hausdorff dimension

Returning to equation (2.1) it is clear that for any given set F and $\delta < 1$, $\mathscr{H}^s_\delta(F)$ is non-increasing with s, so by (2.2) $\mathscr{H}^s(F)$ is also non-increasing. In fact, rather more is true: if $t > s$ and $\{U_i\}$ is a δ-cover of F we have

$$\sum_i |U_i|^t\leqslant \delta^{t-s}\sum_i |U_i|^s \tag{2.10}$$

so, taking infima, $\mathscr{H}^t_\delta(F)\leqslant\delta^{t-s}\mathscr{H}^s_\delta(F)$. Letting $\delta\to 0$ we see that if $\mathscr{H}^s(F)<\infty$ then $\mathscr{H}^t(F) = 0$ for $t > s$. Thus a graph of $\mathscr{H}^s(F)$ against s (figure 2.3) shows that there is a critical value of s at which $\mathscr{H}^s(F)$ 'jumps' from ∞ to 0. This critical value is called the *Hausdorff dimension* of F, and written $\dim_H F$. (Note that

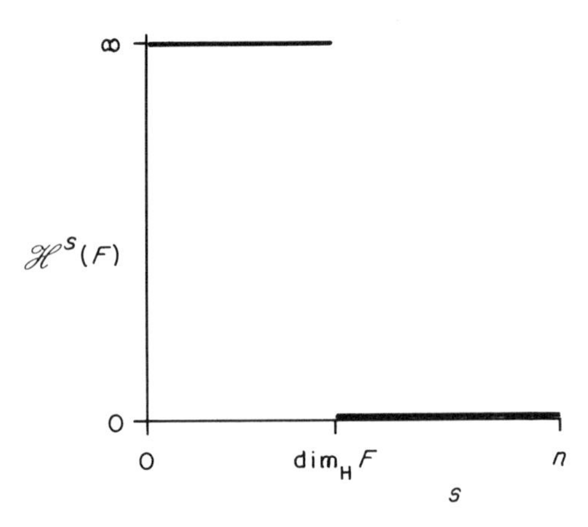

Figure 2.3 Graph of $\mathscr{H}^s(F)$ against s for a set F. The Hausdorff dimension is the value of s at which the 'jump' from ∞ to 0 occurs

some authors refer to Hausdorff dimension as *Hausdorff–Besicovitch dimension.*) Formally

$$\dim_{\mathrm{H}} F = \inf\{s : \mathscr{H}^s(F) = 0\} = \sup\{s : \mathscr{H}^s(F) = \infty\} \tag{2.11}$$

so that

$$\mathscr{H}^s(F) = \begin{cases} \infty & \text{if } s < \dim_{\mathrm{H}} F \\ 0 & \text{if } s > \dim_{\mathrm{H}} F. \end{cases} \tag{2.12}$$

If $s = \dim_{\mathrm{H}} F$, then $\mathscr{H}^s(F)$ may be zero or infinite, or may satisfy

$$0 < \mathscr{H}^s(F) < \infty.$$

A Borel set satisfying this last condition is called an *s-set*. Mathematically, s-sets are by far the most convenient sets to study, and fortunately they occur surprisingly often.

For a very simple example, let F be a flat disc of unit radius in $\mathbb{R}^3$. From familiar proprties of length, area and volume, $\mathscr{H}^1(F) = \text{length}(F) = \infty$, $0 < \mathscr{H}^2(F) = \frac{1}{4}\pi \times \text{area}(F) < \infty$ and $\mathscr{H}^3(F) = \frac{4}{3}\pi \times \text{vol}(F) = 0$. Thus $\dim_{\mathrm{H}} F = 2$, with $\mathscr{H}^s(F) = \infty$ if $s < 2$ and $\mathscr{H}^s(F) = 0$ if $s > 2$.

Hausdorff dimension satisfies the following properties (which might well be expected to hold for any reasonable definition of dimension).

Open sets. If $F \subset \mathbb{R}^n$ is open, then $\dim_{\mathrm{H}} F = n$, since F contains a ball of positive n-dimensional volume.

Smooth sets. If F is a smooth (i.e. continuously differentiable) m-dimensional submanifold (i.e. m-dimensional surface) of $\mathbb{R}^n$ then $\dim_{\mathrm{H}} F = m$. In particular smooth curves have dimension 1 and smooth surfaces have dimension 2. Essentially, this may be deduced from the relationship between Hausdorff and Lebesgue measures.

Monotonicity. If $E \subset F$ then $\dim_{\mathrm{H}} E \leqslant \dim_{\mathrm{H}} F$. This is immediate from the measure property that $\mathscr{H}^s(E) \leqslant \mathscr{H}^s(F)$ for each s.

Countable stability. If $F_1, F_2, \ldots$ is a (countable) sequence of sets then $\dim_{\mathrm{H}} \bigcup_{i=1}^{\infty} F_i = \sup_{1 \leqslant i < \infty} \{\dim_{\mathrm{H}} F_i\}$. Certainly, $\dim_{\mathrm{H}} \bigcup_{i=1}^{\infty} F_i \geqslant \dim_{\mathrm{H}} F_j$ for each j from the monotonicity property. On the other hand, if $s > \dim_{\mathrm{H}} F_i$ for all i, then $\mathscr{H}^s(F_i) = 0$, so that $\mathscr{H}^s(\bigcup_{i=1}^{\infty} F_i) = 0$, giving the opposite inequality.

Countable sets. If F is countable then $\dim_{\mathrm{H}} F = 0$. For if F_i is a single point, $\mathscr{H}^0(F_i) = 1$ and $\dim_{\mathrm{H}} F_i = 0$, so by countable stability $\dim_{\mathrm{H}} \bigcup_{i=1}^{\infty} F_i = 0$.

The transformation properties of Hausdorff dimension follow immediately from the corresponding ones for Hausdorff measures given in Proposition 2.2.

Proposition 2.3

Let $F \subset \mathbb{R}^n$ and suppose that $f : F \to \mathbb{R}^m$ satisfies a Hölder condition

$$|f(x) - f(y)| \leqslant c|x - y|^{\alpha} \qquad (x, y \in F).$$

Then $\dim_{\mathrm{H}} f(F) \leqslant (1/\alpha) \dim_{\mathrm{H}} F$.

Proof. If $s > \dim_{\mathrm{H}} F$ then by Proposition 2.2 $\mathscr{H}^{s/\alpha}(f(F)) \leqslant c^{s/\alpha} \mathscr{H}^s(F) = 0$, implying that $\dim_{\mathrm{H}} f(F) \leqslant s/\alpha$ for all $s > \dim_{\mathrm{H}} F$. $\square$

Corollary 2.4

(*a*) *If* $f:F\to\mathbb{R}^m$ *is a Lipschitz transformation (see (2.8)) then* $\dim_H f(F)\leqslant\dim_H F$.
(*b*) *If* $f:F\to\mathbb{R}^m$ *is a bi-Lipschitz transformation, i.e.*

$$c_1|x-y|\leqslant|f(x)-f(y)|\leqslant c_2|x-y| \qquad (x,y\in F) \tag{2.13}$$

where $0<c_1\leqslant c_2<\infty$, *then* $\dim_H f(F)=\dim_H F$.

Proof. Part (*a*) follows from Proposition 2.3 taking $\alpha=1$. Applying this to $f^{-1}:f(F)\to F$ gives the other inequality required for (*b*). □

This corollary reveals a fundamental property of Hausdorff dimension: *Hausdorff dimension is invariant under bi-Lipschitz transformations.* Thus if two sets have different dimensions there cannot be a bi-Lipschitz mapping from one onto the other. This is reminiscent of the situation in topology where various 'invariants' (such as homotopy or homology groups) are set up to distinguish between sets that are not homeomorphic: if the topological invariants of two sets differ then there cannot be a homeomorphism (continuous one-to-one mapping with continuous inverse) between the two sets.

In topology two sets are regarded as 'the same' if there is a homeomorphism between them. One approach to fractal geometry is to regard two sets as 'the same' if there is a bi-Lipschitz mapping between them. Just as topological invariants are used to distinguish between non-homeomorphic sets, we may seek parameters, including dimension, to distinguish between sets that are not bi-Lipschitz equivalent. Since bi-Lipschitz transformations (2.13) are necessarily continuous, topological parameters provide a start in this direction, and Hausdorff dimension (and other definitions of dimension) provide further distinguishing characteristics between fractals.

In general, the dimension of a set alone tells us little about its topological properties. However, any set of dimension less than 1 is necessarily so sparse as to be totally disconnected; that is, no two of its points lie in the same connected component.

Proposition 2.5

A set $F\subset\mathbb{R}^n$ *with* $\dim_H F<1$ *is totally disconnected.*

Proof. Let x and y be distinct points of F. Define a mapping $f:\mathbb{R}^n\to[0,\infty)$ by $f(z)=|z-x|$. Since f does not increase distances, i.e. $|f(z)-f(w)|\leqslant|z-w|$, we have from Corollary 2.4(*a*) that $\dim_H f(F)\leqslant\dim_H F<1$. Thus $f(F)$ is a subset of $\mathbb{R}$ of $\mathcal{H}^1$-measure or length zero, and so has a dense complement. Choosing r with $r\notin f(F)$ and $0<r<f(y)$ it follows that

$$F=\{z\in F:|z-x|<r\}\cup\{z\in F:|z-x|>r\}.$$

Thus F is contained in two disjoint open sets with x in one set and y in the other, so that x and y lie in different connected components of F. □

2.3 Calculation of Hausdorff dimension—simple examples

This section indicates how to calculate the Hausdorff dimension of some simple fractals such as some of those mentioned in the Introduction. Other methods will be encountered throughout the book.

Example 2.6

Let F be the Cantor dust constructed from the unit square as in figure 0.4. (*At each stage of the construction the squares are divided into* 16 *squares with a quarter of the side length, of which the same pattern of four squares is retained.*) *Then* $1 \leqslant \mathscr{H}^1(F) \leqslant \sqrt{2}$, *so* $\dim_H F = 1$.

Calculation. Taking the obvious covering of F by the 4^k squares of side 4^{-k} (i.e. of diameter $\delta = 4^{-k}\sqrt{2}$) in E_k, the kth stage of construction, we get an estimate $\mathscr{H}^1_\delta(F) \leqslant 4^k 4^{-k}\sqrt{2}$ for the infimum in (2.1). As $k \to \infty$ so $\delta \to 0$ giving $\mathscr{H}^1(F) \leqslant \sqrt{2}$.

For the lower estimate, let proj denote orthogonal projection onto the x-axis. Orthogonal projection does not increase distances, i.e. $|\text{proj}\, x - \text{proj}\, y| \leqslant |x - y|$ if $x, y \in \mathbb{R}^2$, so proj is a Lipschitz mapping. By virtue of the construction of F, the projection or 'shadow' of F on the x-axis, proj F, is the unit interval $[0, 1]$. Using (2.9)

$$1 = \text{length}[0, 1] = \mathscr{H}^1([0, 1]) = \mathscr{H}^1(\text{proj}\, F) \leqslant \mathscr{H}^1(F). \qquad \square$$

Note that the same argument and result hold for a set obtained by repeated division of squares into m^2 squares of side length $1/m$ of which one square in each column is retained.

This trick of using orthogonal projection to get a lower estimate of Hausdorff measure only works in special circumstances and is not the basis of a more general method. Usually we need to work rather harder!

Example 2.7

Let F be the middle third Cantor set (see figure 0.1). *If* $s = \log 2/\log 3 = 0.6309\ldots$ *then* $\dim_H F = s$ and $\frac{1}{2} \leqslant \mathscr{H}^s(F) \leqslant 1$.

Heuristic calculation. The Cantor set F splits into a left part $F_L = F \cup [0, \frac{1}{3}]$ and a right part $F_R = F \cap [\frac{2}{3}, 1]$. Clearly both parts are geometrically similar to F but scaled by a ratio $\frac{1}{3}$, and $F = F_L \cup F_R$ with this union disjoint. Thus for any s

$$\mathscr{H}^s(F) = \mathscr{H}^s(F_L) + \mathscr{H}^s(F_R) = \tfrac{1}{3}^s \mathscr{H}^s(F) + \tfrac{1}{3}^s \mathscr{H}^s(F)$$

by the Scaling property 2.1 of Hausdorff measures. *Assuming that at the critical value* $s = \dim_H F$ *we have* $0 < \mathscr{H}^s(F) < \infty$ (a big assumption, but one that can be justified) we may divide by $\mathscr{H}^s(F)$ to get $1 = 2(\frac{1}{3})^s$ or $s = \log 2/\log 3$.

Rigorous calculation. We call the intervals of length 3^{-k} $(k = 0, 1, 2, \ldots)$ that make up the sets E_k in the construction of F *basic intervals.* The covering $\{U_i\}$ of F

consisting of the 2^k intervals of E_k of length 3^{-k} gives that $\mathscr{H}^s_{3^{-k}}(F) \leqslant \Sigma |U_i|^s = 2^k 3^{-ks} = 1$ if $s = \log 2/\log 3$. Letting $k \to \infty$ gives $\mathscr{H}^s(F) \leqslant 1$.

To prove that $\mathscr{H}^s(F) \geqslant \frac{1}{2}$ we show that

$$\sum |U_i|^s \geqslant \tfrac{1}{2} = 3^{-s} \tag{2.14}$$

for any cover $\{U_i\}$ of F. Clearly, it is enough to assume that the $\{U_i\}$ are intervals, and by expanding them slightly and using the compactness of F, we need only verify (2.14) if $\{U_i\}$ is a finite collection of closed subintervals of $[0, 1]$. For each U_i, let k be the integer such that

$$3^{-(k+1)} \leqslant |U_i| < 3^{-k}. \tag{2.15}$$

Then U_i can intersect at most one basic interval of E_k since the separation of these basic intervals is at least 3^{-k}. If $j \geqslant k$ then, by construction, U_i intersects at most $2^{j-k} = 2^j 3^{-sk} \leqslant 2^j 3^s |U_i|^s$ basic intervals of E_j, using (2.15). If we choose j large enough so that $3^{-(j+1)} \leqslant |U_i|$ for all U_i, then, since the $\{U_i\}$ intersect all 2^j basic intervals of length 3^{-j}, counting intervals gives $2^j \leqslant \sum_i 2^j 3^s |U_i|^s$, which reduces to (2.14). □

With extra effort, the calculation can be adapted to show that $\mathscr{H}^s(F) = 1$.

It is already becoming apparent that calculation of Hausdorff measures and dimensions can be a little involved, even for simple sets. Usually it is the lower estimate that is awkward to obtain.

The 'heuristic' method of calculation used in Example 2.7 gives the right answer for the dimension of many self-similar sets. For example, the von Koch curve is made up of four copies of itself scaled by a factor $\frac{1}{3}$, and hence has dimension $\log 4/\log 3$. More generally, if $F = \bigcup_{i=1}^m F_i$, where each F_i is geometrically similar to F but scaled by a factor c_i then, provided that the F_i do not overlap 'too much', the heuristic argument gives $\dim_H F$ as the number s satisfying $\sum_{i=1}^m c_i^s = 1$. The validity of this formula is discussed fully in Chapter 9.

*2.4 Equivalent definitions of Hausdorff dimension

It is worth pointing out that there are other classes of covering set that define measures leading to Hausdorff dimension. For example, we could use coverings by spherical balls: letting

$$\mathscr{B}^s_\delta(F) = \inf\{\Sigma |B_i|^s : \{B_i\} \text{ is a } \delta\text{-cover of } F \text{ by balls}\} \tag{2.16}$$

we obtain a measure $\mathscr{B}^s(F) = \lim_{\delta \to 0} \mathscr{B}^s_\delta(F)$ and a 'dimension' at which $\mathscr{B}^s(F)$ jumps from ∞ to 0. Clearly $\mathscr{H}^s_\delta(F) \leqslant \mathscr{B}^s_\delta(F)$ since any δ-cover of F by balls is a permissible covering in the definition of $\mathscr{H}^s_\delta$. Also, if $\{U_i\}$ is a δ-cover of F, then so is $\{B_i\}$, where, for each i, B_i is chosen to be some ball containing U_i and of radius $|U_i| \leqslant \delta$. Thus $\Sigma |B_i|^s \leqslant \Sigma (2|U_i|)^s = 2^s \Sigma |U_i|^s$, and taking infima gives $\mathscr{B}^s_{2\delta}(F) \leqslant 2^s \mathscr{H}^s_\delta(F)$. Letting $\delta \to 0$ it follows that $\mathscr{H}^s(F) \leqslant \mathscr{B}^s(F) \leqslant$

$2^s\mathscr{H}^s(F)$. In particular, this implies that the values of s at which $\mathscr{H}^s$ and $\mathscr{B}^s$ jump from ∞ to 0 are equal, so that the dimensions defined by the two measures are equal.

It is easy to check that we get the same values for Hausdorff measure and dimension if in (2.1) we use δ-covers of just open sets or just closed sets. Moreover, if F is compact, then, by expanding the covering sets slightly to open sets, and taking a finite subcover, we get the same value of $\mathscr{H}^s(F)$ if we merely consider δ-covers by finite collections of sets.

Net measures are another useful variant. For the sake of simplicity let F be a subset of the interval $[0, 1)$. A *binary interval* is an interval of the form $[r2^{-k}, (r+1)2^{-k})$ where $k = 0, 1, 2, \ldots$ and $r = 0, 1, \ldots, 2^k - 1$. We define

$$\mathscr{M}^s_\delta(F) = \inf\{\Sigma|U_i|^s : \{U_i\} \text{ is a } \delta\text{-cover of } F \text{ by binary intervals}\} \qquad (2.17)$$

leading to the *net measures*

$$\mathscr{M}^s(F) = \lim_{\delta\to 0} \mathscr{M}^s_\delta(F). \qquad (2.18)$$

Since any interval $U \subset [0, 1)$ is contained in two consecutive binary intervals each of length at most $2|U|$ we see, in just the same way as for the measure $\mathscr{B}^s$, that

$$\mathscr{H}^s(F) \leqslant \mathscr{M}^s(F) \leqslant 2^{s+1}\mathscr{H}^s(F). \qquad (2.19)$$

It follows that the value of s at which $\mathscr{M}^s(F)$ jumps from ∞ to 0 equals the Hausdorff dimension of F, i.e. both definitions of measure give the same dimension.

For certain purposes net measures are much more convenient than Hausdorff measures. This is because two binary intervals are either disjoint or one of them is contained in the other, allowing any cover of binary intervals to be reduced to a cover of *disjoint* binary intervals.

*2.5 Finer definitions of dimension

It is sometimes desirable to have a sharper indication of dimension than just a number. To achieve this let $h: \mathbb{R}^+ \to \mathbb{R}^+$ be a function that is increasing and continuous, which we call a *dimension function*. Analogously to (2.1) we define

$$\mathscr{H}^h_\delta(F) = \inf\{\Sigma h(|U_i|) : \{U_i\} \text{ is a } \delta\text{-cover of } F\} \qquad (2.20)$$

for F a subset of $\mathbb{R}^n$. This leads to a measure, taking $\mathscr{H}^h(F) = \lim_{\delta\to 0}\mathscr{H}^h_\delta(F)$. (If $h(t) = t^s$ this is the usual definition of s-dimensional Hausdorff measure.) If h and g are dimension functions such that $h(t)/g(t) \to 0$ as $t \to 0$ then, by an argument similar to (2.10), we get that $\mathscr{H}^h(F) = 0$ whenever $\mathscr{H}^g(F) < \infty$. Thus partitioning the dimension functions into those for which $\mathscr{H}^h$ is finite and those for which it is infinite gives a more precise indication of the 'dimension' of F than just the number $\dim_H F$.

An important example of this is Brownian motion in $\mathbb{R}^3$ (see Chapter 16 for further details). It may be shown that (with probability 1) a Brownian path has

Hausdorff dimension 2 but with $\mathscr{H}^2$-measure equal to 0. More refined calculations show that such a path has positive and finite $\mathscr{H}^h$-measure, where $h(t) = t^2 \log\log(1/t)$. Although Brownian paths have dimension 2, the dimension is, in a sense, logarithmically smaller than 2.

2.6 Notes and references

The idea of defining measures using covers of sets was introduced by Carathéodory (1914). Hausdorff (1919) used this method to define the measures that now bear his name, and showed that the middle third Cantor set has positive and finite measure of dimension $\log 2/\log 3$. Properties of Hausdorff measures have been developed during the course of this century, largely by Besicovitch and his students.

Technical aspects of Hausdorff measures and dimensions, are discussed in rather more detail in Falconer (1985a), and in greater generality in the books of Rogers (1970) and Federer (1969).

Exercises

2.1 Verify that the value of $\mathscr{H}^s(F)$ is unaltered if, in (2.1), we only consider δ-covers by sets $\{U_i\}$ that are all closed.

2.2 Show that $\mathscr{H}^0(F)$ equals the number of points in the set F.

2.3 Verify from the definition that $\mathscr{H}^s(\bigcup_{i=1}^{\infty} F_i) \leqslant \sum_{i=1}^{\infty} \mathscr{H}^s(F_i)$.

2.4 Let $f:\mathbb{R}\to\mathbb{R}$ be a differentiable function with continuous derivatives. Show that $\dim_{\mathrm{H}} f(F) \leqslant \dim_{\mathrm{H}} F$ for any set F. (Consider the case of F bounded first.)

2.5 Let $f:\mathbb{R}\to\mathbb{R}$ be the function $f(x) = x^2$, and let F be any subset of $\mathbb{R}$. Show that $\dim_{\mathrm{H}} f(F) = \dim_{\mathrm{H}} F$.

2.6 Let F be the set consisting of the numbers between 0 and 1 whose decimal expansions do not contain the digit 5. Use a 'heuristic' argument to show that $\dim_{\mathrm{H}} F = \log 9/\log 10$. Can you prove this by a rigorous argument? Generalize this result.

2.7 Let F consist of the points $(x, y) \in \mathbb{R}^2$ such that the decimal expansions of neither x or y contain the digit 5. Use a 'heuristic' argument to show that $\dim_{\mathrm{H}} F = 2\log 9/\log 10$.

2.8 Use a 'heuristic' argument to show that the Hausdorff dimension of the set depicted in figure 0.5 is given by the solution of the equation $4(\frac{1}{4})^s + (\frac{1}{2})^s = 1$.

2.9 Let F be the set of real numbers with base-3 expansion $b_m b_{m-1} \cdots b_1 \cdot a_1 a_2 \cdots$ with none of the digits b_i or a_i equal to 1. (Thus F is constructed by a Cantor-like process extending outwards as well as inwards.) What is the Hausdorff dimension of F?

2.10 What is the Hausdorff dimension of the set of numbers x with base-3 expansion $0 \cdot a_1 a_2 \cdots$ for which there is a positive integer k (which may depend on x) such that $a_i \neq 1$ for all $i \geqslant k$?

2.11 Show that there is a totally disconnected subset of the plane of Hausdorff dimension s for every $0 \leqslant s \leqslant 2$.

2.12 Let S be the unit circle in the plane, with points on S parametrized by the angle θ subtended at the centre with a fixed axis, so that θ_1 and θ_2 represent the same point if and only if θ_1 and θ_2 differ by a multiple of 2π, in the usual way. Let $F = \{\theta \in S : 0 \leqslant 3^k \theta \leqslant \pi \pmod{2\pi} \text{ for all } k = 1, 2, \ldots\}$. Show that $\dim_{\mathrm{H}} = \log 2/\log 3$.

2.13 Show that if h and g are dimension functions such that $h(t)/g(t) \to 0$ as $t \to 0$ then $\mathscr{H}^h(F) = 0$ whenever $\mathscr{H}^g(F) < \infty$.

Chapter 3 Alternative definitions of dimension

Hausdorff dimension, discussed in the last chapter, is the principal definition of dimension that we shall work with. However, other definitions are in widespread use, and it is appropriate to examine some of these and their inter-relationship. Not all definitions are generally applicable—some only describe particular classes of set, such as curves.

Fundamental to most definitions of dimension is the idea of 'measurement at scale δ'. For each δ, we measure a set in a way that ignores irregularities of size less than δ, and we see how these measurements behave as $\delta \to 0$. For example, if F is a plane curve, then our measurement, $M_\delta(F)$, might be the number of steps required by a pair of dividers set at length δ to traverse F. A dimension of F is then determined by the power law (if any) obeyed by $M_\delta(F)$ as $\delta \to 0$. If

$$M_\delta(F) \sim c\delta^{-s} \tag{3.1}$$

for constants c and s, we might say that F has 'dimension' s, with c regarded as the 's-dimensional length' of F. Taking logarithms

$$\log M_\delta(F) \simeq \log c - s \log \delta \tag{3.2}$$

in the sense that the difference of the two sides tends to 0 with δ, and

$$s = \lim_{\delta \to 0} \frac{\log M_\delta(F)}{-\log \delta}. \tag{3.3}$$

These formulae are appealing for computational or experimental purposes, since s can be estimated as the gradient of a log–log graph plotted over a suitable range of δ; see figure 3.1. (Of course, for real phenomena, we can only work with a finite range of δ; theory and experiment diverge before an atomic scale is reached; see Chapter 18.)

There may be no exact power law for $M_\delta(F)$, and the closest we can get to (3.3) are the lower and upper limits.

For the value of s given by (3.1) to behave like a dimension, the method of measurement needs to scale with the set, so that doubling the size of F and at the same time doubling the scale at which measurement takes place does not affect the answer; that is, we require $M_\delta(\delta F) = M_1(F)$ for all δ. If we modify our example and redefine $M_\delta(F)$ to be the sum of the divider step lengths then $M_\delta(F)$

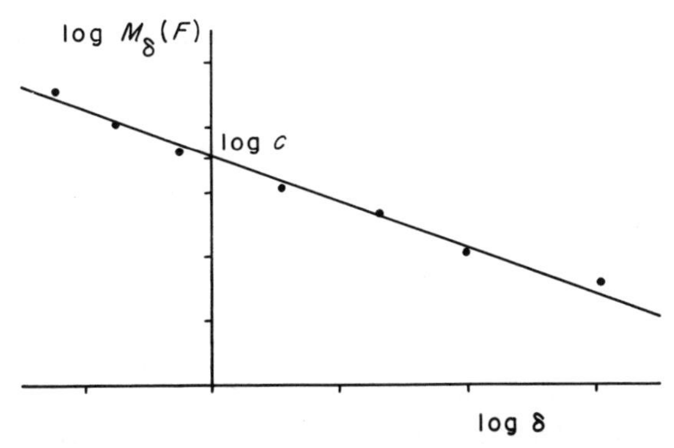

Figure 3.1 Empirical estimation of a dimension of a set F, on the power-law assumption $M_\delta(F) \sim c\delta^{-s}$

is homogeneous of degree 1, i.e. $M_\delta(\delta F) = \delta^1 M_1(F)$ for $\delta > 0$, and this must be taken into account when definining the dimension. In general, if $M_\delta(F)$ is homogeneous of degree d, that is $M_\delta(\delta F) = \delta^d M_1(F)$, then a power law of the form $M_\delta(F) \sim c\delta^{d-s}$ corresponds to a dimension s.

There are no hard and fast rules for deciding whether a quantity may reasonably be regarded as a dimension. There are many definitions that do not fit exactly into the above, rather simplified, scenario. The factors that determine the acceptability of a definition of a dimension are recognized largely by experience and intuition. In general one looks for some sort of scaling behaviour, a naturalness of the definition in the particular context and properties typical of dimensions such as those discussed below.

A word of warning: as we shall see, apparently similar definitions of dimension can have widely differing properties. It should not be assumed that different definitions give the same value of dimension, even for 'nice' sets. Such assumptions have led to major misconceptions and confusion in the past. It is necessary to derive the properties of any 'dimension' from its definition. The properties of Hausdorff dimension (on which we shall largely concentrate in the later chapters of this book) do not necessarily all hold for other definitions.

What are the desirable properties of a 'dimension'? Those derived in the last chapter for Hausdorff dimension are fairly typical.

Monotonicity. If $E \subset F$ then $\dim_{\mathrm{H}} E \leqslant \dim_{\mathrm{H}} F$.

Stability. $\dim_{\mathrm{H}}(E \cup F) = \max(\dim_{\mathrm{H}} E, \dim_{\mathrm{H}} F)$.

Countable stability. $\dim_{\mathrm{H}}(\bigcup_{i=1}^{\infty} F_i) = \sup_{1 \leqslant i < \infty} \dim_{\mathrm{H}} F_i$.

Geometric invariance. $\dim_{\mathrm{H}} f(F) = \dim_{\mathrm{H}} F$ if f is a transformation of $\mathbb{R}^n$ such as a translation, rotation, similarity or affinity.

Lipschitz invariance. $\dim_{\mathrm{H}} f(F) = \dim_{\mathrm{H}} F$ if f is a bi-Lipschitz transformation.

Countable sets. $\dim_{\mathrm{H}} F = 0$ if F is finite or countable.

Open sets. If F is an open subset of $\mathbb{R}^n$ then $\dim_{\mathrm{H}} F = n$.

Smooth manifolds. $\dim_{\mathrm{H}} F = m$ if F is a smooth m-dimensional manifold.

All definitions of dimension are monotonic, most are stable, but, as we shall see, some common definitions fail to exhibit countable stability and may have countable sets of positive dimension. All the usual dimensions are Lipschitz invariant, and, therefore, geometrically invariant. The 'open sets' and 'smooth manifolds' properties ensure that the dimension is an extension of the classical definition. Note that different definitions of dimension can provide different information about which sets are Lipschitz equivalent.

3.1 Box-counting dimensions

Box-counting or box dimension is one of the most widely used dimensions. Its popularity is largely due to its relative ease of mathematical calculation and empirical estimation. The definition goes back at least to the 1930s and it has been variously termed Kolmogorov entropy, entropy dimension, capacity dimension (a term best avoided in view of potential theoretic associations), metric dimension, logarithmic density and information dimension. We shall always refer to box or box counting dimension to avoid confusion.

Let F be any non-empty bounded subset of $\mathbb{R}^n$ and let $N_\delta(F)$ be the smallest number of sets of diameter at most δ which can cover F. The *lower* and *upper box-counting dimensions* of F respectively are defined as

$$\underline{\dim}_{\mathrm{B}} F = \varliminf_{\delta \to 0} \frac{\log N_\delta(F)}{-\log \delta} \tag{3.4}$$

$$\overline{\dim}_{\mathrm{B}} F = \varlimsup_{\delta \to 0} \frac{\log N_\delta(F)}{-\log \delta}. \tag{3.5}$$

If these are equal we refer to the common value as the *box-counting dimension* or *box dimension* of F

$$\dim_{\mathrm{B}} F = \lim_{\delta \to 0} \frac{\log N_\delta(F)}{-\log \delta}. \tag{3.6}$$

There are several equivalent definitions of box dimension that are sometimes more convenient to use. Consider the collection of cubes in the δ-coordinate mesh of $\mathbb{R}^n$, i.e. cubes of the form

$$[m_1\delta, (m_1+1)\delta] \times \cdots \times [m_n\delta, (m_n+1)\delta]$$

where $m_1, \ldots, m_n$ are integers. (Recall that a 'cube' is an interval in $\mathbb{R}^1$ and a square in $\mathbb{R}^2$.) Let $N'_\delta(F)$ be the number of δ-mesh cubes that intersect F. They obviously provide a collection of $N'_\delta(F)$ sets of diameter $\delta\sqrt{n}$ that cover F, so

$$N_{\delta\sqrt{n}}(F) \leqslant N'_\delta(F).$$

If $\delta\sqrt{n} < 1$ then

$$\frac{\log N_{\delta\sqrt{n}}(F)}{-\log(\delta\sqrt{n})} \leqslant \frac{\log N'_\delta(F)}{-\log\sqrt{n} - \log\delta}$$

so taking limits as $\delta \to 0$

$$\underline{\dim}_B F = \varliminf_{\delta\to 0} \frac{\log N'_\delta(F)}{-\log\delta} \tag{3.7}$$

and

$$\overline{\dim}_B F \leqslant \varlimsup_{\delta\to 0} \frac{\log N'_\delta(F)}{-\log\delta}. \tag{3.8}$$

On the other hand, any set of diameter at most δ is contained in 3^n mesh cubes of side δ (by choosing a cube containing some point of the set together with its neighbouring cubes). Thus

$$N'_\delta(F) \leqslant 3^n N_\delta(F)$$

and taking logarithms leads to the opposite inequalities to (3.7) and (3.8). Hence to find the box dimensions (3.4)–(3.6), we can equally well take $N_\delta(F)$ to be the number of mesh cubes of side δ that intersect F.

This version of the definitions is widely used empirically. To find the box dimension of a plane set F we may draw a mesh of squares or boxes of side δ and count the number $N_\delta(F)$ that overlap the set for various small δ (hence the name 'box counting'). The dimension is the logarithmic rate at which $N_\delta(F)$ increases as $\delta \to 0$, and may be estimated by the gradient of the graph of $\log N_\delta(F)$ against $-\log\delta$.

This definition gives an interpretation of the meaning of box dimension. The number of mesh cubes of side δ that intersect a set is an indication of how spread out or irregular the set is when examined at scale δ. The dimension reflects how rapidly the irregularities develop as $\delta \to 0$.

Another frequently used definition of box dimension is obtained by taking $N_\delta(F)$ in (3.4)–(3.6) to be the smallest number of *arbitrary* cubes of side δ required to cover F. The equivalence of this definition follows as in the mesh cube case, noting that any cube of side δ has diameter $\delta\sqrt{n}$, and that any set of diameter of at most δ is contained in a cube of side δ.

Similarly, we get exactly the same values if in (3.4)–(3.6) we take $N_\delta(F)$ as the smallest number of closed balls of radius δ that cover F.

A less obviously equivalent formulation of box dimension involves the *largest* number of *disjoint* balls of radius δ with centres in F. Let this number be $N'_\delta(F)$, and let $B_1, \dots, B_{N'_\delta(F)}$ be disjoint balls centred in F and of radius δ. If x belongs to F then x must be within distance δ of one of the B_i, otherwise the ball of centre x and radius δ can be added to form a larger collection of disjoint balls. Thus the $N'_\delta(F)$ balls concentric with the B_i but of radius 2δ (diameter 4δ) cover F, giving

$$N_{4\delta}(F) \leqslant N'_\delta(F). \tag{3.9}$$

On the other hand, suppose that $B_1, \dots, B_{N'_\delta(F)}$ are disjoint balls of radii δ with centres in F. Let $U_1, \dots, U_k$ be any collection of sets of diameter at most δ which cover F. Since the U_j must cover the centres of the B_i, each B_i must contain at least one of the U_j. As the B_i are disjoint there are at least as many U_j as B_i. Hence

$$N'_\delta(F) \leqslant N_\delta(F). \tag{3.10}$$

Taking logarithms of (3.9) and (3.10) shows that the values of (3.4)–(3.6) are unaltered if $N_\delta(F)$ is replaced by this $N'_\delta(F)$.

These various definitions are summarized below and in figure 3.2.

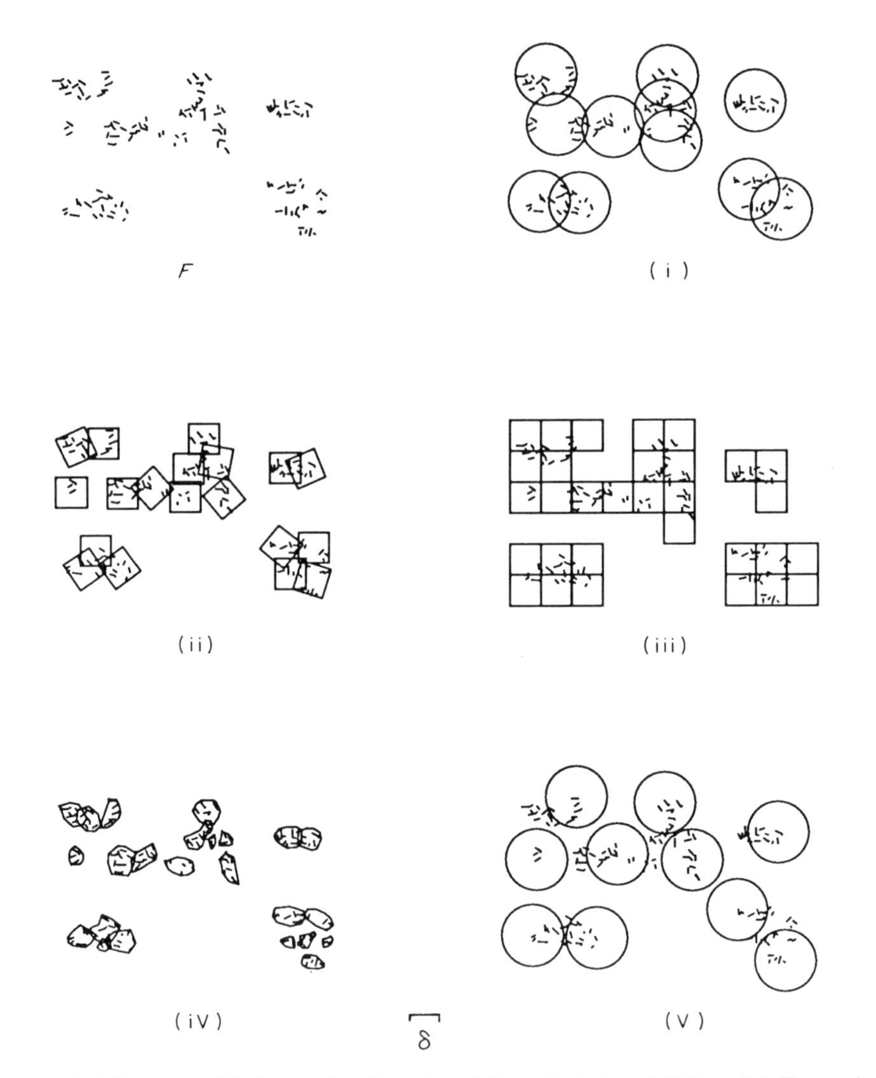

Figure 3.2 Five ways of finding the box dimension of F; see Equivalent definitions 3.1. The number $N_\delta(F)$ is taken to be: (i) the least number of closed balls of radius δ that cover F; (ii) the least number of cubes of side δ that cover F; (iii) the number of δ-mesh cubes that intersect F; (iv) the least number of sets of diameter at most δ that cover F; (v) the greatest number of disjoint balls of radius δ with centres in F

Equivalent definitions 3.1

The lower and upper box-counting dimensions of a subset F of $\mathbb{R}^n$ are given by

$$\underline{\dim}_B F = \underline{\lim}_{\delta \to 0} \frac{\log N_\delta(F)}{-\log \delta} \tag{3.11}$$

$$\overline{\dim}_B F = \overline{\lim}_{\delta \to 0} \frac{\log N_\delta(F)}{-\log \delta} \tag{3.12}$$

and the box-counting dimension of F by

$$\dim_B F = \lim_{\delta \to 0} \frac{\log N_\delta(F)}{-\log \delta} \tag{3.13}$$

(if this limit exists), where $N_\delta(F)$ is any of the following:

(i) *the smallest number of closed balls of radius δ that cover F;*
(ii) *the smallest number of cubes of side δ that cover F;*
(iii) *the number of δ-mesh cubes that intersect F;*
(iv) *the smallest number of sets of diameter at most δ that cover F;*
(v) *the largest number of disjoint balls of radius δ with centres in F.*

This list could be extended further; in practice one adopts the definition most convenient for a particular application.

It is worth noting that, in (3.11)–(3.13), it is enough to consider limits as δ tends to 0 through any decreasing sequence δ_k such that $\delta_{k+1} \geqslant c\delta_k$ for some constant $0 < c < 1$; in particular for $\delta_k = c^k$. To see this, note that if $\delta_{k+1} \leqslant \delta < \delta_k$, then

$$\frac{\log N_\delta(F)}{-\log \delta} \leqslant \frac{\log N_{\delta_{k+1}}(F)}{-\log \delta_k} \leqslant \frac{\log N_{\delta_{k+1}}(F)}{-\log \delta_{k+1} + \log(\delta_{k+1}/\delta_k)} \leqslant \frac{\log N_{\delta_{k+1}}(F)}{-\log \delta_{k+1} + \log c}$$

and so

$$\overline{\lim}_{\delta \to 0} \frac{\log N_\delta(F)}{-\log \delta} \leqslant \overline{\lim}_{k \to \infty} \frac{\log N_{\delta_k}(F)}{-\log \delta_k}. \tag{3.14}$$

The opposite inequality is trivial; the case of lower limits may be dealt with in the same way.

There is an equivalent definition of box dimension of a rather different form that is worth mentioning. Recall that the *δ-parallel body* F_δ of F is

$$F_\delta = \{x \in \mathbb{R}^n : |x - y| \leqslant \delta \text{ for some } y \in F\} \tag{3.15}$$

i.e. the set of points within distance δ of F. We consider the rate at which the n-dimensional volume of F_δ shrinks as $\delta \to 0$. In $\mathbb{R}^3$, if F is a single point then F_δ is a ball with $\mathrm{vol}(F_\delta) = \frac{4}{3}\pi\delta^3$, if F is a segment of length l then F_δ is 'sausage-like' with $\mathrm{vol}(F_\delta) \sim \pi l \delta^2$, and if F is a flat set of area a then F_δ is essentially a thickening of F with $\mathrm{vol}(F_\delta) \sim 2a\delta$. In each case, $\mathrm{vol}(F_\delta) \sim c\delta^{3-s}$ where the integer s is the dimension of F, so that exponent of δ is indicative of the dimension. The coefficient c of δ^{3-s}, known as the *Minkowski content* of F, is a measure of the length, area or volume of the set as appropriate.

This idea extends to fractional dimensions. If F is a subset of $\mathbb{R}^n$ and, for some s, $\mathrm{vol}^n(F_\delta)/\delta^{n-s}$ tends to a positive finite limit as $\delta \to 0$, then it makes sense to regard F as s-dimensional. The limiting value is called the *s-dimensional content* of F—a concept of slightly restricted use since it is not necessarily additive on disjoint subsets, i.e. is not a measure. Even if this limit does not exist, we may be able to extract the critical exponent of δ and this turns out to be related to the box dimension.

Proposition 3.2

If F is a subset of $\mathbb{R}^n$, then

$$\underline{\dim}_B F = n - \overline{\lim_{\delta \to 0}} \frac{\log \mathrm{vol}^n(F_\delta)}{\log \delta}$$

$$\overline{\dim}_B F = n - \underline{\lim_{\delta \to 0}} \frac{\log \mathrm{vol}^n(F_\delta)}{\log \delta}$$

where F_δ is the δ-parallel body to F.

Proof. If F can be covered by $N_\delta(F)$ balls of radius δ then F_δ can be covered by the concentric balls of radius 2δ. Hence

$$\mathrm{vol}^n(F_\delta) \leqslant N_\delta(F) c_n (2\delta)^n$$

where c_n is the volume of the unit ball in $\mathbb{R}^n$. Taking logarithms,

$$\frac{\log \mathrm{vol}^n(F_\delta)}{-\log \delta} \leqslant \frac{\log 2^n c_n + n \log \delta + \log N_\delta(F)}{-\log \delta}$$

gives

$$\underline{\lim_{\delta \to 0}} \frac{\log \mathrm{vol}^n(F_\delta)}{-\log \delta} \leqslant -n + \underline{\dim}_B F \tag{3.16}$$

with a similar inequality for the upper limits. On the other hand if there are $N_\delta(F)$ disjoint balls of radius δ with centres in F, then

$$N_\delta(F) c_n (2\delta)^n \leqslant \mathrm{vol}^n(F_\delta).$$

Taking logarithms gives the opposite inequality to (3.16), using Equivalent definition 3.1(v). □

Because of Proposition 3.2, box dimension is sometimes referred to as *Minkowski dimension.*

It is important to understand the relationship between box-counting dimension and Hausdorff dimension. If F can be covered by $N_\delta(F)$ sets of diameter δ, then, from definition (2.1),

$$\mathcal{H}^s_\delta(F) \leqslant N_\delta(F)\delta^s.$$

If $1 < \mathcal{H}^s(F) = \lim_{\delta\to 0}\mathcal{H}^s_\delta(F)$ then $\log N_\delta(F) + s\log\delta > 0$ if δ is sufficiently small. Thus $s \leqslant \underline{\lim}_{\delta\to 0}\log N_\delta(F)/-\log\delta$ so

$$\dim_{\mathrm{H}} F \leqslant \underline{\dim}_{\mathrm{B}} F \leqslant \overline{\dim}_{\mathrm{B}} F \tag{3.17}$$

for any $F \subset \mathbb{R}^n$. We do *not* in general get equality here. Although Hausdorff and box dimensions are equal for many 'reasonably regular' sets, there are plenty of examples where this inequality is strict.

Roughly speaking (3.6) says that $N_\delta(F) \simeq \delta^{-s}$ for small δ, where $s = \dim_{\mathrm{B}} F$. More precisely, it says that

$$N_\delta(F)\delta^s \to \infty \qquad \text{if } s < \dim_{\mathrm{B}} F$$

and

$$N_\delta(F)\delta^s \to 0 \qquad \text{if } s > \dim_{\mathrm{B}} F.$$

But

$$N_\delta(F)\delta^s = \inf\left\{\sum_i \delta^s : \{U_i\} \text{ is a (finite) } \delta\text{-cover of } F\right\},$$

which should be compared with

$$\mathcal{H}^s_\delta(F) = \inf\left\{\sum_i |U_i|^s : \{U_i\} \text{ is a } \delta\text{-cover of } F\right\},$$

which occurs in the definition of Hausdorff measure and dimension. In calculating Hausdorff dimension, we assign different weights $|U_i|^s$ to the covering sets U_i, whereas for the box dimensions we use the same weight δ^s for each covering set. Box dimensions may be thought of as indicating the efficiency with which a set may be covered by small sets of equal size, whereas Hausdorff dimension involves coverings by sets of small but perhaps widely varying size.

There is a temptation to introduce the quantity $v(F) = \underline{\lim}_{\delta\to 0} N_\delta(F)\delta^s$, but this does *not* give a measure on subsets of $\mathbb{R}^n$. As we shall see, one consequence of this is that box dimensions have a number of unfortunate properties, and can be awkward to handle mathematically.

Since box dimensions are determined by coverings by sets of equal size they tend to be easier to calculate than Hausdorff dimensions.

Example 3.3

Let F be the middle third Cantor set. Then $\underline{\dim}_{\mathrm{B}} F = \overline{\dim}_{\mathrm{B}} F = \log 2/\log 3$.

Calculation. The obvious covering by the 2^k intervals of E_k of length 3^{-k} gives that $N_\delta(F) \leqslant 2^k$ if $3^{-k} < \delta \leqslant 3^{-k+1}$. From (3.5)

$$\overline{\dim}_{\mathrm{B}} F = \overline{\lim_{\delta\to 0}} \frac{\log N_\delta(F)}{-\log\delta} \leqslant \overline{\lim_{k\to\infty}} \frac{\log 2^k}{\log 3^{k-1}} = \frac{\log 2}{\log 3}.$$

On the other hand, any interval of length δ with $3^{-k-1} \leqslant \delta < 3^{-k}$ intersects at most one of the basic intervals of length 3^{-k} used in the construction of F. There are 2^k such intervals so at least 2^k intervals of length δ are required to cover F. Hence $N_\delta(F) \geqslant 2^k$ leading to $\underline{\dim}_B F \geqslant \log 2/\log 3$. □

Thus, at least for the Cantor set, $\dim_H F = \dim_B F$.

3.2 Properties and problems of box-counting dimension

The following elementary properties of box dimension mirror those of Hausdorff dimension, and may be verified in much the same way.

(i) A smooth m-dimensional submanifold of $\mathbb{R}^n$ has $\dim_B F = m$.

(ii) $\underline{\dim}_B$ and $\overline{\dim}_B$ are monotonic.

(iii) $\overline{\dim}_B$ is *finitely* stable, i.e.

$$\overline{\dim}_B (E \cup F) = \max\{\overline{\dim}_B E, \overline{\dim}_B F\}$$

though $\underline{\dim}_B$ is not.

(iv) $\underline{\dim}_B$ and $\overline{\dim}_B$ are Lipschitz invariant. This is so because, if $|f(x) - f(y)| \leqslant c|x - y|$ and F can be covered by $N_\delta(F)$ sets of diameter at most δ, then the $N_\delta(F)$ images of these sets under f form a cover by sets of diameter at most $c\delta$, thus $\dim_B f(F) \leqslant \dim_B F$. Similarly, box dimensions behave just like Hausdorff dimensions under bi-Lipschitz and Hölder transformations.

We now start to encounter the disadvantages of box-counting dimension. The next proposition is at first appealing, but has undesirable consequences.

Proposition 3.4

Let $\bar{F}$ denote the closure of F (i.e. the smallest closed subset of $\mathbb{R}^n$ containing F). Then

$$\underline{\dim}_B \bar{F} = \underline{\dim}_B F$$

and

$$\overline{\dim}_B \bar{F} = \overline{\dim}_B F.$$

Proof. Let $B_1, \ldots, B_k$ be a finite collection of closed balls of radii δ. If the closed set $\bigcup_{i=1}^k B_i$ contains F, it also contains $\bar{F}$. Hence the smallest number of closed balls of radius δ that cover F is enough to cover the larger set $\bar{F}$. The result follows. □

An immediate consequence of this is that if F is a dense subset of an open region of $\mathbb{R}^n$ then $\underline{\dim}_B F = \overline{\dim}_B F = n$. For example, let F be the (countable) set of rational numbers between 0 and 1. Then $\bar{F}$ is the entire interval $[0, 1]$, so that $\underline{\dim}_B F = \overline{\dim}_B F = 1$. Thus countable sets can have non-zero box dimension. Moreover, the box-counting dimension of each rational number regarded as a one-point set is clearly zero, but the countable union of these singleton sets has dimension 1. Consequently, it is not generally true that $\dim_B \bigcup_{i=1}^{\infty} F_i = \sup_i \dim_B F_i$.

This severely limits the usefulness of box dimension—introducing a small, i.e. countable, set of points can play havoc with the dimension. We might hope to salvage something by restricting attention to closed sets, but difficulties still remain.

Example 3.5

$F = \{0, 1, \frac{1}{2}, \frac{1}{3}, \ldots\}$ *is a compact set with* $\dim_B F = \frac{1}{2}$.

Calculation. If $|U| = \delta < \frac{1}{2}$ and k is the integer satisfying $1/(k-1)k > \delta \geqslant 1/k(k+1)$ then U can cover at most one of the points $\{1, \frac{1}{2}, \ldots, 1/k\}$. Thus at least k sets of diameter δ are required to cover F, so

$$\frac{\log N_\delta(F)}{-\log \delta} \geqslant \frac{\log k}{\log k(k+1)}.$$

Letting $\delta \to 0$ gives $\underline{\dim}_B F \geqslant \frac{1}{2}$. On the other hand, if $\frac{1}{2} > \delta > 0$, take k such that $1/(k-1)k > \delta \geqslant 1/k(k+1)$. Then $(k+1)$ intervals of length δ cover $[0, 1/k]$, leaving $k-1$ points of F which can be covered by another $k-1$ intervals. Thus

$$\frac{\log N_\delta(F)}{-\log \delta} \leqslant \frac{\log(2k)}{\log k(k-1)}$$

giving

$$\overline{\dim}_B F \leqslant \tfrac{1}{2}. \qquad \square$$

No-one would regard this set, with all but one of its points isolated, as a fractal, yet it has fractional box dimension.

Nevertheless, as well as being convenient in practice, box dimensions are very useful in theory. If, as often happens, it can be shown that a set has equal box and Hausdorff dimensions, the interplay of these definitions can be used to powerful effect.

*3.3 Modified box-counting dimensions

There are ways of overcoming the difficulties of box dimension outlined in the last section. However, they will probably not appeal to the user since they re-introduce all the difficulties of calculation associated with Hausdorff dimension and more.

If F is a subset of $\mathbb{R}^n$ we can try to decompose F into a countable number of pieces $F_1, F_2, \ldots$ in such a way that the largest piece has as small a dimension as possible. This idea leads to the following *modified box-counting dimensions*:

$$\underline{\dim}_{\text{MB}} F = \inf\left\{\sup_i \underline{\dim}_{\text{B}} F_i : F \subset \bigcup_{i=1}^{\infty} F_i\right\} \tag{3.18}$$

$$\overline{\dim}_{\text{MB}} F = \inf\left\{\sup_i \overline{\dim}_{\text{B}} F_i : F \subset \bigcup_{i=1}^{\infty} F_i\right\}. \tag{3.19}$$

(In both cases the infimum is over all possible countable covers $\{F_i\}$ of F.) Clearly $\underline{\dim}_{\text{MB}} F \leqslant \underline{\dim}_{\text{B}} F$ and $\overline{\dim}_{\text{MB}} F \leqslant \overline{\dim}_{\text{B}} F$. However, we now have that $\underline{\dim}_{\text{MB}} F = \overline{\dim}_{\text{MB}} F = 0$ if F is countable—just take the F_i to be one-point sets. Moreover, for any subset F of $\mathbb{R}^n$,

$$0 \leqslant \dim_{\text{H}} F \leqslant \underline{\dim}_{\text{MB}} F \leqslant \overline{\dim}_{\text{MB}} F \leqslant \overline{\dim}_{\text{B}} F \leqslant n. \tag{3.20}$$

It is easy to see that $\underline{\dim}_{\text{MB}}$ and $\overline{\dim}_{\text{MB}}$ recover all the desirable properties of a dimension, but they can be hard to calculate. However, there is a useful test for compact sets to have equal box and modified box dimensions. It applies to sets that might be described as 'dimensionally homogeneous'.

Proposition 3.6

Let $F \subset \mathbb{R}$ be compact. Suppose that

$$\overline{\dim}_{\text{B}} (F \cap V) = \overline{\dim}_{\text{B}} F \tag{3.21}$$

for all open sets V that intersect F. Then $\overline{\dim}_{\text{B}} F = \overline{\dim}_{\text{MB}} F$. A similar result holds for lower box-counting dimensions.

Proof. Let $F \subset \bigcup_{i=1}^{\infty} F_i$ with each F_i closed. A version of Baire's category theorem (which may be found in any text on basic general topology, and which we quote without proof) states that there is an index i and an open set $V \subset \mathbb{R}^n$ such that $F \cap V \subset F_i$. For this i, $\overline{\dim}_{\text{B}} F_i = \overline{\dim}_{\text{B}} F$. Using (3.19) and Proposition 3.4

$$\begin{aligned}\overline{\dim}_{\text{MB}} F &= \inf\left\{\sup \overline{\dim}_{\text{B}} F_i : F \subset \bigcup_{i=1}^{\infty} F_i \text{ where the } F_i \text{ are closed sets}\right\} \\ &\geqslant \overline{\dim}_{\text{B}} F.\end{aligned}$$

The opposite inequality is contained in (3.20). A similar argument deals with the lower dimensions. □

For an application suppose that F is a compact set with a high degree of self-similarity, for instance the middle third Cantor set or von Koch curve. If V is any open set that intersects F, then $F \cap V$ contains a geometrically similar copy of F which must have upper box dimension equal to that of F, so that (3.21) holds.

*3.4 Packing measures and dimensions

Unlike Hausdorff dimension, neither the box dimensions or modified box dimensions are defined in terms of measures, and this can present difficulties in their theoretical development. Nevertheless, the circle of ideas in the last section may be completed in a way that is, at least mathematically, elegant. Recall that Hausdorff dimension may be defined using economical coverings by small balls (2.16) whilst $\underline{\dim}_B$ may be defined using economical coverings by small balls of equal radius (Equivalent definition 3.1(i)). On the other hand $\overline{\dim}_B$ may be thought of as a dimension that depends on packings by disjoint balls of equal radius that are as dense as possible (Equivalent definition 3.1(v)). It is therefore natural to try to look for a dimension that is defined in terms of dense packings by disjoint balls of differing small radii.

We try to follow the pattern of definition of Hausdorff measure and dimension. Let

$$\mathscr{P}^s_\delta(F) = \sup\left\{\sum_i |B_i|^s : \{B_i\} \text{ is a collection of disjoint balls of radii at most } \delta \text{ with centres in } F\right\}. \tag{3.22}$$

Since $\mathscr{P}^s_\delta(F)$ decreases with δ, the limit

$$\mathscr{P}^s_0(F) = \lim_{\delta\to 0} \mathscr{P}^s_\delta(F) \tag{3.23}$$

exists. At this point we meet the problems encountered with box-counting dimensions. By considering countable dense sets it is easy to see that $\mathscr{P}^s_0(F)$ is not a measure. Hence we modify the definition to

$$\mathscr{P}^s(F) = \inf\left\{\sum_i \mathscr{P}^s_0(F_i) : F \subset \bigcup_{i=1}^\infty F_i\right\}. \tag{3.24}$$

It may be shown that $\mathscr{P}^s(F)$ *is* a measure on $\mathbb{R}^n$, known as the *s-dimensional packing measure*. We may define the *packing dimension* in the usual way:

$$\dim_P F = \sup\{s : \mathscr{P}^s(F) = \infty\} = \inf\{s : \mathscr{P}^s(F) = 0\}. \tag{3.25}$$

The underlying measure structure immediately implies that for a countable collection of sets $\{F_i\}$

$$\dim_P\left(\bigcup_{i=1}^\infty F_i\right) = \sup_i \dim_P F_i \tag{3.26}$$

since if $s > \dim_H F_i$ for all i, then $\mathscr{P}^s(\bigcup_i F_i) \leqslant \sum_i \mathscr{P}^s(F_i) = 0$, and $\dim_P(\bigcup_i F_i) \leqslant s$.

We now investigate the relationship of packing dimension with other definitions of dimension and verify the surprising fact that packing dimension is just the same as the modified upper box dimension.

Lemma 3.7

$$\dim_P F \leqslant \overline{\dim}_B F. \tag{3.27}$$

Proof. Choose any t and s with $t < s < \dim_P F$. Then $\mathscr{P}^s(F) = \infty$, so $\mathscr{P}^s_0(F) = \infty$. Thus, given $0 < \delta \leqslant 1$, there are disjoint balls $\{B_i\}$, of radii at most δ with centres in F, such that $1 < \sum_{i=1}^{\infty} |B_i|^s$. Suppose that, for each k, n_k of these balls satisfy $2^{-k-1} < |B_i| \leqslant 2^{-k}$; then

$$1 < \sum_{k=0}^{\infty} n_k 2^{-ks}. \tag{3.28}$$

There must be some k with $n_k > 2^{kt}(1 - 2^{t-s})$, otherwise (3.28) is contradicted, by summing a geometric series. These n_k balls all contain balls of radii $2^{-k-2} \leqslant \delta$ centred in F. Hence if $N_\delta(F)$ denotes the greatest number of disjoint balls of radius δ with centres in F, then

$$N_{2^{-k-2}}(F)(2^{-k-2})^t \geqslant n_k(2^{-k-2})^t > 2^{-2t}(1 - 2^{t-s})$$

where $2^{-k-2} < \delta$. It follows that $\overline{\lim}_{\delta \to 0} N_\delta(F)\delta^t > 0$, so that $\overline{\dim}_B F \geqslant t$. This is true for any $t < \dim_P F$ so (3.27) follows.

Proposition 3.8

If $F \subset \mathbb{R}^n$ *then* $\dim_P F = \overline{\dim}_{MB} F$.

Proof. If $F \subset \bigcup_{i=1}^{\infty} F_i$ then, by (3.26) and (3.27),

$$\dim_P F \leqslant \sup_i \dim_P F_i \leqslant \sup_i \overline{\dim}_B F_i.$$

Definition (3.19) now gives that $\dim_P F \leqslant \overline{\dim}_{MB} F$. Conversely, if $s > \dim_P F$ then $\mathscr{P}^s(F) = 0$, so that $F \subset \bigcup_i F_i$ for a collection of sets F_i with $\mathscr{P}^s_0(F_i) < \infty$ for each i, by (3.24). Hence, for each i, if δ is small enough, then $\mathscr{P}^s_\delta(F_i) < \infty$, so by (3.22) $N_\delta(F_i)\delta^s$ is bounded as $\delta \to 0$, where $N_\delta(F_i)$ is the largest number of disjoint balls of radius δ with centres in F_i. Thus $\overline{\dim}_B F_i \leqslant s$ for each i, giving that $\overline{\dim}_{MB} F \leqslant s$ by (3.19), as required. □

We have established the following relations:

$$\dim_H F \leqslant \underline{\dim}_{MB} F \leqslant \overline{\dim}_{MB} F = \dim_P F \leqslant \overline{\dim}_B F. \tag{3.29}$$

Suitable examples show that none of the inequalities can be replaced by equality.

As with Hausdorff dimension, packing dimension permits the use of powerful measure theoretic techniques in its study. The recent introduction of packing measures has led to a greater understanding of the geometric measure theory of fractals, with packing measures behaving in a way that is 'dual' to Hausdorff measures in many respects. Nevertheless, one cannot pretend that packing measures and dimensions are easy to work with or to calculate; the extra step (3.24) in their definition makes them more awkward to use than the Hausdorff analogues.

This situation is improved slightly by the equality of packing dimension and the modified upper box dimension. It is improved considerably for compact sets with 'local' dimension constant throughout—a situation that occurs frequently in practice, in particular in sets with some kind of self-similarity.

Corollary 3.9

Let $F \subset \mathbb{R}^n$ be compact and such that

$$\overline{\dim}_B (F \cap V) = \overline{\dim}_B F \tag{3.30}$$

for all open sets V that intersect F. Then $\dim_P F = \overline{\dim}_B F$.

Proof. The is immediate from Propositions 3.6 and 3.8. □

The nicest case, of course, is of fractals with equal Hausdorff and upper box dimensions, in which case equality holds throughout (3.29)—we shall see many such examples later on. However, even the much weaker condition $\dim_H F = \dim_P F$, though sometimes hard to prove, eases analysis of F.

3.5 Some other definitions of dimension

A wide variety of other definitions of dimension have been introduced, many of them only of limited applicability, but nonetheless useful in their context.

The special form of curves gives rise to the several definitions of dimension. We define a *curve* or *Jordan curve* C to be the image of an interval $[a, b]$ under a continuous bijection $f:[a, b] \to \mathbb{R}^n$. (Thus, we restrict attention to curves that are non-self-intersecting.) If C is a curve and $\delta > 0$, we define $M_\delta(C)$ to be the maximum number of points $x_0, x_1, \ldots, x_m$, on the curve C, in that order, such that $|x_k - x_{k-1}| = \delta$ for $k = 1, 2, \ldots, m$. Thus $(M_\delta(C) - 1)\delta$ may be thought of as the 'length' of the curve C measured using a pair of dividers with points set a distance δ apart. The *divider dimension* is defined as

$$\lim_{\delta \to 0} \frac{\log M_\delta(C)}{-\log \delta} \tag{3.31}$$

assuming the limit exists (otherwise we may define upper and lower divider dimensions using upper and lower limits). It is easy to see that the divider dimension of a curve is at least equal to the box dimension (assuming that they both exist) and in simple self-similar examples, such as the von Koch curve, they are equal. The assertion that the coastline of Britain has dimension 1.2 is usually made with the divider dimension in mind—this empirical value comes from estimating the ratio in (3.31) for values of δ between about 20 m and 200 km.

A variation of Hausdorff dimension may be defined for curves by using intervals of the curves themselves as covering sets. Thus we look at $\inf\{\sum_{i=1}^{m}|f[t_{i-1},t_i]|^s\}$ where the infimum is over all dissections $a=t_0<t_1<\cdots<t_m=b$ such that the diameters $|f([t_{i-1},t_i])|$ are all at most δ. We let δ tend to 0 and deem the value of s at which this limit jumps from ∞ to 0 to be the dimension. For self-similar examples such as the von Koch curve, this equals the Hausdorff dimension, but for 'squeezed' curves, such as graphs of certain functions (see Chapter 11) we may get a somewhat larger value.

Sometimes, we are interested in the dimension of a fractal F that is the boundary of a set A. We can define the box dimension of F in the usual way, but sometimes it is useful to take special account of the distinction between A and its complement. Thus the following variation of the 's-dimensional content' definition of box dimension, in which we take the volume of the set of points within δ of F that are contained in A is sometimes useful. We define the *one-sided dimension* of the boundary F of a set A in $\mathbb{R}^n$ as

$$n-\lim_{\delta\to 0}\frac{\log \text{vol}^n(F_\delta\cap A)}{\log\delta} \tag{3.32}$$

where F_δ is the δ-parallel body of F (compare Proposition 3.2). This definition has applications to the surface physics of solids where it is the volume very close to the surface that is important and also to partial differential equations in domains with fractal bounderies.

It is sometimes possible to define dimension in terms of the complement of a set. Suppose F is obtained by removal of a sequence of intervals $I_1, I_2, \ldots$ from, say, the unit interval $[0, 1]$, as, for example, in the Cantor set construction. We may define a dimension as the number s_0 such that the series

$$\sum_{j=1}^{\infty}|I_j|^s \text{ converges if } s<s_0 \text{ and diverges if } s>s_0. \tag{3.33}$$

For the middle third Cantor set, this series is $\sum_{k=1}^{\infty}2^{k-1}3^{-ks}$, giving $s_0=\log 2/\log 3$, equal to the Hausdorff and box dimensions in this case. In general, s_0 equals the upper box dimension of F.

Dimension prints provide an interesting variation on Hausdorff dimension of a rather different nature. Dimension prints may be thought of as a sort of 'fingerprint' that enables sets with differing characteristics to be distinguished, even though they may have the same Hausdorff dimension. In particular they reflect non-isotropic features of a set.

We restrict attention to subsets of the plane, in which case the dimension print will also be planar. The definition of dimension prints is very similar to that of Hausdorff dimension but coverings by rectangles are used with side lengths replacing diameters. Let U be a rectangle (the sides need not be parallel to the coordinate axes) and let $a(U) \geqslant b(U)$ be the lengths of the sides of U. Let s, t be non-negative numbers. For F a subset of $\mathbb{R}^2$, let

$$\mathscr{H}^{s,t}_\delta(F) = \inf\left\{ \sum_i a(U_i)^s b(U_i)^t : \{U_i\} \text{ is a } \delta\text{-cover of } F \text{ by rectangles} \right\}.$$

In the usual way, we get measures of 'Hausdorff type', $\mathscr{H}^{s,t}$, by letting $\delta \to 0$:

$$\mathscr{H}^{s,t}(F) = \lim_{\delta \to 0} \mathscr{H}^{s,t}_\delta(F).$$

(Note that $\mathscr{H}^{s,0}$ is just the s-dimensional Hausdorff measure.) The *dimension*

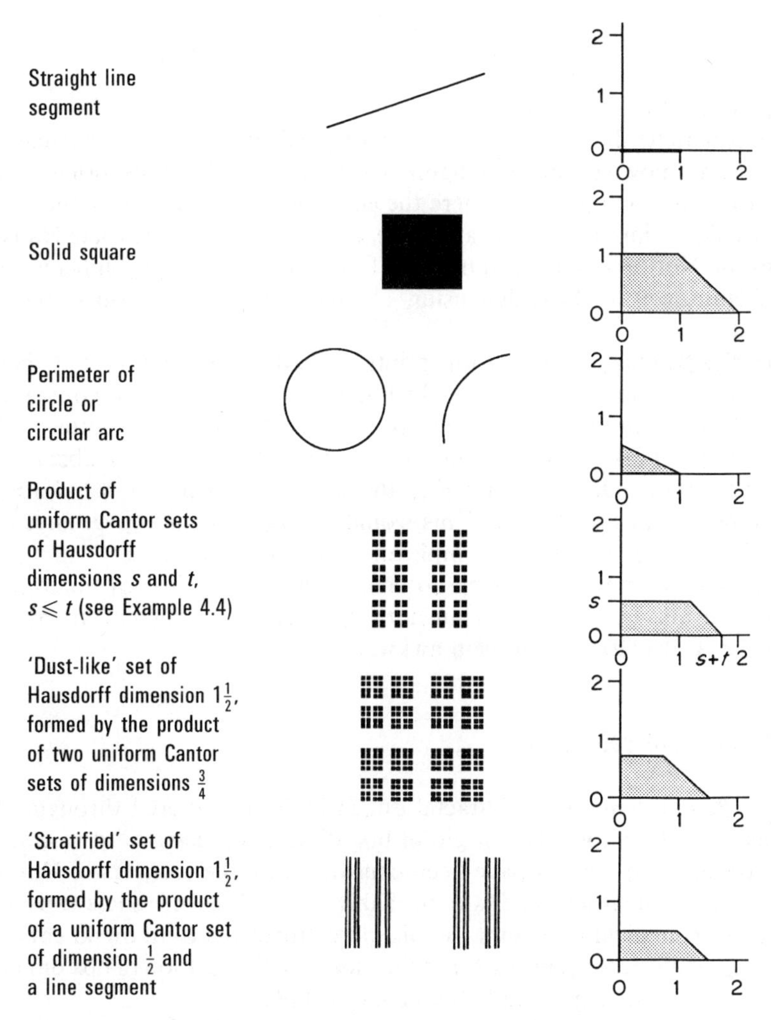

Figure 3.3 A selection of dimension prints of plane sets

print, print F, of F is defined to be the set of non-negative pairs (s,t) for which $\mathscr{H}^{s,t}(F)>0$.

Using standard properties of measures, it is easy to see that we have monotonicity

$$\operatorname{print} F_1 \subset \operatorname{print} F_2 \qquad \text{if } F_1 \subset F_2 \tag{3.34}$$

and countable stability

$$\operatorname{print}\left(\bigcup_{i=1}^{\infty} F_i\right) = \bigcup_{i=1}^{\infty} \operatorname{print} F_i. \tag{3.35}$$

Moreover, if (s,t) is a point in print F and (s',t') satisfies

$$\begin{aligned} s'+t' &\leqslant s+t \\ t' &\leqslant t \end{aligned} \tag{3.36}$$

then (s',t') is also in print F.

Unfortunately, dimension prints are not particularly easy to calculate. We display a few known examples in figure 3.3. Notice that the Hausdorff dimension of a set is given by the point where the edge of its print intersects the x-axis.

Dimension prints are a useful and appealing extension of the idea of Hausdorff dimension. Notice how the prints in the last two cases distinguish between two sets of Hausdorff (or box) dimension $1\frac{1}{2}$, one of which is dust-like, the other stratified.

One disadvantage of dimension prints defined in this way is that they are *not* Lipschitz invariants. The straight line segment and smooth convex curve are bi-Lipschitz equivalent, but their prints are different. In the latter case the dimension print takes into account the curvature. It would be possible to avoid this difficulty by redefining print F as the set of (s,t) such that $\mathscr{H}^{s,t}(F')>0$ for all bi-Lipschitz images F' of F. This would restore Lipschitz invariance of the prints, but would add further complications to their calculation.

Of course, it would be possible to define dimension prints by analogy with box dimensions rather than Hausdorff dimensions, using covers by equal rectangles. Calculations still seem awkward.

3.6 Notes and references

Many different definitions of 'fractal dimension' are scattered throughout the mathematical literature. The origin of box dimension seems hard to trace—it seems certain that it must have been considered by the pioneers of Hausdorff measure and dimension, and was probably rejected as being less satisfactory from a mathematical viewpoint. Bouligand adapted the Minkowski content to non-integral dimensions in 1928, and the more usual definition of box dimension was given by Pontrjagin and Schnirelman in 1932.

Packing measures and dimensions are much more recent, introduced by Tricot

(1982). Their similarities and contrasts to Hausdorff measures and dimensions are proving an important theoretical tool.

Dimension prints are a recent innovation of Rogers (1988).

Exercises

3.1 Let $f: F \to \mathbb{R}^n$ be a Lipschitz function. Show that $\underline{\dim}_B f(F) \leqslant \underline{\dim}_B F$ and $\overline{\dim}_B f(F) \leqslant \overline{\dim}_B F$.

3.2 Verify directly from the definitions that Equivalent definitions 3.1(i) and (iii) give the same values for box dimension.

3.3 Let F consist of those numbers in $[0, 1]$ whose decimal expansions do not contain the digit 5. Find $\dim_B F$, showing that this box dimension exists.

3.4 Verify that the set depicted in figure 0.4 has box dimension 1.

3.5 Use Equivalent definition 3.1(iv) to check that the upper box dimension of the von Koch curve is at most $\log 4/\log 3$ and 3.1(v) to check that the lower box dimension is at least this value.

3.6 Use convenient parts of Equivalent definition 3.1 to find the box dimension of the Sierpiński gasket in figure 0.3.

3.7 Let F be the middle third Cantor set. For $0 < \delta < 1$, find the length of the δ-parallel body F_δ of F, and hence find the box dimension of F using Proposition 3.2.

3.8 Construct a set F for which $\underline{\dim}_B F < \overline{\dim}_B F$. (Hint: consider a variation on the middle third Cantor set construction, with each interval in E_{k-1} containing two equal intervals of E_k. Arrange for these intervals to be 'long' for $k = 1, \ldots, k_1$, 'short' for $k = k_1 + 1, \ldots, k_2$, 'long' for $k = k_2 + 1, \ldots, k_3$, and so on.)

3.9 Find subsets E and F of $\mathbb{R}$ such that $\underline{\dim}_B(E \cup F) > \max\{\underline{\dim}_B E, \underline{\dim}_B F\}$.

3.10 What are the Hausdorff and box dimensions of the set $\{0, 1, \frac{1}{4}, \frac{1}{9}, \frac{1}{16}, \ldots\}$?

3.11 Find two disjoint Borel subsets E and F of $\mathbb{R}$ such that $\mathscr{P}_0^s(E \cup F) \neq \mathscr{P}_0^s(E) + \mathscr{P}_0^s(F)$.

3.12 What is the packing dimension of the von Koch curve?

3.13 Show that the divider dimension (3.31) of a curve is greater than or equal to its box dimension, assuming that they both exist.

3.14 Let $0 < \lambda < 1$ and let F be the 'middle λ Cantor set' obtained by repeated removal of the middle proportion λ from intervals. Show that the dimension of F defined by (3.33) in terms of removed intervals equals the Hausdorff and box dimensions of F.

3.15 Verify properties (3.34)–(3.36) of dimension prints. Given an example of a set with a non-convex dimension print.

Chapter 4 Techniques for calculating dimensions

A direct attempt at calculating the dimensions, in particular the Hausdorff dimension, of almost any set will convince the reader of the practical limitations of working from the definitions. Rigorous dimension calculations often involve pages of complicated manipulations and estimates that provide little intuitive enlightenment.

In this chapter we bring together some of the basic techniques that are available for dimension calculations. Other methods, that are applicable in more specific cases, will be found throughout the book.

4.1 Basic methods

For most fractals 'obvious' upper estimates of dimension may be obtained using natural coverings by small sets.

Proposition 4.1

Suppose F can be covered by n_k sets of diameter at most δ_k with $\delta_k \to 0$ as $k \to \infty$. Then

$$\dim_H F \leqslant \underline{\dim}_B F \leqslant \underline{\lim}_{k\to\infty} \frac{\log n_k}{-\log \delta_k}$$

and, if $\delta_{k+1} \geqslant c\delta_k$ for some $0 < c < 1$,

$$\overline{\dim}_B F \leqslant \overline{\lim}_{k\to\infty} \frac{\log n_k}{-\log \delta_k}.$$

Moreover, if $n_k \delta_k^s$ remains bounded as $k \to \infty$, then $\mathscr{H}^s(F) < \infty$.

Proof. The inequalities for the box-counting dimension are immediate from the definitions. For the last part, $\mathscr{H}^s_{\delta_k}(F) \leqslant n_k \delta_k^s$, so $\mathscr{H}^s_{\delta_k}(F)$ tends to a finite limit $\mathscr{H}^s(F)$ as $k \to \infty$. □

Thus, as we have seen already (Example 2.7), in the case of the middle third Cantor set the natural coverings by 2^k intervals of length 3^{-k} give $\dim_H F \leqslant \log 2/\log 3$.

Surprisingly often, the 'obvious' upper bound for the Hausdorff dimension of a set turns out to be the actual value. However, demonstrating this can be difficult. To obtain an upper bound it is enough to evaluate sums of the form $\sum|U_i|^s$ for *specific* coverings $\{U_i\}$ of F, whereas for a lower bound we must show that $\sum|U_i|^s$ is greater than some positive constant for *all* δ-coverings of F. Clearly an enormous number of such coverings are available. In particular, when working with Hausdorff dimension as opposed to box dimension, consideration must be given to covers where some of the U_i are very small and others have relatively large diameter—this prohibits sweeping estimates for $\sum|U_i|^s$ such as those available for upper bounds.

One way of getting around these difficulties is to show that no *individual* set U can cover too much of F compared with its size measured as $|U|^s$. Then if $\{U_i\}$ covers the whole of F the sum $\sum|U_i|^s$ cannot be too small. The usual way to do this is to concentrate a suitable mass distribution μ on F and compare the mass $\mu(U)$ covered by U with $|U|^s$ for each U. (Recall that a mass distribution on F is a measure with support contained in F such that $0<\mu(F)<\infty$, see Section 1.3.)

Mass distribution principle 4.2

Let μ be a mass distribution on F and suppose that for some s there are numbers $c>0$ and $\delta>0$ such that

$$\mu(U)\leqslant c|U|^s \tag{4.1}$$

for all sets U with $|U|\leqslant\delta$. Then $\mathscr{H}^s(F)\geqslant\mu(F)/c$ and

$$s\leqslant\dim_{\mathrm{H}}F\leqslant\underline{\dim}_{\mathrm{B}}\leqslant\overline{\dim}_{\mathrm{B}}F.$$

Proof. If $\{U_i\}$ is any cover of F then

$$0<\mu(F)=\mu\left(\bigcup_i U_i\right)\leqslant\sum_i\mu(U_i)\leqslant c\sum_i|U_i|^s. \tag{4.2}$$

Taking infima, $\mathscr{H}^s_\delta(F)\geqslant\mu(F)/c$ if δ is small enough, so $\mathscr{H}^s(F)\geqslant\mu(F)/c$. □

Notice that the conclusion $\mathscr{H}^s(F)\geqslant\mu(F)/c$ remains true if μ is a mass distribution on $\mathbb{R}^n$ and F is any subset.

The mass distribution principle 4.2 gives a quick lower estimate for the Hausdorff dimension of the middle third Cantor set F (figure 0.1). Let μ be the natural mass distribution on F, so that each of the 2^k basic intervals of length 3^{-k} in E_k in the construction of F, carry a mass 2^{-k}. (We imagine that we start with unit mass on E_0 and repeatedly divide the mass on each interval of E_k between its two subintervals in E_{k+1}; see Proposition 1.7.) Let U be a set with $|U|<1$ and let k be the integer such that $3^{-(k+1)}\leqslant|U|<3^{-k}$. Then U can intersect at most one of the intervals of E_k, so

$$\mu(U)\leqslant 2^{-k}=(3^{-k})^{\log 2/\log 3}\leqslant(3|U|)^{\log 2/\log 3}$$

and hence $\mathscr{H}^{\log 2/\log 3}(F)>0$ by the mass distribution principle giving $\dim_{\mathrm{H}}F\geqslant\log 2/\log 3$.

Example 4.3

Let $F_1 = F \times [0,1] \subset \mathbb{R}^2$ *be the product of the middle third Cantor set* F *and the unit interval. Then* $\dim_B F_1 = \dim_H F_1 = 1 + \log 2/\log 3 = s$, *with* $0 < \mathscr{H}^s(F_1) < \infty$.

Calculation. For each k, there is a covering of F by 2^k intervals of length 3^{-k}. A column of 3^k squares of side 3^{-k} (diameter $3^{-k}\sqrt{2}$) covers the part of F_1 above each such interval, so taking these all together, F_1 may be covered by $2^k 3^k$ squares of side 3^{-k}. Thus $\mathscr{H}^s_{3^{-k}\sqrt{2}}(F_1) \leqslant 3^k 2^k (3^{-k}\sqrt{2})^s = 2^{s/2}$, so $\mathscr{H}^s(F_1) \leqslant 2^{s/2}$ and $\dim_H F_1 \leqslant \underline{\dim}_B F_1 \leqslant \overline{\dim}_B F_1 \leqslant s$.

We define a mass distribution μ on F_1 by taking the natural mass distribution on F described above (each basic interval of F of side 3^{-k} having mass 2^{-k}) and 'spreading it' uniformly along the intervals above F. Thus if U is a rectangle, with sides parallel to the coordinate axes, of height h, above a basic interval of F of side 3^{-k}, then $\mu(U) = h 2^{-k}$. Any set U is contained in a square of side $|U|$ with sides parallel to the coordinate axes. If $3^{-(k+1)} \leqslant |U| < 3^{-k}$ then U lies above at most one basic interval of F of side 3^{-k}, so

$$\mu(U) \leqslant |U| 2^{-k} \leqslant |U| 3^{-k \log 2/\log 3} \leqslant |U| (3|U|)^{\log 2/\log 3} \leqslant 3^{\log 2/\log 3} |U|^s.$$

By the Mass distribution principle 4.2, $\mathscr{H}^s(F_1) > 0$. □

Notice that in this example the dimension of the product of two sets equals the sum of the dimensions of the sets. We study this is greater depth in Chapter 7.

The following *general construction* of a subset of $\mathbb{R}$ may be thought of as a generalization of the Cantor set construction. Let $[0,1] = E_0 \supset E_1 \supset E_2 \supset \ldots$ be a decreasing sequence of sets, with each E_k a union of a finite number of disjoint closed intervals (called *basic* intervals), with each interval of E_k containing at least two intervals of E_{k+1}, and the maximum length of intervals in E_k tending to 0 as $k \to 0$. Then the set

$$F = \bigcap_{k=0}^{\infty} E_k \tag{4.3}$$

is a totally disconnected subset of $[0,1]$ which is generally a fractal (figure 4.1).

Obvious upper bounds for the dimension of F are available by taking the intervals of E_k as covering intervals, for each k, but, as usual, lower bounds are harder to find. Note that, in the following examples, the upper estimates for

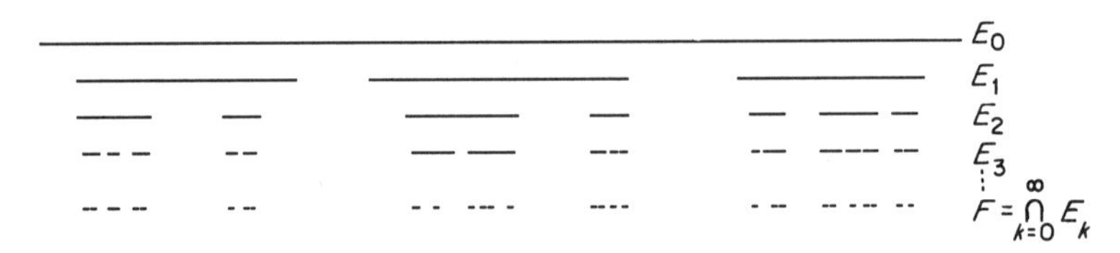

Figure 4.1 An example of the general construction of a subset of $\mathbb{R}$

$\dim_H F$ depend on the number and size of the basic intervals, whilst the lower estimates depend on their spacing. For these to be equal, the intervals of E_{k+1} must be 'nearly uniformly distributed' inside the intervals of E_k.

Example 4.4

Let s be a number strictly between 0 and 1. Assume that the E_k in the general construction (4.3) have the following property: for each basic interval I of E_k, the intervals $I_1, \ldots, I_m$ $(m \geqslant 2)$ of E_{k+1} contained in I are of equal length and equally spaced, the lengths being given by

$$|I_i|^s = \frac{1}{m}|I|^s \qquad (1 \leqslant i \leqslant m) \tag{4.4}$$

with the left-hand ends of I_1 and I coinciding, and the right-hand ends of I_m and I coinciding. Then $\dim_H F = s$ and $0 < \mathscr{H}^s(F) < \infty$. (Notice that m may be different for different intervals I in the construction, so that the intervals of E_k may have widely differing lengths.)

Calculation. With I, I_i, as above,

$$|I|^s = \sum_{i=1}^{m} |I_i|^s. \tag{4.5}$$

Applying this inductively to the intervals of E_k for successive k, it follows that, for each k, $1 = \sum |I_i|^s$, where the sum is over all the intervals in E_k. The intervals of E_k cover F; since the maximum interval length tends to 0 as $k \to \infty$, we have $\mathscr{H}^s_\delta(F) \leqslant 1$ for sufficiently small δ giving $\mathscr{H}^s(F) \leqslant 1$.

Now distribute a mass μ on F in such a way that $\mu(I) = |I|^s$ whenever I is a basic interval. Thus, starting with unit mass on $[0, 1]$ we divide this equally between each interval of E_1, the mass on each of these intervals being divided equally between each subinterval of E_2, and so on; see Proposition 1.7. Equation (4.5) ensures that we get a mass distribution on F with $\mu(I) = |I|^s$ for every basic interval. We estimate $\mu(U)$ for any interval U with endpoints in F. Let I be the smallest basic interval that contains U; suppose that I is an interval of E_k, and let $I_i, \ldots, I_m$ be the intervals of E_{k+1} contained in I. Then U intersects a number $j \geqslant 2$ of the I_i, otherwise U would be contained in a smaller basic interval. The spacing between consecutive I_i is

$$\begin{aligned}(|I| - m|I_i|)/(m-1) &= |I|(1 - m|I_i|/|I|)/(m-1)\\ &= |I|(1 - m^{1-1/s})/(m-1)\\ &\geqslant c_s|I|/m\end{aligned}$$

using (4.4), where $c_s = (1 - 2^{1-1/s})$. Thus

$$|U| \geqslant \frac{j-1}{m} c_s|I| \geqslant \frac{j}{2m} c_s|I|.$$

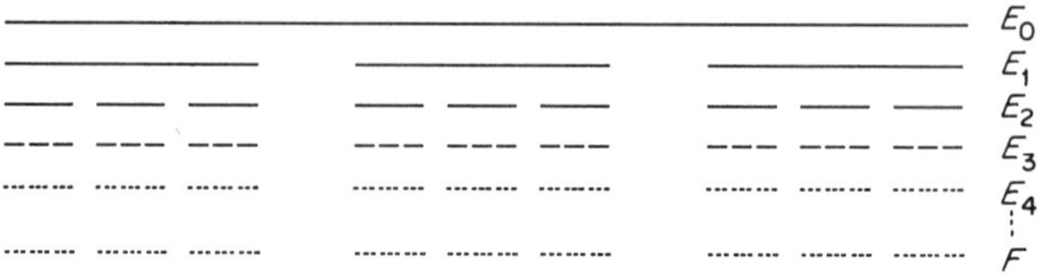

Figure 4.2 A uniform Cantor set (Example 4.5) with $m = 3$, $\lambda = \frac{4}{15}$, $\dim_H F = \dim_B F = \log 3 / -\log \frac{4}{15} = 0.831$

By (4.5)

$$\mu(U) \leqslant j\mu(I_i) = j|I_i|^s = \frac{j}{m}|I|^s$$

$$\leqslant 2^s c_s^{-s} \left(\frac{j}{m}\right)^{1-s} |U|^s \leqslant 2^s c_s^{-s} |U|^s. \tag{4.6}$$

This is true for any interval U with endpoints in F, and so for any set U (by applying (4.6) to the smallest interval containing $U \cap F$). By the Mass distribution principle 4.2, $\mathscr{H}^s(F) > 0$. $\square$

A more careful estimate of $\mu(U)$ in Example 4.4 leads to $\mathscr{H}^s(F) = 1$.

We call the sets obtained when m is kept constant throughout the construction of Example 4.4 *uniform Cantor sets*; see figure 4.2. These provide a natural generalisation of the middle third Cantor set.

Example 4.5 Uniform Cantor sets

Let $m \geqslant 2$ be an integer and $0 < \lambda < 1/m$. Let F be the set obtained by the construction in which each basic interval I is replaced by m equally spaced subintervals of lengths $\lambda|I|$, the ends of I coinciding with the ends of the extreme subintervals. Then $\dim_H F = \dim_B F = \log m / -\log \lambda$, and $0 < \mathscr{H}^{\log m / -\log \lambda}(F) < \infty$.

Calculation. The set F is obtained on taking m constant and $s = \log m/(-\log \lambda)$ in Example 4.4. Equation (4.4) becomes $(\lambda|I|)^s = (1/m)|I|^s$, which is satisfied identically, so $\dim_H F = s$. For the box dimension, note that F is covered by the m^k basic intervals of length λ^{-k} in E_k for each k, leading to $\overline{\dim}_B F \leqslant \log m / -\log \lambda$ in the usual way. $\square$

The next example is another case of the general construction.

Example 4.6

Suppose in the general construction (4.3) each interval of E_{k-1} contains at least m_k intervals of $E_k (k = 1, 2, \ldots)$ which are separated by gaps of at least ε_k, where

$0<\varepsilon_{k+1}<\varepsilon_k$ for each k. Then

$$\dim_{\mathrm{H}} F \geqslant \varliminf_{k\to\infty} \frac{\log(m_1\cdots m_{k-1})}{-\log(m_k\varepsilon_k)}. \tag{4.7}$$

Calculation. We may assume that each set E_{k-1} contains exactly m_k intervals of E_k; if not we may throw out excess intervals to get smaller sets E_k and F for which this is so. We may define a mass distribution μ on F by assigning a mass of $(m_1\cdots m_k)^{-1}$ to each of the $m_1\cdots m_k$ basic intervals of E_k.

Let U be an interval with $0<|U|<\varepsilon_1$; we estimate $\mu(U)$. Let k be the integer such that $\varepsilon_k\leqslant|U|<\varepsilon_{k-1}$. The number of intervals of E_k that intersect U is

(i) at most m_k since U intersects at most one interval of E_{k-1}

(ii) at most $|U|/\varepsilon_k+1\leqslant 2|U|/\varepsilon_k$ since the intervals of E_k have gaps of at least ε_k between them. Each interval of E_k supports mass $(m_1\cdots m_k)^{-1}$ so that

$$\begin{aligned}\mu(U) &\leqslant (m_1\cdots m_k)^{-1}\min\{2|U|/\varepsilon_k, m_k\}\\ &\leqslant (m_1\cdots m_k)^{-1}(2|U|/\varepsilon_k)^s m_k^{1-s}\end{aligned}$$

for any $0\leqslant s\leqslant 1$.

Hence

$$\frac{\mu(U)}{|U|^s}\leqslant\frac{2^s}{(m_1\cdots m_{k-1})m_k^s\varepsilon_k^s}$$

which is bounded above by a constant provided that

$$s<\varliminf_{k\to\infty}\log(m_1\cdots m_{k-1})/-\log(m_k\varepsilon_k).$$

The result follows by Principle 4.2. $\square$

Now suppose that in Example 4.6 the intervals of E_k are all of length δ_k, and that each interval of E_{k-1} contains exactly m_k intervals of E_k, which are 'roughly equally spaced' in the sense that $m_k\varepsilon_k\geqslant c\delta_{k-1}$, where $c>0$ is a constant. Then (4.7) becomes

$$\dim_{\mathrm{H}} F\geqslant\varliminf_{k\to\infty}\frac{\log(m_1\cdots m_{k-1})}{-\log c-\log\delta_{k-1}}=\varliminf_{k\to\infty}\frac{\log(m_1\cdots m_{k-1})}{-\log\delta_{k-1}}.$$

But E_{k-1} comprises $m_1\cdots m_{k-1}$ intervals of length δ_{k-1}, so this expression equals the upper bound for $\dim_{\mathrm{H}} F$ given by Proposition 4.1. Thus in the situation where the intervals are well spaced, we get equality in (4.7).

Examples of the following form occur in number theory; see Section 10.3.

Example 4.7

Fix $0<s<1$ and let $n_1, n_2,\ldots$ be a rapidly increasing sequence of integers, say $n_{k+1}\geqslant\max\{n_k^k, 3n_k^{1/s}\}$ for each k. For each k let $H_k\subset\mathbb{R}$ consist of equally spaced *equal intervals of lengths $n_k^{-1/s}$ with the midpoints of consecutive intervals distance n_k^{-1} apart. Then* $\dim_H F = \dim_B F = s$, *where* $F = \bigcap_{k=1}^{\infty} H_k$.

Calculation. Since $F \subset H_k$ for each k, the set $F \cap [0,1]$ is contained in at most $n_k + 1$ intervals of length $n_k^{-1/s}$, so Proposition 4.1 gives $\overline{\dim}_B (F \cap [0,1]) \leqslant \overline{\lim}_{k\to\infty} \log(n_k + 1)/-\log n_k^{-1/s} = s$. Similarly, $\overline{\dim}_B (F \cap [n, n+1]) \leqslant s$ for any $n \in \mathbb{Z}$, so F, as a countable union of such sets, has $\underline{\dim}_B F \leqslant s$.

Now let $E_0 = [0,1]$ and, for $k \geqslant 1$, let E_k consist of the intervals of H_k that are completely contained in E_{k-1}. Then each interval I of E_{k-1} contains at least $n_k|I| - 1 \geqslant n_k n_{k-1}^{-1/s} - 1 \geqslant 2$ intervals of E_k, which are separated by gaps of at least $n_k^{-1} - n_k^{-1/s} \geqslant \frac{1}{2} n_k^{-1}$ if k is large enough. Using Example 4.6, and noting that replacing $n_k n_{k-1}^{-1/s} - 1$ by $n_k n_{k-1}^{-1/s}$ does not affect the limit,

$$\dim_H F \geqslant \dim_H \bigcap_{k=1}^{\infty} E_k = \varliminf_{k\to\infty} \frac{\log((n_1 \cdots n_{k-2})^{1-1/s} n_{k-1})}{-\log(n_k n_{k-1}^{-1/s} \frac{1}{2} n_k^{-1})}$$

$$= \varliminf_{k\to\infty} \frac{\log(n_1 \cdots n_{k-2})^{1-1/s} + \log n_{k-1}}{\log 2 + (\log n_{k-1})/s}.$$

Provided that n_k is sufficiently rapidly increasing, the terms in $\log n_{k-1}$ in the numerator and denominator of this expression are dominant, so that $\dim_H F \geqslant s$, as required. □

Although the Mass distribution principle 4.2 is based on a simple idea, we have seen that it can be very useful in finding Hausdorff and box dimensions. We now develop some important variations of the method.

It is enough for condition (4.1) to hold just for sufficiently small balls centred at each point of F. This is expressed in Proposition 4.9(*a*). Although mass distribution methods for upper bounds are required for less frequently, we include part (*b*) because it is, in a sense, dual to (*a*). Note that density expressions, such as $\lim_{r\to 0} \mu(B_r(x))/r^s$ play a major role in the study of local properties of fractals—see Chapter 5. (Recall that $B_r(x)$ is the closed ball of centre x and radius r.)

We require the following covering lemma in the proof of Proposition 4.9(*b*).

Covering lemma 4.8

Let $\mathscr{C}$ be a family of balls contained in some bounded region of $\mathbb{R}^n$. Then there is a (finite or countable) disjoint subcollection $\{B_i\}$ such that

$$\bigcup_{B \in \mathscr{C}} B \subset \bigcup_i \tilde{B}_i \tag{4.8}$$

where $\tilde{B}_i$ is the closed ball concentric with B_i and of four times the radius.

Proof. For simplicity, we give the proof when $\mathscr{C}$ is a finite family; the basic idea is the same in the general case. We select the $\{B_i\}$ inductively. Let B_1 be a ball in $\mathscr{C}$ of maximum radius. Suppose that $B_1, \ldots, B_{k-1}$ have been chosen. We take B_k to be the largest ball in $\mathscr{C}$ (or one of the largest) that does not intersect $B_1, \ldots, B_{k-1}$. The process terminates when no such ball remains. Clearly the balls selected are disjoint; we must check that (4.8) holds. If $B \in \mathscr{C}$, then either $B = B_i$ for some i, or B intersects one of the B_i with $|B_i| \geqslant |B|$; if this were not the case, then B would have been chosen instead of the first ball B_k with $|B_k| < |B|$. Either way, $B \subset \widetilde{B}_i$, so we have (4.8). (It is easy to see that the result remains true taking $\widetilde{B}_i$ as the ball concentric with B_i and of $3+\varepsilon$ times the radius, for any $\varepsilon > 0$; if $\mathscr{C}$ is finite we may take $\varepsilon = 0$.) □

Proposition 4.9

Let μ be a mass distribution on $\mathbb{R}^n$, let $F \subset \mathbb{R}^n$ be a Borel set and let $0 < c < \infty$ be a constant.

(a) *If $\overline{\lim}_{r\to 0} \mu(B_r(x))/r^s < c$ for all $x \in F$ then $\mathscr{H}^s(F) \geqslant \mu(F)/c$*
(b) *If $\overline{\lim}_{r\to 0} \mu(B_r(x))/r^s > c$ for all $x \in F$ then $\mathscr{H}^s(F) \leqslant 2^s \mu(\mathbb{R}^n)/c$.*

Proof

(a) For each $\delta > 0$ let

$$F_\delta = \{x \in F : \mu(B_r(x)) < (c-\varepsilon) r^s \text{ for all } 0 < r \leqslant \delta \text{ for some } \varepsilon > 0\}.$$

Let $\{U_i\}$ be a δ-cover of F and thus of F_δ. For each U_i containing a point x of F_δ, the ball B with centre x and radius $|U_i|$ certainly contains U_i. By definition of F_δ,

$$\mu(U_i) \leqslant \mu(B) < c|U_i|^s$$

so that

$$\mu(F_\delta) \leqslant \sum_i \{\mu(U_i) : U_i \text{ intersects } F_\delta\} \leqslant c \sum_i |U_i|^s.$$

Since $\{U_i\}$ is any δ-cover of F, it follows that $\mu(F_\delta) \leqslant c\mathscr{H}^s_\delta(F) \leqslant c\mathscr{H}^s(F)$. But F_δ increases to F as δ decreases to 0, so $\mu(F) \leqslant c\mathscr{H}^s(F)$ by (1.7).

(b) For simplicity, we prove a weaker version of (*b*) with 2^s replaced by 8^s, but the basic idea is similar. Suppose first that F is bounded. Fix $\delta > 0$ and let $\mathscr{C}$ be the collection of balls

$$\{B_r(x) : x \in F, 0 < r \leqslant \delta \text{ and } \mu(B_r(x)) > cr^s\}.$$

Then by the hypothesis of (*b*) $F \subset \bigcup_{B\in\mathscr{C}} B$. Applying Covering lemma 4.8 to the collection $\mathscr{C}$, there is a sequence of disjoint balls $B_i \in \mathscr{C}$ such that $\bigcup_{B\in\mathscr{C}} B \subset \bigcup_i \widetilde{B}_i$ where $\widetilde{B}_i$ is the ball concentric with B_i but of four times the radius. Thus $\{\widetilde{B}_i\}$ is an 8δ-cover of F, so

$$\begin{aligned}\mathscr{H}^s_{8\delta}(F) &\leqslant \sum_i |\widetilde{B}_i|^s \leqslant 4^s \sum_i |B_i|^s \\ &\leqslant 8^s c^{-1} \sum_i \mu(B_i) \leqslant 8^s c^{-1} \mu(\mathbb{R}^n).\end{aligned}$$

Letting $\delta \to 0$, we get $\mathcal{H}^s(F) \leqslant 8^s c^{-1} \mu(\mathbb{R}^n) < \infty$. Finally, if F is unbounded and $\mathcal{H}^s(F) > 8^s c^{-1} \mu(\mathbb{R}^n)$, the $\mathcal{H}^s$-measure of some bounded subset of F will also exceed this value, contrary to the above. □

Note that it is immediate from this Proposition 4.9 that $\dim_{\mathrm{H}} F = \lim_{r \to 0} \log \mu(B_r(x))/\log r$ if this limit exists.

Applications of Proposition 4.9 will occur throughout the book.

The densities $\lim_{r \to 0} \mu(B_r(x))/r^s$ that occur in Proposition 4.9 are sometimes used (often rather imprecisely) to define the dimension of a set. Often a fractal F is naturally endowed with a mass distribution μ, for example, an invariant measure on the attractor of a dynamical system; see Section 13.7. If the mass of small balls obeys a law $\log \mu(F \cap B_r(x))/\log r \to s$ as $r \to 0$ for *all* x in F, then the Hausdorff dimension of F equals s. This is sometimes used as a practical method for estimating a 'dimension' of a set that carries a natural mass distribution. For a 'typical' point x, we might estimate $\mu(F \cap B_r(x))$ for a series of small values of r, and read off the dimension as the gradient of the graph of $\log \mu(F \cap B_r(x))$ against $\log r$.

4.2 Subsets of finite measure

This section may seem out of place in a chapter about finding dimensions. However, Theorem 4.10 is required for the important potential theoretic methods developed in the following section. It also permits a simplification which can be very useful in fractal analysis.

Theorem 4.10 guarantees that any (Borel) set F with $\mathcal{H}^s(F) = \infty$ contains a subset E with $0 < \mathcal{H}^s(E) < \infty$ (i.e. with E an s-set). At first, this might seem obvious—just shave pieces off F until what remains has positive finite measure. Unfortunately it is not quite this simple—it is possible to jump from infinite measure to zero measure without passing through any intermediate value. Stating this in mathematical terms, it is possible to have a decreasing sequence of sets $E_1 \supset E_2 \supset \ldots$ with $\mathcal{H}^s(E_k) = \infty$ for all k, but with $\mathcal{H}^s(\bigcap_{k=1}^{\infty} E_k) = 0$. (For a simple example, take $E_k = [0, 1/k] \subset \mathbb{R}$ and $0 < s < 1$.) To prove the theorem we need to look rather more closely at the structure of Hausdorff measures. Readers mainly concerned with applications may prefer to omit the proof!

Theorem 4.10

Let F be a Borel subset of $\mathbb{R}^n$ with $\mathcal{H}^s(F) = \infty$. Then there is a compact set $E \subset F$ such that $0 < \mathcal{H}^s(E) < \infty$.

****Sketch of proof.*** The complete proof of this is complicated. We indicate the ideas involved in the case where F is a compact subset of $[0, 1) \subset \mathbb{R}$ and $0 < s < 1$.

We work with the net measures $\mathcal{M}^s$ (2.17)–(2.18) which are defined using the binary intervals $[r2^{-k}, (r+1)2^{-k})$ and are related to Hausdorff measure by (2.19).

We define inductively a decreasing sequence $E_0 \supset E_1 \supset E_2 \supset \ldots$ of compact subsets of F. Let $E_0 = F$. For $k \geqslant 0$ we define E_{k+1} by specifying its intersection with each binary interval I of length 2^{-k}. If $\mathcal{M}^s_{2^{-(k+1)}}(E_k \cap I) \leqslant 2^{-sk}$ we let $E_{k+1} \cap I = E_k \cap I$. Then

$$\mathcal{M}^s_{2^{-(k+1)}}(E_{k+1} \cap I) = \mathcal{M}^s_{2^{-k}}(E_k \cap I) \tag{4.9}$$

since using I itself as a covering interval in calculating $\mathcal{M}^s_{2^{-k}}$ gives an estimate at least as large as using shorter binary intervals. On the other hand, if $\mathcal{M}^s_{2^{-(k+1)}}(E_{k+1} \cap I) > 2^{-sk}$ we take $E_{k+1} \cap I$ to be a compact subset of $E_k \cap I$ with $\mathcal{M}^s_{2^{-(k+1)}}(E_{k+1} \cap I) = 2^{-sk}$. Such a subset exists since $\mathcal{M}^s_{2^{-(k+1)}}(E_k \cap I \cap [0, u])$ is finite and continuous in u. (This is why we need to work with the $\mathcal{M}^s_\delta$ rather than $\mathcal{M}^s$.) Since $\mathcal{M}^s_{2^{-k}}(E_k \cap I) = 2^{-sk}$, (4.9) again holds. Summing (4.9) over all binary intervals of length 2^{-k} we get

$$\mathcal{M}^s_{2^{-(k+1)}}(E_{k+1}) = \mathcal{M}^s_{2^{-k}}(E_k). \tag{4.10}$$

Repeated application of (4.10) gives $\mathcal{M}^s_{2^{-k}}(E_k) = \mathcal{M}^s_1(E_0)$ for all k. Let E be the compact set $\bigcap_{k=0}^{\infty} E_k$. Taking the limit as $k \to \infty$ gives $\mathcal{M}^s(E) = \mathcal{M}^s_1(E_0)$ (this step needs some justification). The covering of $E_0 = F$ by the single interval $[0, 1)$ gives $\mathcal{M}^s_1(E_0) \leqslant 1$. Since $\mathcal{M}^s(E_0) \geqslant \mathcal{H}^s(E_0) = \infty$ we have $\mathcal{M}^s_{2^{-k}}(E_0) > 0$ if k is large enough, so $\mathcal{M}^s_1(E_0) \geqslant \min\{\mathcal{M}^s_{2^{-k}}(E_0), 2^{-ks}\} > 0$. Thus $0 < \mathcal{M}^s(E) < \infty$, and the theorem follows from (2.19). $\square$

A number of results, for example those in Chapter 5, apply only to s-sets, i.e. sets with $0 < \mathcal{H}^s(F) < \infty$. One way of approaching s-dimensional sets with $\mathcal{H}^s(F) = \infty$ is to use Theorem 4.10 to extract a subset of positive finite measure, to study its properties as an s-set, and then to interpret these properties in the context of the larger set F. Similarly, any set F of Hausdorff dimension $t > 0$ has $\mathcal{H}^s(F) = \infty$ if $0 < s < t$, and so contains an s-set.

The following proposition, really a corollary of Proposition 4.9, allows us to strengthen Theorem 4.10 even further.

Proposition 4.11

Let F be a Borel set satisfying $0 < \mathcal{H}^s(F) < \infty$. There is a constant b and a compact set $E \subset F$ with $\mathcal{H}^s(E) > 0$ such that

$$\mathcal{H}^s(E \cap B_r(x)) \leqslant br^s \tag{4.11}$$

for all $x \in \mathbb{R}^n$ and $r > 0$.

Proof. In Proposition 4.9(*b*) take μ as the restriction of $\mathcal{H}^s$ to F, i.e. $\mu(A) = \mathcal{H}^s(F \cap A)$. Then, if

$$F_1 = \left\{ x \in \mathbb{R}^n : \overline{\lim_{r \to 0}}\, \mathcal{H}^s(F \cap B_r(x))/r^s > 2^{1+s} \right\}$$

it follows that $\mathcal{H}^s(F_1) \leqslant 2^s 2^{-(1+s)} \mathcal{H}^s(F) \leqslant \frac{1}{2}\mathcal{H}^s(F)$. Thus $\mathcal{H}^s(F \backslash F_1) \geqslant \frac{1}{2}\mathcal{H}^s(F) >$

0, so if $E_1 = F\backslash F_1$ then $\mathscr{H}^s(E_1) > 0$ and $\overline{\lim}_{r\to 0}\mathscr{H}^s(F\cap B_r(x))/r^s \leqslant 2^{1+s}$ for $x\in E_1$. By Egoroff's theorem (see page 16) it follows that there is a compact set $E\subset E_1$ with $\mathscr{H}^s(E)>0$ and a number $r_0>0$ such that $\mathscr{H}^s(F\cap B_r(x))/r^s \leqslant 2^{2+s}$ for all $x\in E$ and all $0<r\leqslant r_0$. But $\mathscr{H}^s(F\cap B_r(x))/r^s \leqslant \mathscr{H}^s(F)/r_0^s$ if $r\geqslant r_0$ so (4.11) follows. □

Corollary 4.12

Let F be a Borel subset of $\mathbb{R}^n$ with $\mathscr{H}^s(F)=\infty$. Then there is a compact set $E\subset F$ such that $0<\mathscr{H}^s(E)<\infty$ and such that for some constant b

$$\mathscr{H}^s(E\cap B_r(x)) \leqslant br^s$$

for all $x\in\mathbb{R}^n$ and $r\geqslant 0$.

Proof. Theorem 4.10 provides us with a subset of F of positive finite measure, and applying Proposition 4.11 to this gives the result. □

4.3 Potential theoretic methods

In this section we introduce a technique for calculating Hausdorff dimensions that is important both in theory and in practice. This replaces the need for estimating the mass of a large number of small sets by a single check for the convergence of a certain integral.

The ideas of potential and energy will be familiar to readers with a knowledge of gravitation or electrostatics. For $s\geqslant 0$ the *s-potential* at a point x of $\mathbb{R}^n$ due to the mass distribution μ on $\mathbb{R}^n$ is defined as

$$\phi_s(x) = \int \frac{\mathrm{d}\mu(y)}{|x-y|^s}. \tag{4.12}$$

(If we are working in $\mathbb{R}^3$ and $s=1$ then this is essentially the familiar Newtonian gravitational potential.) The *s-energy* of μ is

$$I_s(\mu) = \int \phi_s(x)\,\mathrm{d}\mu(x) = \iint \frac{\mathrm{d}\mu(x)\,\mathrm{d}\mu(y)}{|x-y|^s}. \tag{4.13}$$

The following theorem relates Hausdorff dimension to seemingly unconnected potential theoretic ideas. In particular, if there is a mass distribution on a set F which has finite s-energy, then F has dimension at least s.

Theorem 4.13

Let F be a subset of $\mathbb{R}^n$.

(a) *If there is a mass distribution μ on F with $I_s(\mu)<\infty$ then $\mathscr{H}^s(F)=\infty$ and $\dim_{\mathrm{H}} F\geqslant s$.*

(*b*) *If F is a Borel set with $\mathscr{H}^s(F) > 0$ then there exists a mass distribution μ on F with $I_t(\mu) < \infty$ for all $t < s$.*

Proof

(*a*) Suppose that $I_s(\mu) < \infty$ for some mass distribution μ with support contained in F. Define

$$F_1 = \left\{ x \in F : \overline{\lim_{r \to 0}}\, \mu(B_r(x))/r^s > 0 \right\}.$$

If $x \in F_1$ we may find $\varepsilon > 0$ and a sequence of numbers $\{r_i\}$ decreasing to 0 such that $\mu(B_r(x)) \geqslant \varepsilon r_i^s$. Unless $\mu(\{x\}) > 0$ (in which case it is clear that $I_s(\mu) = \infty$) it follows from the continuity of μ that, by taking q_i $(0 < q_i < r_i)$ small enough, we get $\mu(A_i) \geqslant \frac{1}{4}\varepsilon r_i^s (i = 1, 2, \ldots)$, where A_i is the annulus $B_{r_i}(x) \backslash B_{q_i}(x)$. Taking subsequences if necessary, we may assume that $r_{i+1} < q_i$ for all i, so that the A_i are disjoint annuli centred on x. Hence for $x \in F_1$

$$\phi_s(x) = \int \frac{\mathrm{d}\mu(y)}{|x - y|^s} \geqslant \sum_{i=1}^{\infty} \int_{A_i} \frac{\mathrm{d}\mu(y)}{|x - y|^s}$$

$$\geqslant \sum_{i=1}^{\infty} \tfrac{1}{4} \varepsilon r_i^s r_i^{-s} = \infty$$

since $|x - y|^{-s} \geqslant r_i^{-s}$ on A_i. But $I_s(\mu) = \int \phi_s(x)\,\mathrm{d}\mu(x) < \infty$, so $\phi_s(x) < \infty$ for μ-almost all x. We conclude that $\mu(F_1) = 0$. Since $\overline{\lim}_{r \to 0} \mu(B_r(x))/r^s = 0$ if $x \in F \backslash F_1$, the Proposition 4.9(*a*) tells us that, for all $c > 0$, we have

$$\mathscr{H}^s(F) \geqslant \mathscr{H}^s(F \backslash F_1) \geqslant \mu(F \backslash F_1)/c \geqslant (\mu(F) - \mu(F_1))/c = \mu(F)/c.$$

Hence $\mathscr{H}^s(F) = \infty$.

(*b*) Suppose that $\mathscr{H}^s(F) > 0$. We use $\mathscr{H}^s$ to construct a mass distribution μ on F with $I_t(\mu) < \infty$ for any $t < s$.

By Corollary 4.12 there exists a compact set $E \subset F$ with $0 < \mathscr{H}^s(E) < \infty$ such that

$$\mathscr{H}^s(E \cap B_r(x)) \leqslant b r^s \qquad (x \in \mathbb{R}^n)$$

for some constant b. Let μ be the restriction of $\mathscr{H}^s$ to E, so that $\mu(A) = \mathscr{H}^s(E \cap A)$; then μ is a mass distribution on F. Fix $x \in \mathbb{R}^n$ and write

$$m(r) = \mu(B_r(x)) = \mathscr{H}^s(E \cap B_r(x)) \leqslant b r^s. \tag{4.14}$$

Then, if $0 \leqslant t < s$

$$\phi_t(x) = \int_{|x-y| \leqslant 1} \frac{\mathrm{d}\mu(y)}{|x - y|^t} + \int_{|x-y| > 1} \frac{\mathrm{d}\mu(y)}{|x - y|^t}$$

$$\leqslant \int_0^1 r^{-t} \mathrm{d}m(r) + \mu(\mathbb{R}^n)$$

$$= [r^{-t}m(r)]_{0^+}^{1} + t\int_0^1 r^{-(t+1)}m(r)\,\mathrm{d}r + \mu(\mathbb{R}^n)$$

$$\leqslant b + bt\int_0^1 r^{s-t-1}\,\mathrm{d}r + \mu(\mathbb{R}^n)$$

$$= b\left(1 + \frac{t}{s-t}\right) + \mathscr{H}^s(F)$$

after integrating by parts and using (4.14). Thus $\phi_t(x) \leqslant c$, say, so that $I_t(\mu) = \int \phi_t(x)\,\mathrm{d}\mu(x) \leqslant c\mu(\mathbb{R}^n) < \infty$. □

Important applications of Theorem 4.13 will be given later in the book; for example, in the proof of the projection theorems in Chapter 6 and in the determination of the dimension of Brownian paths in Chapter 16. The theorem is often used to find the dimension of fractals F_θ which depend on a parameter θ. There may be a natural way to define a mass distribution μ_θ on F_θ for each θ. If we can show, that for some s,

$$\int I_s(\mu_\theta)\,\mathrm{d}\theta = \iiint \frac{\mathrm{d}\mu_\theta(x)\,\mathrm{d}\mu_\theta(y)\,\mathrm{d}\theta}{|x-y|^s} < \infty$$

then $I_s(\mu_\theta) < \infty$ for almost all θ, so that $\dim_{\mathrm{H}} F_\theta \geqslant s$ for almost all θ.

Readers familiar with potential theory will have encountered the definition of the *s-capacity* of a set F:

$$C_s(F) = \sup_\mu \{1/I_s(\mu) : \mu \text{ is a mass distribution on } F \text{ with } \mu(F) = 1\}$$

(with the convention that $1/\infty = 0$). Thus another way of expressing Theorem 4.13 is

$$\dim_{\mathrm{H}} F = \inf\{s : C_s(F) = 0\} = \sup\{s : C_s(F) > 0\}.$$

Whilst this is reminiscent of the definition (2.11) of Hausdorff dimension in terms of Hausdorff measures, it should be noted that capacities behave very differently from measures. In particular, they are not generally additive.

*4.4 Fourier transform methods

In this section, we do no more than indicate that Fourier transforms can be a powerful tool for analysing dimensions.

The n-dimensional Fourier transforms of an integrable function f and a mass distribution μ on $\mathbb{R}^n$ are defined by

$$\hat{f}(u) = \int_{\mathbb{R}^n} f(x)\exp(\mathrm{i}x\cdot u)\,\mathrm{d}x \qquad (u \in \mathbb{R}^n) \tag{4.15}$$

$$\hat{\mu}(u) = \int_{\mathbb{R}^n} \exp(\mathrm{i}x \cdot u)\, \mathrm{d}\mu(x) \qquad (u \in \mathbb{R}^n) \tag{4.16}$$

where $x \cdot u$ represents the usual scalar product. (Fourier transformation extends to a much wider class of function using the theory of distributions.)

The s-potential (4.12) of a mass distribution μ is just the convolution

$$\phi_s(x) = (|\cdot|^{-s} * \mu)(x) \equiv \int |x - y|^{-s} \mathrm{d}\mu(y).$$

Formally, the transform of $|x|^{-s}$ may be shown to be $c|u|^{s-n}$, where c depends on n and s, so the convolution theorem, which states that the transform of the convolution of two function equals the product of the transforms of the functions, gives

$$\hat{\phi}_s(u) = c|u|^{s-n}\hat{\mu}(u).$$

Parseval's theorem tells us that

$$\int \phi_s(x)\, \mathrm{d}\mu(x) = (2\pi)^n \int \hat{\phi}_s(u)\overline{\hat{\mu}(u)}\, \mathrm{d}u$$

where the bar denotes complex conjugation, so

$$I_s(\mu) = (2\pi)^n c \int |u|^{s-n} |\hat{\mu}(u)|^2\, \mathrm{d}u. \tag{4.17}$$

This expression for $I_s(\mu)$, which may be established rather more rigorously, is sometimes a convenient way of expressing the energy (4.13) required in Theorem 4.13. Thus if there is a mass distribution μ on a set F for which the integral (4.17) is finite, then $\dim_{\mathrm{H}} F \geqslant s$. In particular, if

$$|\hat{\mu}(u)| \leqslant b|u|^{-t/2} \tag{4.18}$$

for some constant b, then, noting that, by (4.16), $|\hat{\mu}(u)| \leqslant \mu(\mathbb{R}^n)$ for all u, we have

$$I_s(\mu) \leqslant c_1 \int_{|u| \leqslant 1} |u|^{s-n}\, \mathrm{d}u + c_2 \int_{|u| > 1} |u|^{s-n} |u|^{-t}\, \mathrm{d}u$$

which converges if $s < t$. Thus if (4.18) holds, any set F which supports μ has dimension at least t. The greatest value of t for which there is a mass distribution μ on F satisfying (4.18) is sometimes called the *Fourier dimension* of F.

4.5 Notes and references

I know of no systematic general account on calculating dimensions, although there are a number that discuss particular cases in detail. Among these are the papers of Eggleston (1952), Beardon (1965) and Peyrière (1977).

The potential theoretic approach was, essentially, due to Frostman (1935);

see Taylor (1961), Hayman and Kennedy (1976) or Carleson (1967) for more recent accounts. For an introduction to Fourier transforms see Papoulis (1962).

The work on subsets of finite measure originates from Besicovitch (1952) and a very general treatment is given in Rogers (1970). A complete proof of Theorem 4.10 may be found in Falconer (1985a).

Exercises

4.1 What is the Hausdorff dimension of the 'Cantor tartan' given by $\{(x, y) \in \mathbb{R}^2 : \text{either } x \in F \text{ or } y \in F\}$ where F is the middle third Cantor set?

4.2 Fix $0 < \lambda \leqslant \frac{1}{2}$, and let F be the set of real numbers

$$F = \left\{ \sum_{k=1}^{\infty} a_k \lambda^k : a_k = 0 \text{ or } 1 \text{ for } k = 1, 2, \ldots \right\}.$$

Find the Hausdorff and box dimensions of F.

4.3 Let F be the middle third Cantor set. What is the Hausdorff dimension of the plane set given by $\{(x, y) \in \mathbb{R}^2 : x \in F \text{ and } 0 \leqslant y \leqslant x^2\}$?

4.4 Use the mass distribution method to show that the 'Cantor dust' depicted in figure 0.4 has Hausdorff dimension 1. (Hint: note that any two squares in the set E_k of the construction are separated by a distance of at least 4^{-k}.)

4.5 Use a mass distribution method to obtain the result of Example 4.5 directly rather than via Example 4.4.

4.6 Show that every number $x \geqslant 0$ may be expressed in the form

$$x = m + \frac{a_2}{2!} + \frac{a_3}{3!} + \cdots$$

where $m \geqslant 0$ is an integer and a_k is an integer with $0 \leqslant a_k \leqslant k - 1$ for each k. Let $F = \{x \geqslant 0 : m = 0 \text{ and } a_k \text{ is even for } k = 2, 3, \ldots\}$. Find $\dim_H F$.

4.7 Show that there is a compact subset F of $[0, 1]$ of Hausdorff dimension 1 but with $\mathcal{H}^1(F) = 0$. (Hint: try a 'Cantor set' construction, but reduce the proportion of intervals removed at each stage.)

4.8 Deduce from Theorem 4.10 that if F is a Borel subset of $\mathbb{R}^n$ with $\mathcal{H}^s(F) = \infty$ and c is a positive number, then there is a Borel subset E of F with $\mathcal{H}^s(E) = c$.

4.9 Let μ be the natural mass distribution on the middle third Cantor set F (see after Principle 4.2). Estimate the s-energy of μ for $s < \log 2/\log 3$ and deduce from Theorem 4.13 that $\dim_H F \geqslant \log 2/\log 3$.

Chapter 5 Local structure of fractals

Classical calculus involves finding local approximations to curves and surfaces by tangent lines and planes. Viewed on a large scale the neighbourhood of a point on a smooth curve appears close to a line segment. Can we say anything about the local structure of as diverse a class of objects as fractals? Surprisingly, the answer in many cases is yes. We can go some way to establishing the form of fractals in a neighbourhood of a general point. In particular, we can study the concentration of fractals about typical points; in other words, their local densities, and the directional distribution of fractals around points including the question of whether tangents exist. A knowledge of the local form of a fractal is useful both in developing theory and in applications.

In order to realize the power of Hausdorff measures, it is necessary to restrict attention to *s-sets*, i.e. Borel sets of Hausdorff dimension s with positive finite s-dimensional Hausdorff measure. (More generally, it is possible to work with s-sets of positive finite $\mathscr{H}^h$-measure for some dimension function h; see Section 2.5—we do not consider this generalization here.) This is not so restrictive as it first appears. Many fractals encountered in practice are s-sets, but even if $\mathscr{H}^s(F) = \infty$ then, by Theorem 4.10, F has subsets that are s-sets to which this theory can be applied. Alternatively, it sometimes happens that a set F of dimension s is a countable union of s-sets, and the properties of these component sets can often be transferred to F.

The material outlined in this chapter lies at the heart of geometric measure theory, a subject where rigorous proofs are often intricate and difficult. We omit the harder proofs here; it is hoped that those included will be found instructive and will give the flavour of the subject. We generally restrict attention to subsets of the plane—the higher-dimensional analogues, though valid, are often an order of magnitude harder.

5.1 Densities

Let F be a subset of the plane. The *density* of F at x is

$$\lim_{r\to 0} \frac{\text{area}\,(F\cap B_r(x))}{\text{area}\,(B_r(x))} = \lim_{r\to 0} \frac{\text{area}\,(F\cap B_r(x))}{\pi r^2} \tag{5.1}$$

where $B_r(x)$ is the closed disc of radius r and centre x. The classical Lebesgue density theorem tells us that, for a Borel set F, this limit exists and equals 1 when $x \in F$ and 0 when $x \notin F$, except for a set of x of area 0. In other words, for a typical point x of F, small discs centred at x are almost entirely filled by F, but if x is outside F then small discs centred at x generally contain very little of F; see figure 5.1.

Similarly, if F is a smooth curve in the plane and x is a point of F (other than an endpoint), then, $F \cap B_r(x)$ is close to a diametrical chord of F for small r and

$$\lim_{r \to 0} \frac{\text{length}\,(F \cap B_r(x))}{2r} = 1.$$

If $x \notin F$ then this limit is clearly 0.

Density theorems such as these tell us how much of the set F, in the sense of area or length, is concentrated near x. In the same way it is natural to investigate densities of fractals—if F has dimension s, how does the s-dimensional Hausdorff measure of $F \cap B_r(x)$ behave as $r \to 0$? We look at this question when F is an s-set in $\mathbb{R}^2$ with $0 < s < 2$ (0-sets are just finite sets of points, and there is little to say, and $\mathscr{H}^2$ is essentially area, so if $s = 2$ we are in the Lebesgue density situation (5.1)).

We define the *lower* and *upper densities* of an s-set F at a point $x \in \mathbb{R}^n$ as

$$\underline{D}^s(F, x) = \varliminf_{r \to 0} \frac{\mathscr{H}^s(F \cap B_r(x))}{(2r)^s} \tag{5.2}$$

and

$$\overline{D}^s(F, x) = \varlimsup_{r \to 0} \frac{\mathscr{H}^s(F \cap B_r(x))}{(2r)^s} \tag{5.3}$$

respectively (note that $|B_r(x)| = 2r$). If $\underline{D}^s(F, x) = \overline{D}^s(F, x)$ we say that the density of F at x exists and we write $D^s(F, x)$ for the common value.

A point x at which $\underline{D}^s(F, x) = \overline{D}^s(F, x) = 1$ is called a *regular* point of F, otherwise x is an *irregular* point. An s-set is termed *regular* if $\mathscr{H}^s$-almost all of its points

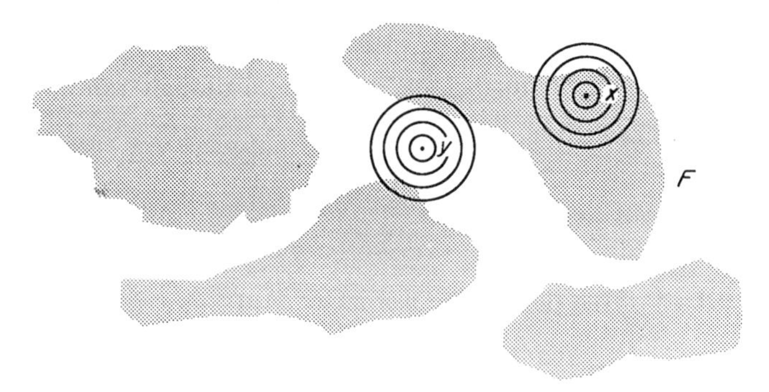

Figure 5.1 The Lebesgue density theorem. The point x is in F, and area $(F \cap B_r(x))$/area $(B_r(x))$ is close to 1 if r is small. The point y is outside F, and area $(F \cap B_r(y))$/area $(B_r(y))$ is close to 0 if r is small

are regular (i.e. all of its points except for a set of $\mathscr{H}^s$-measure 0), and *irregular* if $\mathscr{H}^s$-almost all of its points are irregular. As we shall see a fundamental result is that an s-set F must be irregular unless s is an integer. However, if s is integral an s-set decomposes into a regular and an irregular part. Roughly speaking, a regular 1-set consists of portions of rectifiable curves of finite length, whereas an irregular 1-set is totally disconnected and dust-like, and typically of fractal form.

By definition, a regular set is one for which the direct analogue of the Lebesgue density theorem holds. However, even the densities of irregular sets cannot behave too erratically.

Proposition 5.1

Let F be an s-set in $\mathbb{R}^n$. Then

(*a*) $\underline{D}^s(F,x)=\overline{D}^s(F,x)=0$ *for $\mathscr{H}^s$-almost all $x\notin F$*
(*b*) $2^{-s}\leqslant\overline{D}^s(F,x)\leqslant 1$ *for $\mathscr{H}^s$-almost all $x\in F$.*

Partial proof

(*a*) If F is closed and $x\notin F$, then $B_r(x)\cap F=\varnothing$ if r is small enough. Hence $\lim_{r\to 0}\mathscr{H}^s(F\cap B_r(x))/(2r)^s=0$. If F is not closed the proof is a little more involved and we omit it here.
(*b*) This follows quickly from Proposition 4.9(*a*) by taking μ as the restriction of $\mathscr{H}^s$ to F, i.e. $\mu(A)=\mathscr{H}^s(F\cap A)$: if

$$F_1=\left\{x\in F:\overline{D}^s(F,x)=\overline{\lim_{r\to 0}}\frac{\mathscr{H}^s(F\cap B_r(x))}{(2r)^s}<2^{-s}c\right\}$$

then $\mathscr{H}^s(F_1)\geqslant\mathscr{H}^s(F)/c$. If $c<1$ this is only possible if $\mathscr{H}^s(F_1)=0$; thus for almost all $x\in F$ we have $\overline{D}^s(F,x)\geqslant 2^{-s}$. The upper bound follows in essentially the same way using Proposition 4.9(*b*). □

Note that an immediate consequence of Proposition 5.1(*b*) is that an irregular set has a lower density which is strictly less than 1 almost everywhere.

We will sometimes need to relate the densities of a set to those of certain subsets. Let F be an s-set and let E be a Borel subset of F. Then

$$\frac{\mathscr{H}^s(F\cap B_r(x))}{(2r)^s}=\frac{\mathscr{H}^s(E\cap B_r(x))}{(2r)^s}+\frac{\mathscr{H}^s((F\backslash E)\cap B_r(x))}{(2r)^s}.$$

For almost all x in E, we have

$$\frac{\mathscr{H}^s((F\backslash E)\cap B_r(x))}{(2r)^s}\to 0\qquad\text{as } r\to 0$$

by Proposition 5.1(*a*), so letting $r\to 0$ gives

$$\underline{D}^s(F,x)=\underline{D}^s(E,x);\qquad \overline{D}^s(F,x)=\overline{D}^s(E,x)\tag{5.4}$$

for $\mathcal{H}^s$-almost all x in E. Thus, from the definitions of regularity, if E is a subset of an s-set F with $\mathcal{H}^s(E) > 0$, then E is regular if F is regular and irregular if F is irregular. In particular, the intersection of a regular and an irregular set, being a subset of both, has measure zero.

Estimates for lower densities are altogether harder to obtain, and we do not pursue them here.

In general it is quite hard to show that s-sets of non-integral dimension are irregular, but in the case $0 < s < 1$ the following 'annulus' proof is appealing.

Theorem 5.2

Let F be an s-set in $\mathbb{R}^2$. Then F is irregular unless s is an integer.

Partial proof. We show that F is irregular if $0 < s < 1$ by showing that the density $D(F, x)$ fails to exist almost everywhere in F. Suppose to the contrary: then there is a set $F_1 \subset F$ of positive measure where the density exists and therefore where $\frac{1}{2} < 2^{-s} \leqslant D^s(F, x)$, by Proposition 5.1 (*b*). By Egoroff's theorem (see page 16) we may find $r_0 > 0$ and a Borel set $E \subset F_1 \subset F$ with $\mathcal{H}^s(E) > 0$ such that

$$\mathcal{H}^s(F \cap B_r(x)) > \tfrac{1}{2}(2r)^s \tag{5.5}$$

for all $x \in E$ and $r < r_0$. Let $y \in E$ be a cluster point of E (i.e. a point y with other points of E arbitrarily close). Let η be a number with $0 < \eta < 1$ and let $A_{r,\eta}$ be

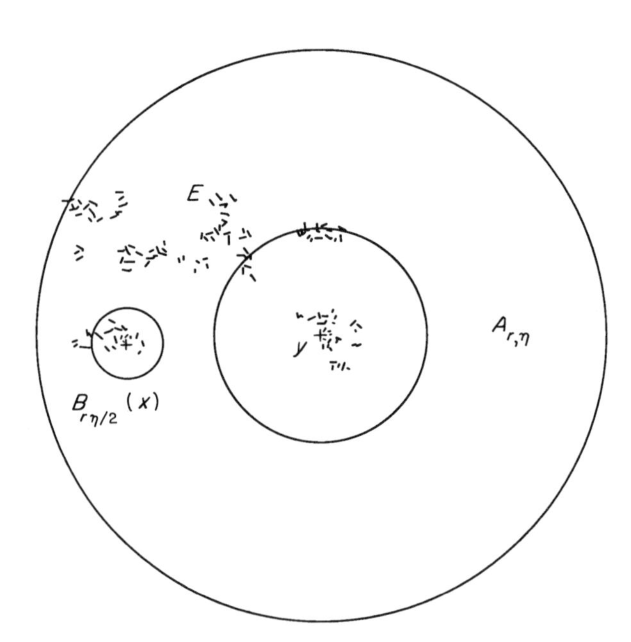

Figure 5.2 The 'annulus' proof of Theorem 5.2

the annulus $B_{r(1+\eta)}(y)\backslash B_{r(1-\eta)}(y)$; see figure 5.2. Then

$$(2r)^{-s}\mathscr{H}^s(F\cap A_{r,\eta})=(2r)^{-s}\mathscr{H}^s(F\cap B_{r(1+\eta)}(y))-(2r)^{-s}\mathscr{H}^s(F\cap B_{r(1-\eta)}(y))$$
$$\to D^s(F,y)((1+\eta)^s-(1-\eta)^s) \tag{5.6}$$

as $r\to 0$. For a sequence of values of r tending to 0, we may find $x\in E$ with $|x-y|=r$. Then $B_{\frac{1}{2}r\eta}(x)\subset A_{r,\eta}$ so by (5.5)

$$\tfrac{1}{2}r^s\eta^s<\mathscr{H}^s(F\cap B_{\frac{1}{2}r\eta}(x))\leqslant \mathscr{H}^s(F\cap A_{r,\eta}).$$

Combining with (5.6) this implies that

$$\begin{aligned}2^{-s-1}\eta^s&\leqslant D^s(F,y)((1+\eta)^s-(1-\eta)^s)\\&=D^s(F,y)(2s\eta+\text{terms in }\eta^2\text{ or higher}).\end{aligned}$$

Letting $\eta\to 0$ we see that this is impossible when $s<1$ and the result follows by contradiction. □

5.2 Structure of 1-sets

As we have pointed out, sets of non-integral dimension must be irregular. The situation for sets of integral dimension is more complicated. The following decomposition theorem, indicated in figure 5.3, enables us to split a 1-set into a regular and an irregular part, so that we can analyse each separately, and recombine them without affecting density properties.

Decomposition theorem 5.3

Let F be a 1-set. The set of regular points of F forms a regular set, the set of irregular points forms an irregular set.

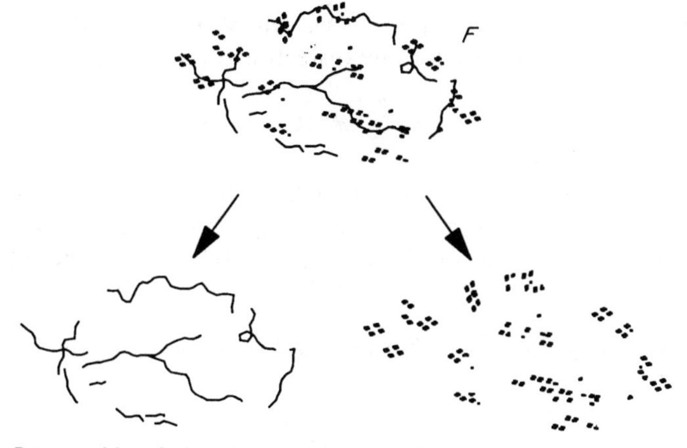

Figure 5.3 Decomposition of a 1-set into a regular 'curve-like' part and an irregular 'curve-free' part

Proof. This is immediate from (5.4), taking E as the set of regular and irregular points respectively. □

Examples of regular and irregular 1-sets abound. Smooth curves are regular, and provide us with the shapes of classical geometry such as the perimeters of circles or ellipses. On the other hand the iterated construction of figure 0.4 gives an irregular 1-set which is a totally disconnected fractal. This is typical—as we shall see, regular 1-sets are made up from pieces of curve, whereas irregular 1-sets are dust-like and 'curve-free', i.e. intersect any (finite length) curve in length zero.

To study 1-sets we need a few facts about curves. For our purposes a *curve* or *Jordan curve* C is the image of a continuous injection (one-to-one function) $\psi: [a, b] \to \mathbb{R}^2$, where $[a, b] \subset \mathbb{R}$ is a proper closed interval. According to our definition, curves are not self-intersecting, have two ends, and are compact connected subsets of the plane. The length $\mathscr{L}(C)$ of the curve C is given by polygonal approximation:

$$\mathscr{L}(C) = \sup \sum_{i=1}^{m} |x_i - x_{i-1}|$$

where the supremum is taken over all dissections of C by points $x_0, x_1, \ldots, x_m$ in that order along the curve. If the length $\mathscr{L}(C)$ is finite we call C a *rectifiable curve*.

As one might expect, the length of a curve equals its 1-dimensional Hausdorff measure.

Lemma 5.4

If C is a rectifiable curve then $\mathscr{H}^1(C) = \mathscr{L}(C)$.

Proof. For $x, y \in C$, let $C_{x,y}$ denote that part of C between x and y. Since orthogonal projection onto the line through x and y does not increase distances, (2.9) gives $\mathscr{H}^1(C_{x,y}) \geqslant \mathscr{H}^1[x, y] = |x - y|$, where $[x, y]$ is the straight-line segment joining x to y. Hence for any dissection $x_0, x_1, \ldots, x_m$ of C,

$$\sum_{i=1}^{m} |x_i - x_{i-1}| \leqslant \sum_{i=1}^{m} \mathscr{H}^1(C_{x_i, x_{i-1}}) \leqslant \mathscr{H}^1(C)$$

so that $\mathscr{L}(C) \leqslant \mathscr{H}^1(C)$. On the other hand, let $f: [0, \mathscr{L}(C)] \to C$ be the mapping that takes t to the point on C at distance t along the curve from one of its ends. Clearly $|f(t) - f(u)| \leqslant |t - u|$ for $0 \leqslant t, u \leqslant \mathscr{L}(C)$, so $\mathscr{H}^1(C) \leqslant \mathscr{L}(C)$ by (2.9) as required.

It is straightforward to show that rectifiable curves are regular.

Lemma 5.5

A rectifiable curve is a regular 1-set.

Proof. If C is rectifiable, $\mathscr{L}(C) < \infty$, and since C has distinct endpoints p and q, we get $\mathscr{L}(C) \geqslant |p-q| > 0$. By Lemma 5.4, $0 < \mathscr{H}^1(C) < \infty$, so C is a 1-set.

A point x of C that is not an endpoint, divides C into two parts $C_{p,x}$ and $C_{x,q}$. If r is sufficiently small, then moving away from x along the curve $C_{x,q}$ we reach a first point y on C with $|x-y| = r$. Then $C_{x,y} \subset B_r(x)$ and

$$r = |x-y| \leqslant \mathscr{L}(C_{x,y}) = \mathscr{H}^1(C_{x,y}) \leqslant \mathscr{H}^1(C_{x,q} \cap B_r(x)).$$

Similarly, $r \leqslant \mathscr{H}^1(C_{p,x} \cap B_r(x))$, so, adding, $2r \leqslant \mathscr{H}^1(C \cap B_r(x))$, if r is small enough. Thus

$$\underline{D}^1(C,x) = \lim_{r \to 0} \frac{\mathscr{H}^1(C \cap B_r(x))}{2r} \geqslant 1.$$

By Proposition 5.1(*b*) $\underline{D}^1(C,x) \leqslant \bar{D}^1(C,x) \leqslant 1$, so $D^1(C,x)$ exists and equals 1 for all $x \in C$ other than the two endpoints, so C is regular. □

Other regular sets are easily constructed. By (5.4), subsets of regular sets, and unions of regular sets should also be regular. With this in mind we define a 1-set to be *curve-like* if it is contained in a countable union of rectifiable curves.

Proposition 5.6

A curve-like set is a regular 1-set.

Proof. If F is curve-like then $F \subset \bigcup_{i=1}^{\infty} C_i$ where the C_i are rectifiable curves. For each i and $\mathscr{H}^1$-almost all $x \in F \cap C_i$ we have, using Lemma 5.5 and equation (5.4),

$$1 = \underline{D}^1(C_i, x) = \underline{D}^1(F \cap C_i, x) \leqslant \underline{D}^1(F, x)$$

and hence $1 \leqslant \underline{D}^1(F,x)$ for almost all $x \in F$. But for almost all $x \in F$ we have $\underline{D}^1(F,x) \leqslant \bar{D}^1(F,x) \leqslant 1$ so $D^1(F,x) = 1$ almost everywhere, and F is regular. □

It is natural to introduce a complementary definition: a 1-set is called *curve-free* if its intersection with every rectifiable curve has $\mathscr{H}^1$-measure-zero.

Proposition 5.7

An irregular 1-set is curve-free.

Proof. If F is irregular and C is a rectifiable curve then $F \cap C$ is a subset of both a regular and an irregular set, so has zero $\mathscr{H}^1$-measure. □

These two propositions begin to suggest that regular and irregular sets might be characterized as curve-like and curve-free respectively. This is indeed the case, but it is far from easy to prove. The crux of the matter is the following lower-density estimate, which depends on an intricate investigation of the properties of curves and connected sets and some ingenious geometrical arguments.

Proposition 5.8

Let F be a curve-free 1-set in $\mathbb{R}^2$. Then $\underline{D}^1(F,x) \leqslant \frac{3}{4}$ at almost all $x \in E$.

Proof. Omitted. □

Assuming this proposition, a complete characterization of regular and irregular sets is relatively easy.

Theorem 5.9

(*a*) *A 1-set in $\mathbb{R}^2$ is irregular if and only if it is curve-free.*
(*b*) *A 1-set in $\mathbb{R}^2$ is regular if and only if it is the union of a curve-like set and a set of $\mathscr{H}^1$-measure zero.*

Proof

(*a*) A curve-free set must be irregular by Proposition 5.8. Proposition 5.7 provides the converse implication.
(*b*) By Proposition 5.6 a curve-like set is regular, and adding in a set of measure zero does not affect densities or, therefore, regularity.

If F is regular, then any Borel subset E of positive measure is regular with $\underline{D}^1(E,x)=1$ for almost all $x \in E$. By Proposition 5.8 the set E cannot be curve-free, so some rectifiable curve intersects E in a set of positive length. We use this fact to define inductively a sequence of rectifiable curves $\{C_i\}$. We choose C_1 to cover a reasonably large part of F, say

$$\mathscr{H}^1(F \cap C_1) \geqslant \tfrac{1}{2}\sup\{\mathscr{H}^1(F \cap C) : C \text{ is rectifiable}\} > 0.$$

If $C_1, \ldots, C_k$ have been selected and $F_k = F \backslash \bigcup_{i=1}^k C_i$ has positive measure, let C_{k+1} be a rectifiable curve for which

$$\mathscr{H}^1(F_k \cap C_{k+1}) \geqslant \tfrac{1}{2}\sup\{\mathscr{H}^1(F_k \cap C) : C \text{ is rectifiable}\} > 0. \tag{5.7}$$

If the process terminates then for some k the curves $C_1, \ldots, C_k$ cover almost all of F and F is curve-like. Otherwise,

$$\infty > \mathscr{H}^1(F) \geqslant \sum_k \mathscr{H}^1(F_k \cap C_{k+1})$$

since the $F_k \cap C_{k+1}$ are disjoint, so that $\mathscr{H}^1(F_k \cap C_{k+1}) \to 0$ as $k \to \infty$. If $\mathscr{H}^1(F \backslash \bigcup_{i=1}^\infty C_i) > 0$ there is a rectifiable curve C such that $\mathscr{H}^1((F \backslash \bigcup_{i=1}^\infty C_i) \cap C) = d$ for some $d > 0$. But $\mathscr{H}^1(F_k \cap C_{k+1}) < \frac{1}{2}d$ for some k, so, according to (5.7), C would have been selected in preference to C_{k+1}. Hence $\mathscr{H}^1(F\backslash\bigcup_{i=1}^{\infty} C_i) = 0$, and F consists of the curve-like set $F \cap \bigcup_{i=1}^{\infty} C_i$ together with $F\backslash\bigcup_{i=1}^{\infty} C_i$, which is of measure zero. □

Thus regular 1-sets are essentially unions of subsets of rectifiable curves, but irregular 1-sets contain no pieces of rectifiable curves at all. This dichotomy is remarkable in that the definition of regularity is purely in terms of densities and makes no reference to curves. Propositions 5.6 and 5.8 provide a further contrast. Almost everywhere, a regular set has lower density 1, whereas an irregular set has lower density at most $\frac{3}{4}$. Thus in any 1-set F the set of points for which $\frac{3}{4} < \underline{D}^1(F, x) < 1$ has $\mathscr{H}^1$-measure zero.

Regular 1-sets may be connected but, like sets of dimension less than 1, irregular 1-sets must be totally disconnected. We know at least that distinct points cannot be joined by a rectifiable curve in an irregular set, and further investigation shows that no two points can lie in the same connected component.

Further differences between regular and irregular sets include the existence of tangents (see Section 5.3) and projection properties (see Chapter 6). In all these ways, the classes of regular and irregular 1-sets are distanced from each other. For the special case of 1-sets, it would make sense mathematically to define fractals to be those sets which are irregular.

5.3 Tangents to *s*-sets

Suppose that a smooth curve C has a tangent (in the classical sense) at x. This means that close to x the set C is concentrated in two diametrically opposite directions. What can be said about the directional distribution of an s-set about a typical point? Is there a meaningful definition of a tangent to an s-set, and when do such tangents exist?

Any generalization of the definition of tangents should reflect the directional distribution of sets of positive measure—for sets of the complexity that we have in mind, there is no hope of a definition involving *all* nearby points; we must be content with a condition on *almost all* points. We say that an s-set F in $\mathbb{R}^n$ has a *tangent at x in direction $\boldsymbol{\theta}$* ($\boldsymbol{\theta}$ a unit vector) if

$$\overline{D}^s(F, x) > 0 \tag{5.8}$$

and, for every angle $\varphi > 0$,

$$\lim_{r\to 0} r^{-s}\mathscr{H}^s(F \cap (B_r(x)\backslash S(x, \boldsymbol{\theta}, \varphi))) = 0 \tag{5.9}$$

where $S(x, \boldsymbol{\theta}, \varphi)$ is the *double sector* with vertex x, consisting of those y such that the line segment $[x, y]$ makes an angle at most φ with $\boldsymbol{\theta}$ or $-\boldsymbol{\theta}$; see figure 5.4. Thus, for a tangent, (5.8) requires that a significant part of F lies near x, of which, by (5.9), a negligible amount lies outside any double sector $S(x, \boldsymbol{\theta}, \varphi)$; see figure 5.5.

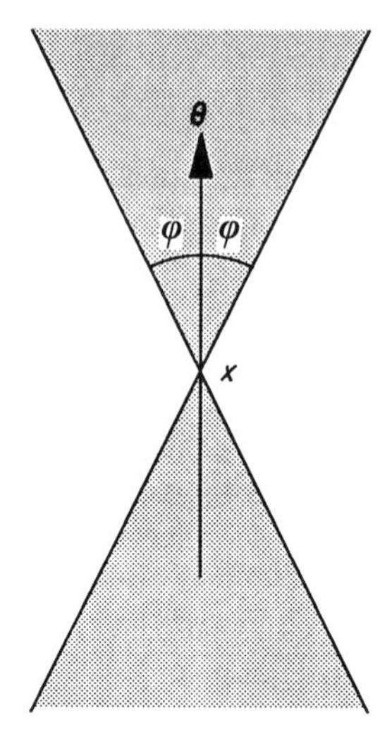

Figure 5.4 The double sector $S(x, \boldsymbol{\theta}, \varphi)$

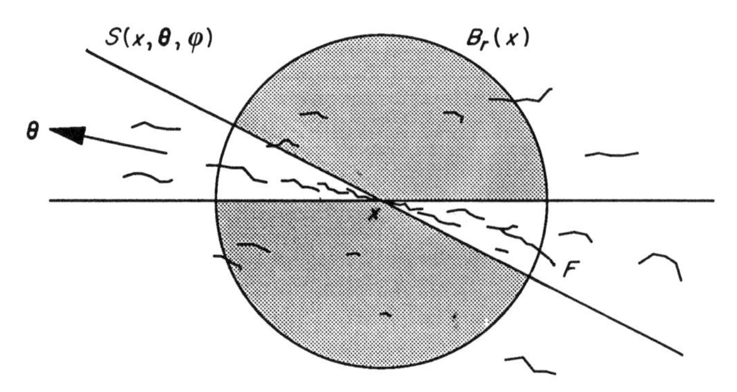

Figure 5.5 For F to have a tangent in direction $\boldsymbol{\theta}$ at x, there must be a negligible part of F in $B_r(x)\backslash S(x, \boldsymbol{\theta}, \varphi)$ (shaded) for small r

We first discuss tangents to regular 1-sets in the plane, a situation not far removed from the classical calculus of curves.

Proposition 5.10

A rectifiable curve C has a tangent at almost all of its points.

Proof. By Lemma 5.5 the upper density $\bar{D}^1(C, x) = 1 > 0$ for almost all $x \in C$. We may reparametrize the defining function of C by arc length, so that $\psi: [0, \mathscr{L}(C)] \to \mathbb{R}^2$ gives $\psi(t)$ as the point distance t along C from the endpoint $\psi(0)$. To say that $\mathscr{L}(C) < \infty$ simply means that ψ has bounded variation, in other words $\sup \sum_{i=1}^{m} |\psi(t_i) - \psi(t_{i-1})| < \infty$ where the supremum is over dissections $0 = t_0 < t_1 < \cdots < t_m = \mathscr{L}(C)$. We quote a standard result from the theory of functions, that functions of bounded variation are differentiable almost everywhere, so $\psi'(t)$ exists as a vector for almost all t. Because of the arc-length parametrization, $|\psi'(t)|=1$ for such t. Hence at almost all points $\psi(t)$ on C, there exists a unit vector $\boldsymbol{\theta}$ such that $\lim_{u\to t}(\psi(u)-\psi(t))/(u-t)=\boldsymbol{\theta}$. Thus, given $\varphi>0$, there is a number $\varepsilon>0$ such that $\psi(u)\in S(\psi(t),\boldsymbol{\theta},\varphi)$ whenever $|u-t|<\varepsilon$. Since C has no double points we may find r such that $\psi(u)\notin B_r(\psi(t))$ if $|u-t|\geqslant\varepsilon$, so $C\cap(B_r(\psi(t))\backslash S(\psi(t),\boldsymbol{\theta},\varphi))$ is empty. By the definition (5.8) and (5.9), the curve C has a tangent at $\psi(t)$. Such points account for almost all points on C. □

Just as with densities, we can transfer tangency properties from curves to curve-like sets.

Proposition 5.11

A regular 1-set F in $\mathbb{R}^2$ has a tangent at almost all of its points.

Proof. By definition of regularity, $\underline{D}^1(F,x)=1>0$ at almost all $x\in F$.

If C is any rectifiable curve, then for almost all x in C there exists $\boldsymbol{\theta}$ such that if $\varphi>0$

$$\begin{aligned}&\lim_{r\to\infty} r^{-1}\mathscr{H}^1((F\cap C)\cap(B_r(x)\backslash S(x,\boldsymbol{\theta},\varphi)))\\ &\quad\leqslant \lim_{r\to\infty} r^{-1}\mathscr{H}^1(C\cap(B_r(x)\backslash S(x,\boldsymbol{\theta},\varphi)))=0\end{aligned}$$

by Proposition 5.10. Moreover

$$\lim_{r\to\infty} r^{-1}\mathscr{H}^1((F\backslash C)\cap(B_r(x)\backslash S(x,\boldsymbol{\theta},\varphi)))\leqslant \lim_{r\to\infty} r^{-1}\mathscr{H}^1((F\backslash C)\cap B_r(x))=0$$

for almost all $x\in C$ by Property 5.1(*a*). Adding these inequalities

$$\lim_{r\to\infty} r^{-1}\mathscr{H}^1(F\cap(B_r(x)\backslash S(x,\boldsymbol{\theta},\varphi)))=0$$

for almost all $x\in C$ and so for almost all $x\in F\cap C$. Since a countable collection of such curves covers almost all of F, the result follows. □

In contrast to regular sets, irregular 1-sets do not generally support tangents.

Proposition 5.12

At almost all points of an irregular 1-set, no tangent exists.

Proof. The proof, which depends on the characterization of irregular sets as curve-free sets, is too involved to include here. □

We turn now to s-sets in $\mathbb{R}^2$ for non-integral s, which, as we have seen, are necessarily irregular. For $0<s<1$ tangency questions are not particularly interesting, since any set contained in a smooth curve will automatically satisfy (5.9) with $\boldsymbol{\theta}$ the direction of the tangent to the curve at x. For example, the middle third Cantor set F regarded as a subset of the plane is a ($\log 2/\log 3$)-set that satisfies (5.8) and (5.9) for all x in F and $\varphi > 0$, where $\boldsymbol{\theta}$ is a vector pointing along the set. On the other hand, if F, say, is a Cartesian product of two uniform Cantor sets, each formed by repeated removal of a proportion $\alpha > \frac{1}{2}$ from the centre of intervals, then a little calculation (see Chapter 7) shows that F is an s-set with $s = 2\log 2/\log(2/(1-\alpha)) < 1$ with no tangents at any of its points.

It is at least plausible that s-sets in $\mathbb{R}^2$ with $1 < s < 2$ do not have tangents—such sets are so large that they radiate in many directions from a typical point, so that (5.9) cannot hold. This is made precise in the following proposition.

Proposition 5.12

If F is an s-set in $\mathbb{R}^2$ with $1 < s < 2$, then at almost all points of F, no tangent exists.

Proof. For $r_0 > 0$ let

$$E = \{ y \in F : \mathcal{H}^s(F \cap B_r(y)) < 2(2r)^s \text{ for all } r < r_0 \}. \tag{5.10}$$

For any $x \in F$, any unit vector $\boldsymbol{\theta}$ and any angle φ with $0 < \varphi < \frac{1}{2}\pi$, we estimate how much of E lies in $B_r(x) \cap S(x, \boldsymbol{\theta}, \varphi)$. For $r < r_0/20$ and $i = 1, 2, \ldots$ let A_i be the intersection of the annulus and the double sector given by

$$A_i = (B_{ir\varphi}(x) \backslash B_{(i-1)r\varphi}(x)) \cap S(x, \boldsymbol{\theta}, \varphi).$$

Then $B_r(x) \cap S(x, \boldsymbol{\theta}, \varphi) \subset \bigcup_{i=1}^m A_i \cup \{x\}$ for some integer $m < 2/\varphi$. Each A_i comprises two parts, both of diameter at most $10r\varphi < r_0$, so applying (5.10) to the parts that contain points of E, and summing,

$$\mathcal{H}^s(E \cap B_r(x) \cap S(x, \boldsymbol{\theta}, \varphi)) \leqslant (4\varphi^{-1}) 2(20r\varphi)^s$$

so that

$$(2r)^{-s} \mathcal{H}^s(E \cap B_r(x) \cap S(x, \boldsymbol{\theta}, \varphi)) \leqslant 8.10^s \varphi^{s-1} \tag{5.11}$$

if $r < r_0/20$.

Now, almost all $x \in E$ satisfy $\bar{D}^s(F \backslash E, x) = 0$ by Proposition 5.1(*a*). Decomposing $F \cap B_r(x)$ into three parts we get

$$\begin{aligned}\mathcal{H}^s(F \cap B_r(x)) = \mathcal{H}^s((F \backslash E) \cap B_r(x)) + \mathcal{H}^s(E \cap B_r(x) \cap S(x, \boldsymbol{\theta}, \varphi)) \\ + \mathcal{H}^s(E \cap (B_r(x) \backslash S(x, \boldsymbol{\theta}, \varphi))).\end{aligned}$$

Dividing by $(2r)^s$ and taking upper limits as $r \to 0$,

$$\bar{D}^s(F, x) \leqslant 0 + 8.10^s \varphi^{s-1} + \overline{\lim_{r \to 0}}\, (2r)^{-s} \mathcal{H}^s(F \cap (B_r(x) \backslash S(x, \boldsymbol{\theta}, \varphi)))$$

for almost all $x \in E$, using (5.11). Choosing φ sufficiently small, it follows that

(5.8) and (5.9) cannot both hold for any $\boldsymbol{\theta}$, so no tangent exists at x. To complete the proof, we note that almost all $x \in F$ belong to the set E defined in (5.10) for some $r_0 > 0$, by Proposition 5.1(*b*). □

The results of this chapter begin to provide a local picture of fractals that are s-sets. By taking these methods rather further, it is possible to obtain much more precise estimates of densities and also of the directional distributions of s-sets about typical points. For example, it may be shown that if $s > 1$, almost every line through $\mathcal{H}^s$-almost every point of an s-set F intersects F in a set of dimension $s - 1$.

Recently, packing measures (see Section 3.4) have been employed in the study of local properties, and it has been shown that regularity of a set corresponds closely to the equality of its packing measure and (slightly modified) Hausdorff measure.

These ideas extend, albeit with considerable effort, to higher dimensions. Regular s-sets in $\mathbb{R}^n$ may be defined using densities and, again, s-sets can only be regular if s is an integer. Regular s-sets have tangents almost everywhere, and are 's-dimensional-surface-like' in the sense that, except for a subset of $\mathcal{H}^s$-measure zero, they may be covered by a countable collection of Lipschitz images of subsets of $\mathbb{R}^s$.

5.4 Notes and references

This chapter touches the surface of a deep area of mathematics known as geometric measure theory. It has its origins in the fundamental papers of Besicovitch (1928, 1938) which contain a remarkably complete analysis of 1-sets in the plane. The results on s-sets in the plane for non-integral s are due to Marstrand (1954a). A succession of writers have extended this work to subsets of higher-dimensional space, culminating in the paper of Priess (1987) which solved many of the outstanding problems. A more detailed discussion of s-sets in the plane may be found in Falconer (1985a), see also Federer (1969).

Exercises

5.1 By applying Proposition 5.1 with $s = n = 2$, deduce the Lebesgue density theorem (5.1).

5.2 Let $f:\mathbb{R} \to \mathbb{R}$ be a continuously differentiable function such that $0 < c_1 \leqslant f'(x) \leqslant c_2$ for all x. Show that, if F is an s-set in $\mathbb{R}$, then $\underline{D}^s(f(F), f(x)) = \underline{D}^s(F, x)$ for all x in $\mathbb{R}$, with a similar result for upper densities.

5.3 Let F be the middle third Cantor set. Show that $\underline{D}^s(F, x) \leqslant 2^{-s}$ for all x, where $s = \log 2/\log 3$. Deduce that F is irregular.

5.4 Estimate the upper and lower densities at points of the 1-set depicted in figure 0.4 and show that it is irregular.

5.5 Adapt the proof of Theorem 5.2 to show that if F is an s-set with $0 < s < 1$, then $\underline{D}^s(F, x) \leqslant (1 + 2^{s/(s-1)})^{s-1}$ for almost all x.

5.6 Construct a regular 1-set that is totally disconnected. (Hint: start with a rectifiable curve.)

5.7 Let E and F be s-sets in $\mathbb{R}^2$ such that for every disc $B_r(x)$ we have that $\mathscr{H}^s(B_r(x) \cap E) \leqslant \mathscr{H}^s(B_r(x) \cap F)$. Show that $\mathscr{H}^s(E \setminus F) = 0$. Need we have $E \subset F$?

5.8 Let $F_1, F_2, \ldots$ be 1-sets in the plane such that $F = \bigcup_{k=1}^{\infty} F_k$ is a 1-set. Show that if F_k is regular for all k then F is regular, and if F_k is irregular for all k then F is irregular.

Chapter 6 Projections of fractals

In this chapter we consider the orthogonal projection or 'shadow' of fractals in $\mathbb{R}^n$ onto lower-dimensional subspaces. A smooth (1-dimensional) curve in $\mathbb{R}^3$ generally has a (1-dimensional) curve as its shadow on a plane, but a (2-dimensional) surface or (3-dimensional) solid object generally has a 2-dimensional shadow, as in the upper part of figure 6.1. We examine analogues of this for fractals. Intuitively, one would expect a set F in $\mathbb{R}^3$ to have plane projections of dimension 2 if $\dim_H F > 2$ and of dimension $\dim_H F$ if $\dim_H F < 2$, as in the lower part of figure 6.1. Roughly speaking this is correct, but a precise formulation of the projection properties requires some care.

We prove the projection theorems in the simplest case, for projection of subsets of the plane onto lines, and then state the higher-dimensional analogues.

6.1 Projections of arbitrary sets

Let L_θ be the line through the origin of $\mathbb{R}^2$ that makes an angle θ with the horizontal axis. We denote orthogonal projection onto L_θ by proj_θ, so that if F is a subset of $\mathbb{R}^2$, then $\text{proj}_\theta F$ is the projection of F onto L_θ; see figure 6.2. Clearly, $|\text{proj}_\theta x - \text{proj}_\theta y| \leqslant |x-y|$ if $x, y \in \mathbb{R}^2$, i.e. proj_θ is a Lipschitz mapping. Thus

$$\dim_H(\text{proj}_\theta F) \leqslant \min\{\dim_H F, 1\} \tag{6.1}$$

for any F and θ, by Corollary 2.4(*a*). (As $\text{proj}_\theta F$ is a subset of the line L_θ, its dimension cannot be more than 1.) The interesting question is whether the opposite inequality is valid. The projection theorems tell us that this is so for almost all $\theta \in [0, \pi)$; that is, the exceptional values of θ for which inequality (6.1) is strict form a set of zero length (1-dimensional Lebesgue measure).

Projection theorem 6.1

Let $F \subset \mathbb{R}^2$ be a Borel set.

(*a*) *If* $\dim_H F \leqslant 1$ *then* $\dim_H(\text{proj}_\theta F) = \dim_H F$ *for almost all* $\theta \in [0, \pi)$.

(*b*) *If* $\dim_H F > 1$ *then* $\text{proj}_\theta F$ *has positive length (as a subset of* L_θ*) and so has dimension* 1 *for almost all* $\theta \in [0, \pi)$.

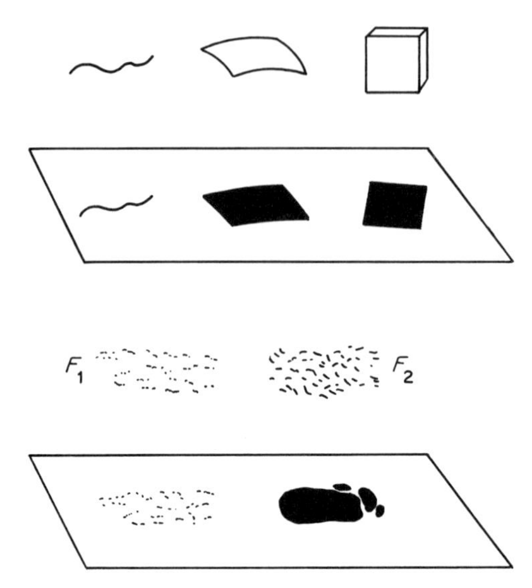

Figure 6.1 Top: projections of classical sets onto a plane—a curve 'typically' has projection of dimension 1, but the surface and cube have projections of dimension 2 and of positive area. Bottom: projections of fractal sets onto a plane. If $\dim_H F_1 < 1$ and $\dim_H F_2 > 1$ then 'typically' the projection of F_1 has dimension equal to $\dim_H F_1$ (and zero area) and the projection of F_2 has dimension 2 and positive area

Proof. We give a proof that uses the potential theoretic characterization of Hausdorff dimension in a very effective way. If $s < \dim F \leqslant 1$ then by Theorem 4.13(*b*) there exists a mass distribution μ on (a compact subset of) F with $0 < \mu(F) < \infty$ and

$$\int_F \int_F \frac{\mathrm{d}\mu(x)\,\mathrm{d}\mu(y)}{|x-y|^s} < \infty. \tag{6.2}$$

For each θ we 'project' the mass distribution μ onto the line L_θ to get a mass

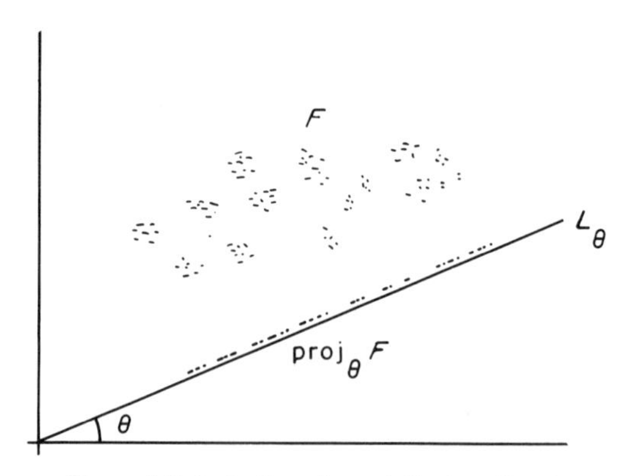

Figure 6.2 Projection of a set F onto a line L_θ

distribution μ_θ on $\text{proj}_\theta F$. Thus μ_θ is defined by the requirement that

$$\mu_\theta([a,b]) = \mu\{x : a \leqslant x\cdot\boldsymbol{\theta} \leqslant b\}$$

for each interval $[a,b]$, or equivalently,

$$\int_{-\infty}^{\infty} f(t)\,\mathrm{d}\mu_\theta(t) = \int_F f(x\cdot\boldsymbol{\theta})\,\mathrm{d}\mu(x)$$

for each non-negative function f. (Here $\boldsymbol{\theta}$ is the unit vector in the direction θ, x is identified with its position vector and $x\cdot\boldsymbol{\theta}$ is the usual scalar product.) Then

$$\begin{aligned}\int_0^\pi\left[\int_{-\infty}^{\infty}\int_{-\infty}^{\infty}\frac{\mathrm{d}\mu_\theta(u)\,\mathrm{d}\mu_\theta(v)}{|u-v|^s}\right]\mathrm{d}\theta &= \int_0^\pi\left[\int_F\int_F\frac{\mathrm{d}\mu(x)\,\mathrm{d}\mu(y)}{|x\cdot\boldsymbol{\theta}-y\cdot\boldsymbol{\theta}|^s}\right]\mathrm{d}\theta\\ &= \int_0^\pi\left[\int_F\int_F\frac{\mathrm{d}\mu(x)\,\mathrm{d}\mu(y)}{|(x-y)\cdot\boldsymbol{\theta}|^s}\right]\mathrm{d}\theta\\ &= \int_0^\pi\frac{\mathrm{d}\theta}{|\boldsymbol{\tau}\cdot\boldsymbol{\theta}|^s}\int_F\int_F\frac{\mathrm{d}\mu(x)\,\mathrm{d}\mu(y)}{|x-y|^s} \qquad (6.3)\end{aligned}$$

for any fixed unit vector $\boldsymbol{\tau}$. (Note that the integral of $|(x-y)\cdot\boldsymbol{\theta}|^{-s}$ with respect to θ depends only on $|x-y|$.) If $s<1$ then (6.3) is finite by virtue of (6.2) and that

$$\int_0^\pi\frac{\mathrm{d}\theta}{|\boldsymbol{\tau}\cdot\boldsymbol{\theta}|^s} = \int_0^\pi\frac{\mathrm{d}\theta}{|\cos(\tau-\theta)|^s} < \infty.$$

Hence

$$\int_F\int_F\frac{\mathrm{d}\mu_\theta(u)\,\mathrm{d}\mu_\theta(v)}{|u-v|^s} < \infty$$

for almost all $\theta\in[0,\pi)$. By Theorem 4.13(*a*) the existence of such a mass distribution μ_θ on $\text{proj}_\theta F$ implies that $\dim_{\mathrm{H}}(\text{proj}_\theta F) > s$. This is true for all $s < \dim_{\mathrm{H}} F$, so part (*a*) of the result follows.

The proof of (*b*) follows similar lines, though Fourier transforms need to be introduced to show that the projections have positive length. □

These projection theorems generalize to higher dimensions in the natural way. Let $G_{n,k}$ be the set of k-dimensional subspaces or 'k-planes through the origin' in $\mathbb{R}^n$. These subspaces are naturally parametrized by $k(n-k)$ coordinates ('generalized direction cosines') so that we may refer to 'almost all' subspaces in a consistent way in terms of $k(n-k)$-dimensional Lebesgue measure. We write proj_Π for orthogonal projection onto the k-plane Π.

Theorem 6.2. Higher-dimensional projection theorems.

Let $F\subset\mathbb{R}^n$ be a Borel set.

(*a*) *If* $\dim_{\mathrm{H}} F \leqslant k$ *then* $\dim_{\mathrm{H}}(\text{proj}_\Pi F) = \dim_{\mathrm{H}} F$ *for almost all* $\Pi\in G_{n,k}$

(*b*) *If* $\dim_{\mathrm{H}} F > k$ *then* $\text{proj}_\Pi F$ *has positive k-dimensional measure and so has dimension k for almost all* $\Pi\in G_{n,k}$.

Proof. The proof of Theorem 6.1 extends to higher dimensions without difficulty. □

Thus if F is a subset of $\mathbb{R}^3$, the plane projections of F are, in general, of dimension $\min\{2, \dim_H F\}$. This result has important practical implications. We can estimate the dimension of an object in space by estimating the dimension of a photograph taken from a random direction. Provided this is less than 2, it may be assumed to equal the dimension of the object. Such a reduction can make dimension estimates of spatial objects tractable—box-counting methods are difficult to apply in 3 dimensions but can be applied with reasonable success in the plane.

6.2 Projections of *s*-sets of integral dimension

If a subset F of $\mathbb{R}^2$ has Hausdorff dimension exactly 1, then Theorem 6.1 tells us that the projections of F onto almost every L_θ have dimension 1. However, in this critical case, no information is given as to whether these projections have zero or positive length. In the special case where F is a 1-set, i.e. with $0 < \mathscr{H}^s(F) < \infty$, an analysis is possible. Recall from Theorem 5.3 that a 1-set may be decomposed into a regular curve-like part and an irregular curve-free part. The following two theorems provide another sharp contrast between these types of set.

Theorem 6.3

Let F be a regular 1-set in $\mathbb{R}^2$. Then $\text{proj}_\theta F$ has positive length except for at most one $\theta \in [0, \pi)$.

Sketch of proof. By Theorem 5.9(*b*) it is enough to prove the result if F is a subset of positive length of a rectifiable curve C. Using the Lebesgue density theorem to approximate to such an F by short continuous subcurves of C, essentially all we need to consider is the case when F is itself a rectifiable curve C_1 joining distinct points x and y. But clearly, the projection onto L_θ of such a curve is an interval of positive length, except possibly for the one value of θ for which L_θ is perpendicular to the straight line through x and y. □

(In general $\text{proj}_\theta F$ will have positive length for all θ; there is an exceptional value of θ only if F is contained in a set of parallel line segments.)

Theorem 6.4

Let F be an irregular 1-set in $\mathbb{R}^2$. Then $\text{proj}_\theta F$ has length zero for almost all $\theta \in [0, \pi)$.

Proof. The proof is complicated, depending on the intricate density and angular density structure of irregular sets. We omit it! □

These theorems may be combined in several ways.

Corollary 6.5

Let F be a 1-set in $\mathbb{R}^2$. If the regular part of F has $\mathscr{H}^1$-measure zero, then $\text{proj}_\theta F$ *has length zero for almost all θ; otherwise it has positive length for all but at most one value of θ.*

The following characterization of irregular sets is also useful.

Corollary 6.6

A 1-set in $\mathbb{R}^2$ is irregular if and only if it has projections of zero length in at least two directions.

Example 6.7

The set F of figure 0.4 *is an irregular* 1-set.

Calculation. In Example 2.6 we showed that F is a 1-set. It is easy to see that the projections of F onto lines L_θ with $\tan\theta = \frac{1}{2}$ and $\tan\theta = -2$ have zero length (look at the first few iterations), so F is irregular by Corollary 6.6. □

The results of this section have been stated for sets for which $0 < \mathscr{H}^1(F) < \infty$, which is rather a strong property for 1-dimensional sets to have, although one which occurs surprisingly often. However, the theorems can be applied rather more widely. If F is any set that intersects some rectifiable curve in a set of positive length, so that F contains a regular subset, then $\text{proj}_\theta F$ has positive length for almost all θ. Again, if F is a σ-finite irregular set, i.e. one which may be expressed as a countable union of irregular 1-sets each of finite measure, then $\text{proj}_\theta F$ has zero length for almost all θ; this follows by taking countable unions of the projections of these component 1-sets.

For the record, we state the higher-dimensional analogue of Theorems 6.3 and 6.4, though the proofs are even more complicated than in the plane case.

Theorem 6.8

Let F be a k-set in $\mathbb{R}^n$, where k is an integer.

(*a*) *If F is regular then* $\text{proj}_\Pi F$ *has positive k-dimensional measure for almost all $\Pi \in G_{n,k}$.*

(*b*) *If F is irregular then* $\text{proj}_\Pi F$ *has zero k-dimensional measure for almost all $\Pi \in G_{n,k}$.*

6.3 Projections of arbitrary sets of integral dimension

The theorems of the last section, although mathematically elegant and sophisticated, do not provide a complete answer to the question of whether projections of plane sets onto lines have zero or positive length. A subset F of $\mathbb{R}^2$ of Hausdorff dimension 1 need not be a 1-set or even be of σ-finite $\mathscr{H}^1$-measure (i.e. a countable union of sets of finite $\mathscr{H}^1$-measure). Moreover there need not be any dimension function h (see Section 2.5) for which $0 < \mathscr{H}^h(F) < \infty$, in which case mathematical analysis is extremely difficult. What can be said about the projections of such sets? The surprising answer is that, by working in this rather delicate zone of sets of Hausdorff dimension 1 but of non-σ-finite $\mathscr{H}^1$-measure, we can construct sets with projections more or less what we please. For example, there is a set F in $\mathbb{R}^2$ such that $\text{proj}_\theta F$ contains an interval of length 1 for almost all θ with $0 \leqslant \theta < \frac{1}{2}\pi$ but with $\text{proj}_\theta F$ of length zero for $\frac{1}{2}\pi \leqslant \theta < \pi$. More generally, we have the following result which says that there exist sets for which the projections in almost all directions are, to within length zero, anything that we care to prescribe. The measurability condition in square brackets is included for completeness, but is best ignored by non-specialists!

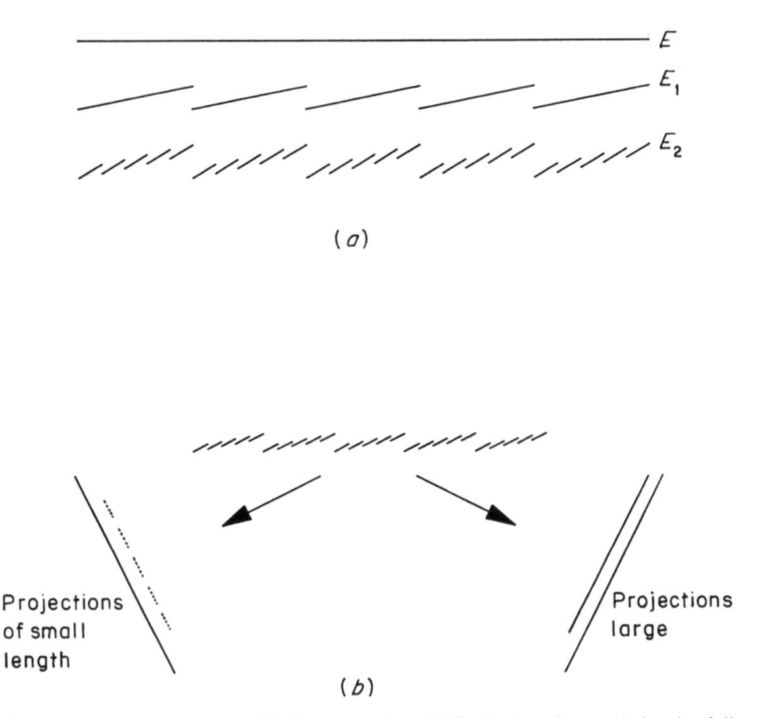

Figure 6.3 (*a*) The 'iterated Venetian blind' construction. (*b*) Projections in certain bands of directions have large lengths, whilst projections in other bands of directions have very small lengths

Theorem 6.9

Let G_θ be a subset of L_θ for each $\theta \in [0, \pi)$ [such that the set $\bigcup_{0 \leqslant \theta < \pi} G_\theta$ is plane Lebesgue measurable]. Then there exists a Borel set $F \subset \mathbb{R}^2$ such that

(*a*) $\text{proj}_\theta F \supset G_\theta$ *for all* θ, *and*

(*b*) $\text{length}\,(\text{proj}_\theta F \backslash G_\theta) = 0$ *for almost all* θ.

In particular, for almost all θ, the set of points of L_θ belonging to either G_θ or $\text{proj}_\theta F$, *but not both, has zero length.*

Idea of proof. Without going into much detail, we indicate the basic building block for such sets, which has been termed the 'iterated Venetian blind' construction. This is shown in figure 6.3. Let E be a line segment of length λ. Let ε be a small angle and k a large number. We replace E by k line segments of lengths roughly λ/k, each at an angle ε to E and with endpoints equally spaced along E to form a new set, E_1. We repeat this process with each segment of E_1 to form a set E_2 consisting of k^2 line segments all of lengths about λ/k^2 and at angle 2ε to E. We continue in this way, to get E_r, a set of k^r segments all of lengths about λ/k^r and at angle $r\varepsilon$ to E. We stop when r is such that $r\varepsilon$ is, say, about $\frac{1}{4}\pi$. Comparing the projections of E_r with that of the original line segment E, we see that if $0 \leqslant \theta < \frac{1}{2}\pi$ then $\text{proj}_\theta E$ and $\text{proj}_\theta E_r$ are nearly the same (since lines perpendicular to L_θ that cut E also cut E_r). However, if $-\frac{1}{4}\pi < \theta < 0$ then $\text{proj}_\theta E_r$ will have very small length, since most lines perpendicular to L_θ will pass straight between appropriately angled 'slats' of the construction. Thus the projections of E_r are very similar to those of E in certain directions, but are almost negligible in other directions. This idea may be adapted to obtain sets with projections very close to G_θ in a narrow band

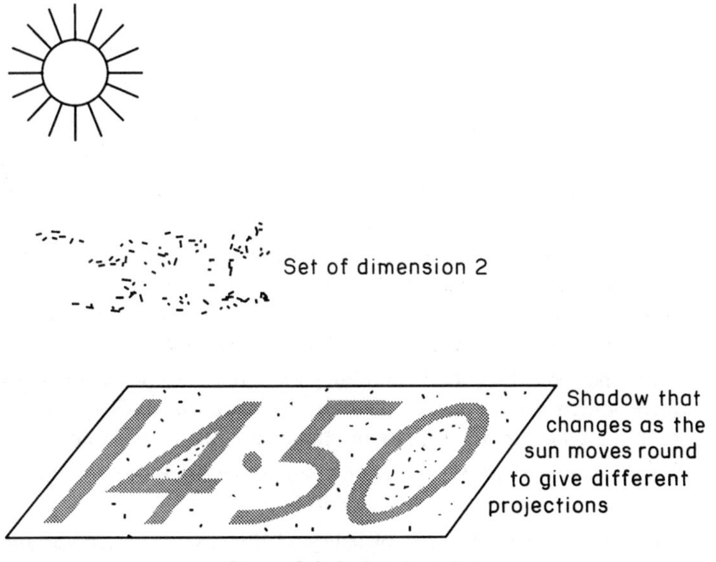

Figure 6.4 A digital sundial

of directions but with almost null projections in other directions. Taking unions of such sets for various small bands of directions gives a set with approximately the required property. Taking a limit of a sequence of sets which give increasingly accurate approximations leads to a set with the properties stated. □

This construction may be extended to higher dimensions: there exists a set F in $\mathbb{R}^n$ such that almost all projections of F onto k-dimensional subspaces differ from prescribed sets by zero k-dimensional measure. In particular there exists a set in 3-dimensional space with almost all of its plane shadows anything we care to prescribe to within zero area. By specifying the shadows to be the thickened digits of the time when the sun is shining from a perpendicular direction, we obtain, at least in theory, a digital sundial; see figure 6.4. As the sun moves across the sky we get different projections of the set. It is perhaps better to regard this as providing an intuitive view of the result, rather than as a feasible method of chronography!

6.4 Notes and references

A geometric proof of the projection theorems for arbitrary subsets of the plane was given by Marstrand (1954a); the potential theoretic proof was due to Kaufman (1968). Mattila (1975) obtained various generalizations including extensions to higher dimensions. The projection results for regular and irregular 1-sets in the plane are, surprisingly, older, dating back to Besicovitch (1939), with the analogous results for s-sets in $\mathbb{R}^n$ in the mammoth paper of Federer (1947). A dual version of Theorem 6.9 was given by Davies (1952) and a direct proof, with the higher-dimensional generalizations, by Falconer (1986a).

Exercises

6.1 For $0 < s < 1$, give an example of an s-set F in $\mathbb{R}^2$ such that $\text{proj}_\theta F$ is an s-set for all θ.

6.2 Let E and F be subsets of $\mathbb{R}$. Show that, for almost all real numbers λ, $\dim_H(E + \lambda F) = \min\{1, \dim_H(E \times F)\}$, where $E + \lambda F$ denotes the set of real numbers $\{x + \lambda y : x \in E, y \in F\}$.

6.3 Let E and F be subsets of $\mathbb{R}$ with Hausdorff dimension strictly between 0 and 1. You are given that the subset $E \times F$ of $\mathbb{R}^2$ has Hausdorff dimension at least $\dim_H E + \dim_H F$ (see Chapter 7). Show that the projections of $E \times F$ onto the coordinate axes are always 'exceptional' as far Projection theorem 6.1 is concerned.

6.4 Let F be a connected subset of $\mathbb{R}^2$ containing more than one point. Show that $\text{proj}_\theta F$ has positive length for all except possibly one value of θ. (Thus the projection theorems in the plane are only really of interest for sets that are not connected.)

6.5 Show that the conclusions of Theorem 6.4 remain true if F is a countable union of irregular 1-sets.

6.6 Let E and F be any subsets of $\mathbb{R}$ of length (1-dimensional Lebesgue measure) 0. Show that any rectifiable curve in $\mathbb{R}^2$ intersects the product $E \times F$ in a set of length 0.

6.7 If F is a set and x is a point in $\mathbb{R}^2$, the projection of F at x, denoted by $\text{proj}_x F$, is defined as the set of θ in $[0, 2\pi)$ such that the half-line emanating from x in direction θ intersects F. Let L be a line. Show that if $\dim_H F \leqslant 1$ then $\dim_H \text{proj}_x F = \dim_H F$ for almost all x on L (in the sense of Lebesgue measure) and if $\dim_H F > 1$ then $\text{proj}_x F$ has positive length for almost all x on L. (Hint: consider a sphere tangential to the plane and a transformation that maps a point x on the plane to the point on the sphere intersected by the line joining x to the centre of the sphere.)

Chapter 7 Products of fractals

One way of constructing 'new fractals from old' is by forming Cartesian products. Indeed, many fractals that occur in practice are products or, at least, are locally product-like. In this chapter we develop dimension formulae for products.

7.1 Product formulae

Recall that if E is a subset of $\mathbb{R}^n$ and F is a subset of $\mathbb{R}^m$, the *Cartesian product*, or just *product*, $E \times F$ is defined as the set of points with first coordinate in E and second coordinate in F, i.e.

$$E \times F = \{(x, y) \in \mathbb{R}^{n+m} : x \in E, y \in F\}. \tag{7.1}$$

Thus if E is a unit interval in $\mathbb{R}$, and F is a unit interval in $\mathbb{R}^2$, then $E \times F$ is a unit square in $\mathbb{R}^3$ (figure 7.1). Again, if F is the middle third Cantor set, then $F \times F$ is the 'Cantor product' (figure 7.2) consisting of those points in the plane with both coordinates in F.

In the first example above it is obvious that

$$\dim(E \times F) = \dim E + \dim F$$

using the classical defnintion of dimension. This holds more generally, in the 'smooth' situation, where E and F are smooth curves, surfaces or higher-dimensional manifolds. Unfortunately, this equation is not always valid for 'fractal' dimensions. For Hausdorff dimensions the best general result possible is an inequality $\dim_H(E \times F) \leqslant \dim_H E + \dim_H F$. Nevertheless, as we shall see, in many situations equality does hold.

The proof of the product rule uses the Hausdorff measures on E and F to define a mass distribution μ on $E \times F$. Density bounds on E and F lead to estimates for μ suitable for a mass distribution method.

Proposition 7.1

If $E \subset \mathbb{R}^n$, $F \subset \mathbb{R}^m$ are Borel sets with $\mathcal{H}^s(E), \mathcal{H}^t(F) < \infty$, then

$$\mathcal{H}^{s+t}(E \times F) \geqslant c\mathcal{H}^s(E)\mathcal{H}^t(F) \tag{7.2}$$

where c depends only on s and t.

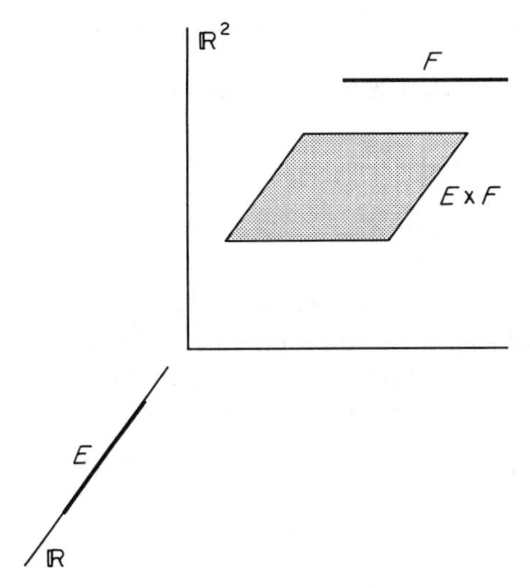

Figure 7.1 The Cartesian product of a unit interval in $\mathbb{R}$ and a unit interval in $\mathbb{R}^2$

Proof. For simplicity we assume that $E, F \subset \mathbb{R}$, so that $E \times F \subset \mathbb{R}^2$; the general proof is almost identical. If either $\mathscr{H}^s(E)$ or $\mathscr{H}^t(F)$ is zero, then (7.2) is trivial, so suppose that E is an s-set and F is a t-set, i.e. $0 < \mathscr{H}^s(E), \mathscr{H}^t(F) < \infty$. We may define a mass distribution μ on $E \times F$ by utilizing the 'product measure' of $\mathscr{H}^s$ and $\mathscr{H}^t$. Thus if $I, J \subset \mathbb{R}$, we define μ on the 'rectangle' $I \times J$ by

$$\mu(I \times J) = \mathscr{H}^s(E \cap I)\mathscr{H}^t(F \cap J). \tag{7.3}$$

It may be shown that this defines a mass distribution μ on $E \times F$ with $\mu(\mathbb{R}^2) = \mathscr{H}^s(E)\mathscr{H}^t(F)$.

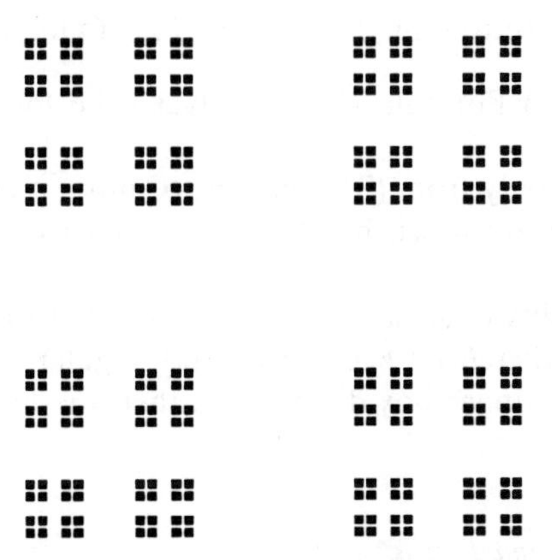

Figure 7.2 The product $F \times F$, where F is the middle third Cantor set. In this case, $\dim_H F \times F = 2 \dim_H F = 2 \log 2/\log 3$

By the density estimate Proposition 5.1(*b*) we have that

$$\overline{\lim_{r\to 0}}\, \mathscr{H}^s(E\cap B_r(x))(2r)^{-s} \leqslant 1 \tag{7.4}$$

for $\mathscr{H}^s$-almost all $x\in E$ and

$$\overline{\lim_{r\to 0}}\, \mathscr{H}^t(F\cap B_r(y))(2r)^{-t} \leqslant 1 \tag{7.5}$$

for $\mathscr{H}^t$-almost all $y\in F$. (Of course, since we are concerned with subsets of $\mathbb{R}$, $B_r(x)$ is just the interval of length $2r$ with midpoint x.) From the definition of μ, both (7.4) and (7.5) hold for μ-almost all (x, y) in $E\times F$. Since the disc $B_r(x, y)$ is contained in the square $B_r(x)\times B_r(y)$ we have that

$$\mu(B_r(x, y)) \leqslant \mu(B_r(x)\times B_r(y)) = \mathscr{H}^s(E\cap B_r(x))\mathscr{H}^t(F\cap B_r(y))$$

so

$$\frac{\mu(B_r(x, y))}{(2r)^{s+t}} \leqslant \frac{\mathscr{H}^s(E\cap B_r(x))}{(2r)^s}\frac{\mathscr{H}^t(F\cap B_r(y))}{(2r)^t}.$$

It follows, using (7.4) and (7.5), that $\overline{\lim}_{r\to\infty}\, \mu(B_r(x, y))(2r)^{-(s+t)} \leqslant 1$ for μ-almost all $(x, y)\in E\times F$. By Proposition 4.9(*a*)

$$\mathscr{H}^s(E\times F) \geqslant 2^{-(s+t)}\mu(E\times F) = 2^{-(s+t)}\mathscr{H}^s(E)\mathscr{H}^t(F). \qquad \square$$

Product formula 7.2

If $E\subset\mathbb{R}^n$, $F\subset\mathbb{R}^m$ are any Borel sets then

$$\dim_{\mathrm{H}}(E\times F) \geqslant \dim_{\mathrm{H}} E + \dim_{\mathrm{H}} F. \tag{7.6}$$

Proof. If s, t are any numbers with $s<\dim_{\mathrm{H}} E$ and $t<\dim_{\mathrm{H}} F$, then $\mathscr{H}^s(E) = \mathscr{H}^t(F) = \infty$. Theorem 4.10 implies that there are Borel sets $F_0\subset E$ and $F_0\subset F$ with $0<\mathscr{H}^s(E_0), \mathscr{H}^t(F_0)<\infty$. By Proposition 7.1 $\mathscr{H}^{s+t}(E\times F) \geqslant \mathscr{H}^{s+t}(E_0\times F_0) \geqslant c\mathscr{H}^s(E_0)\mathscr{H}^t(F_0) > 0$. Hence $\dim_{\mathrm{H}}(E\times F) \geqslant s+t$. By choosing s and t arbitrarily close to $\dim_{\mathrm{H}} E$ and $\dim_{\mathrm{H}} F$, (7.6) follows. $\square$

Proposition 7.1 and Formula 7.2 are in fact valid for arbitrary (non-Borel) sets.

It follows immediately from (7.6) that the 'Cantor product' $F\times F$, where F is the middle third Cantor set, has Hausdorff dimension at least $2\log 2/\log 3$ (see figure 7.2).

In general, inequality (7.6) cannot be reversed; see Example 7.8. However, if, as often happens, either E or F is 'reasonably regular' in the sense of having equal Hausdorff and upper box dimensions, then we do get equality.

Product formula 7.3

For any sets $E\subset\mathbb{R}^n$ and $F\subset\mathbb{R}^m$

$$\dim_{\mathrm{H}}(E\times F) \leqslant \dim_{\mathrm{H}} E + \overline{\dim}_{\mathrm{B}} F. \tag{7.7}$$

Proof. For simplicity take $E \subset \mathbb{R}$ and $F \subset \mathbb{R}$. Choose numbers $s > \dim_{\mathrm{H}} E$ and $t > \overline{\dim}_{\mathrm{B}} F$. Then there is a number $\delta_0 > 0$ such that F may be covered by $N_\delta(F) \leqslant \delta^{-t}$ intervals of length δ for all $\delta \leqslant \delta_0$. Let $\{U_i\}$ be any δ-cover of E by intervals with $\sum_i |U_i|^s < 1$. For each i, let $U_{i,j}$ be a cover of F by $N_{|U_i|}(F)$ intervals of length $|U_i|$. Then $U_i \times F$ is covered by $N_{|U_i|}(F)$ squares $\{U_i \times U_{i,j}\}$ of side $|U_i|$. Thus $E \times F \subset \bigcup_i \bigcup_j (U_i \times U_{i,j})$, so that

$$\mathcal{H}^{s+t}_{\delta\sqrt{2}}(E \times F) \leqslant \sum_i \sum_j |U_i \times U_{i,j}|^{s+t} \leqslant \sum_i N_{|U_i|}(F) 2^{\frac{1}{2}(s+t)} |U_i|^{s+t}$$

$$\leqslant 2^{\frac{1}{2}(s+t)} \sum_i |U_i|^{-t} |U_i|^{s+t} < 2^{\frac{1}{2}(s+t)}.$$

Hence $\mathcal{H}^{s+t}(E \times F) < \infty$ whenever $s > \dim_{\mathrm{H}} E$ and $t > \overline{\dim}_{\mathrm{B}} F$, giving $\dim_{\mathrm{H}}(E \times F) \leqslant s + t$. □

Corollary 7.4

If $\dim_{\mathrm{H}} F = \overline{\dim}_{\mathrm{B}} F$ *then*

$$\dim_{\mathrm{H}}(E \times F) = \dim_{\mathrm{H}} E + \dim_{\mathrm{H}} F.$$

Proof. Note that combining Product formulae 7.2 and 7.3 gives

$$\dim_{\mathrm{H}} E + \dim_{\mathrm{H}} F \leqslant \dim_{\mathrm{H}}(E \times F) \leqslant \dim_{\mathrm{H}} E + \overline{\dim}_{\mathrm{B}} F. \quad \square \tag{7.8}$$

It is worth noting that the basic product inequality for upper box dimensions is opposite to that for Hausdorff dimensions.

Product formula 7.5

For any sets $E \subset \mathbb{R}^n$ *and* $F \subset \mathbb{R}^m$

$$\overline{\dim}_{\mathrm{B}}(E \times F) \leqslant \overline{\dim}_{\mathrm{B}} E + \overline{\dim}_{\mathrm{B}} F. \tag{7.9}$$

Proof. This is left as an exercise. The idea is just as in Formula 7.3—note that if E and F can be covered by $N_\delta(E)$ and $N_\delta(F)$ intervals of side δ, then $E \times F$ is covered by the $N_\delta(E) N_\delta(F)$ squares formed by products of these intervals. □

Example 7.6. Product with uniform Cantor sets

Let E, F be subsets of $\mathbb{R}$ with F a uniform Cantor set (see Example 4.5). Then $\dim_{\mathrm{H}}(E \times F) = \dim_{\mathrm{H}} E + \dim_{\mathrm{H}} F$.

Calculation. Example 4.5 shows that uniform Cantor sets have equal Hausdorff and upper box dimensions, so the result follows from Corollary 7.4. □

Thus the 'Cantor product' of the middle third Cantor set with itself has dimension exactly $2\log 2/\log 3$. Similarly, if E is a subset of $\mathbb{R}$ and F is a straight line segment, then $\dim_H(E \times F) = \dim_H E + 1$.

Many fractals encountered in practice are not actually products, but are locally product-like. For example, the Hénon attractor (see (13.5)) looks locally like a product of a line segment and a Cantor-like set F. More precisely, there are smooth bijections from $[0, 1] \times F$ to small neighbourhoods of the attractor. Such sets may be analysed as the image of a product under a suitable Lipschitz transformation.

Example 7.7

The 'Cantor target' is the plane set given in polar coordinates by $F' = \{(r, \theta): r \in F, 0 \leqslant \theta \leqslant 2\pi\}$ *where F is the middle third Cantor set; see figure 7.3. Then* $\dim_H F' = 1 + \log 2/\log 3$.

Calculation. Let $f: \mathbb{R}^2 \to \mathbb{R}^2$ be given by $f(x, y) = (x \cos y, x \sin y)$. It is easy to see that f is a Lipschitz mapping and $F' = f(F \times [0, 2\pi])$. Thus

$$\begin{aligned}\dim_H F' &= \dim_H f(F \times [0, 2\pi]) \leqslant \dim_H(F \times [0, 2\pi]) \\ &= \dim_H F + \dim_H [0, 2\pi] = (\log 2/\log 3) + 1\end{aligned}$$

by Corollary 2.4(*a*) and Example 7.6. On the other hand, if we restrict f to $[\frac{2}{3}, 1] \times [0, \pi]$ then f is a bi-Lipschitz function on this domain. Since $F' \supset f((F \cap [\frac{2}{3}, 1]) \times [0, \pi])$ we have

$$\begin{aligned}\dim_H F' &\geqslant \dim_H f((F \cap [\tfrac{2}{3}, 1]) \times [0, \pi]) \\ &= \dim_H((F \cap [\tfrac{2}{3}, 1]) \times [0, \pi]) \\ &= \dim_H(F \cap [\tfrac{2}{3}, 1]) + \dim_H [0, \pi] \\ &= (\log 2/\log 3) + 1\end{aligned}$$

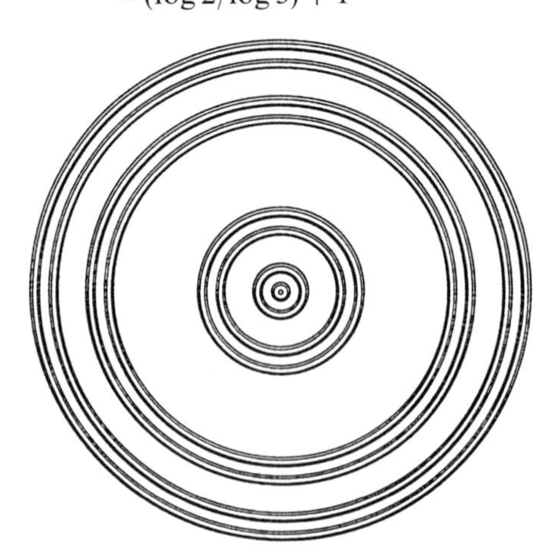

Figure 7.3 The 'Cantor target'—the set swept out by rotating the middle third Cantor set about an endpoint

by Corollary 2.4(*b*) and Example 7.6. This argument requires only minor modification to show that F' is an s-set for this value of s. □

The following example demonstrates that we do not in general get equality in the product formula (7.6) for Hausdorff measures.

Example 7.8

There exist sets $E, F \subset \mathbb{R}$ *with* $\dim_H E = \dim_H F = 0$ *and* $\dim_H(E \times F) \geqslant 1$.

Calculation. Let $0 = m_0 < m_1 < \cdots$ be a rapidly increasing sequence of integers satisfying a condition to be specified shortly. Let E consist of those numbers in $[0, 1]$ with a zero in the rth decimal place whenever $m_k + 1 \leqslant r \leqslant m_{k+1}$ and k is even, and let F consist of those numbers with zero in the rth decimal place if $m_k + 1 \leqslant r \leqslant m_{k+1}$ and k is odd. Looking at the first m_{k+1} decimal places for even k, there is an obvious cover of E by 10^{j_k} intervals of length $10^{-m_{k+1}}$, where $j_k = (m_2 - m_1) + (m_4 - m_3) + \cdots + (m_k - m_{k-2})$. Then $\log 10^{j_k}/-\log 10^{-m_{k+1}} = j_k/m_{k+1}$ which tends to 0 as $k \to \infty$ provided that the m_k are chosen to increase sufficiently rapidly. Thus $\dim_H E \leqslant \underline{\dim}_B E = 0$. Similarly $\dim_H F = 0$.

If $0 < w < 1$ then we can write $w = x + y$ where $x \in E$ and $y \in F$; just take the rth decimal digit of w from E if $m_k + 1 \leqslant r \leqslant m_{k+1}$ and k is odd and from F if k is even. The mapping $f: \mathbb{R}^2 \to \mathbb{R}$ given by $f(x, y) = x + y$ is easily seen to be Lipschitz, so

$$\dim_H(E \times F) \geqslant \dim_H f(E \times F) \geqslant \dim_H(0, 1) = 1$$

by Corollary 2.4(*a*). □

A useful generalization of the product formula relates the dimension of a set to the dimensions of parallel sections. We work in the (x, y)-plane and let L_x denote the line parallel to the y-axis through the point $(x, 0)$.

Proposition 7.9

Let F be a Borel subset of $\mathbb{R}^2$. *If* $1 \leqslant s \leqslant 2$ *then*

$$\int_{-\infty}^{\infty} \mathscr{H}^{s-1}(F \cap L_x)\,\mathrm{d}x \leqslant \mathscr{H}^s(F). \tag{7.10}$$

Proof. Given $\varepsilon > 0$, let $\{U_i\}$ be a δ-cover of F such that

$$\sum_i |U_i|^s \leqslant \mathscr{H}^s_\delta(F) + \varepsilon.$$

Each U_i is contained in a square S_i of side $|U_i|$ with sides parallel to the coordinate axes.

Let χ_i be the indicator function of S_i (i.e. $\chi_i(x, y) = 1$ if $(x, y) \in S_i$ and $\chi_i(x, y) = 0$ if $(x, y) \notin S_i$). For each x, the sets $\{S_i \cap L_x\}$ from a δ-cover of $F \cap L_x$, so

$$\begin{aligned}\mathscr{H}_\delta^{s-1}(F \cap L_x) &\leqslant \sum_i |S_i \cap L_x|^{s-1} \\ &= \sum_i |U_i|^{s-2} |S_i \cap L_x| \\ &= \sum_i |U_i|^{s-2} \int \chi_i(x, y)\, \mathrm{d}y.\end{aligned}$$

Hence

$$\begin{aligned}\int \mathscr{H}_\delta^{s-1}(F \cap L_x)\, \mathrm{d}x &\leqslant \sum_i |U_i|^{s-2} \int\int \chi_i(x, y)\, \mathrm{d}x\, \mathrm{d}y \\ &= \sum_i |U_i|^s \\ &\leqslant \mathscr{H}_\delta^s(F) + \varepsilon.\end{aligned}$$

Since $\varepsilon > 0$ is arbitrary, $\int \mathscr{H}_\delta^{s-1}(F \cap L_x)\, \mathrm{d}x \leqslant \mathscr{H}_\delta^s(F)$. Letting $\delta \to 0$ gives (7.10). □

Corollary 7.10

Let F be a Borel subset of $\mathbb{R}^2$. Then, for almost all x (in the sense of 1-dimensional Lebesgue measure), $\dim_H(F \cap L_x) \leqslant \max\{0, \dim_H F - 1\}$.

Proof. Take $s > \dim_H F$, then $\mathscr{H}^s(F) = 0$. If $s > 1$, formula (7.10) gives $\mathscr{H}^{s-1}(F \cap L_x) = 0$ and so $\dim_H(F \cap L_x) \leqslant s - 1$ for almost all x. □

We state, without proof, a further useful generalization.

Proposition 7.11

Let F be any subset of $\mathbb{R}^2$, and let E be any subset of the x-axis. Suppose that there is a constant c such that $\mathscr{H}^t(F \cap L_x) \geqslant c$ for all $x \in E$. Then

$$\mathscr{H}^{s+t}(F) \geqslant bc\,\mathscr{H}^s(E) \tag{7.11}$$

where b depends only on s and t.

This result may be phrased in terms of dimensions.

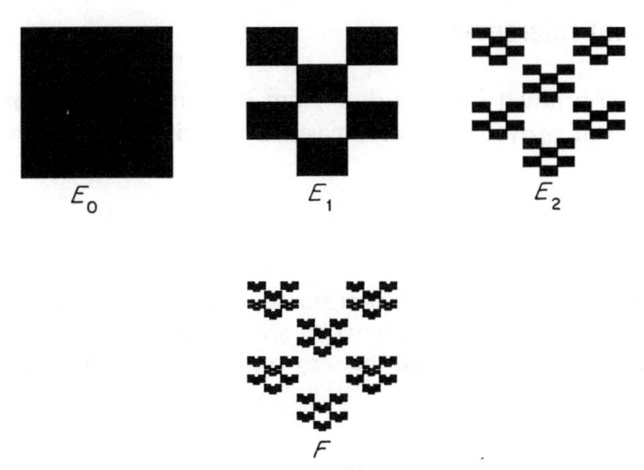

Figure 7.4 Construction of a self-affine set, $\dim_H F = 1\frac{1}{2}$

Corollary 7.12

Let F be any subset of $\mathbb{R}^2$, and let E be a subset of the x-axis. If $\dim_H (F \cap L_x) \geqslant t$ *for all* $x \in E$, *then* $\dim_H F \geqslant t + \dim_H E$.

The obvious higher-dimensional analogues of these results are all valid.

The following illustration of Proposition 7.9 is an example of a self-affine set, a class of sets which will be discussed in detail in Section 9.4.

Example 7.13. A self-affine set

Let F be the set with iterated construction indicated in figure 7.4. (At the kth stage each rectangle of E_k is replaced with an affine copy of the rectangles in E_1. Thus the contraction is greater in the 'y' direction than in the 'x' direction, with the width to height ratio of the rectangles in E_k tending to infinity as $k \to \infty$.) Then $\dim_H F = \dim_B F = 1\frac{1}{2}$.

Calculation. E_k consists of 6^k rectangles of size $3^{-k} \times 4^{-k}$. Each of these rectangles may be covered by at most $(4/3)^k + 1$ squares of side 4^{-k}, by dividing the rectangles using a series of vertical cuts. Hence E_k may be covered by $6^k \times 2 \times 4^k \times 3^{-k} = 2 \times 8^k$ squares of diameter $4^{-k}\sqrt{2}$. In the usual way this gives $\dim_H F \leqslant \overline{\dim}_B F \leqslant 1\frac{1}{2}$.

On the other hand, except for x of the form $j3^{-k}$ where j and k are integers, we have that $E_k \cap L_x$ consists of 2^k intervals of length 4^{-k}. A standard application of the mass distribution method shows that $\mathcal{H}^{\frac{1}{2}}(E_k \cap L_x) \geqslant \frac{1}{2}$ for each such x (Exercise). By Proposition 7.9, $\mathcal{H}^{1\frac{1}{2}}(F) \geqslant \frac{1}{2}$. Hence $\dim_H F = \dim_B F = 1\frac{1}{2}$. $\square$

7.2 Notes and references

Versions of the product formula date from Besicovitch and Moran (1945). A very general result, proved using net measures, was given by Marstrand (1954b).

Exercises

7.1 Show that there is a subset F of $\mathbb{R}^2$ of Hausdorff dimension 2 with projections onto both coordinate axes of length 0. (Hint: see Exercise 4.7.) Deduce that any 1-set contained in F is irregular, and that any rectifiable curve intersects F in a set of length 0.

7.2 Derive Product formula 7.5.

7.3 What are the Hausdorff and box dimensions of the plane set $\{(x, y) \in \mathbb{R}^2 : x + y \in F$ and $x - y \in F\}$, where F is the middle third Cantor set?

7.4 Let $F \subset \mathbb{R}$ have equal Hausdorff and upper box dimensions. Let D be the set $\{x - y : x, y \in F\}$, known as the difference set of F. Show that $\dim_H D \leqslant \min\{1, 2\dim_H F\}$. (Hint: consider the set $F \times F$.)

7.5 Let F be any subset of $[0, \infty)$ and let F' be the 'target' in $\mathbb{R}^2$ given in polar coordinates by $\{(r, \theta) : r \in F, 0 \leqslant \theta < 2\pi\}$. Show that $\dim_H F' = 1 + \dim_H F$.

7.6 Find the Hausdorff and box dimensions of the plane set $\{(x, y) : y - x^2 \in F\}$ where F is the middle third Cantor set.

7.7 Let L_x be as in Proposition 7.9. Let F be a subset of $\mathbb{R}^2$ and let $E_s = \{x \in \mathbb{R} : \dim_H(F \cap L_x) \geqslant s\}$ for $0 \leqslant s \leqslant 1$. Show that $\dim_H F \geqslant \sup\{s + \dim_H E_s\}$.

7.8 Divide the unit square E_0 into a three column, five row array of rectangles of sides $\frac{1}{3}$ and $\frac{1}{5}$, and let E_1 be a set obtained by choosing some four of the five rectangles from each column. Let F be the self-affine set formed by repeatedly replacing rectangles by affine copies of E_1 (compare Example 7.13). Adapt the method of Example 7.13 to show that $\dim_H F = 1 + \log 4/\log 5$.

7.9 Suppose that the construction of the previous exercise is modified so that E_1 contains four rectangles from each of the first and third columns but none from the middle column. Show that $\dim_H F = \log 2/\log 3 + \log 4/\log 5$.

Chapter 8 Intersections of fractals

The intersection of two fractals is often a fractal; it is natural to try to relate the dimension of this intersection to that of the original sets. It is immediately apparent that we can say almost nothing in the general case. For if F is bounded, there is a congruent copy F_1 of F such that $\dim_H(F \cap F_1) = \dim_H F$ (take $F_1 = F$) and another congruent copy with $\dim_H(F \cap F_1) = \varnothing$ (take F and F_1 disjoint). However, if we consider the intersection of F and a congruent copy in a 'typical' relative position, then some progress is possible.

To illustrate this, let F be a unit line segment in the plane. If F_1 is a congruent copy of F, then $F \cap F_1$ can be a line segment, but only in the exceptional situation when F and F_1 are collinear. If F and F_1 cross at an angle, then $F \cap F_1$ is a single point, but now the set $F \cap F_2$ is also a point for all congruent copies F_2 of F close enough to F_1. Thus, whilst 'in general' $F \cap F_1$ contains at most one point, this situation occurs 'frequently'.

We can make this rather more precise. Recall that a rigid motion or direct congruence transformation σ of the plane transforms any set E to a congruent copy $\sigma(E)$ without reflection. The rigid motions may be parametrized by three coordinates (x, y, θ) where the origin is transformed to (x, y) and θ is the angle of rotation. Such a parametrization provides a natural measure on the space of rigid motions, with the measure of a set A of rigid motions given by the 3-dimensional Lebesgue measure of the (x, y, θ) parametrizing the motions in A. For example, the set of all rigid motions which map the origin to a point of the rectangle $[1, 2] \times [0, 3]$ has measure $1 \times 3 \times 2\pi$.

In the example with F a unit line segment, the set of transformations σ for which $F \cap \sigma(F)$ is a line segment has measure 0. However, $F \cap \sigma(F)$ is a single point for a set of transformations of positive measure, in fact a set of measure 4.

Similar results hold in higher dimensions. 'Typically', in $\mathbb{R}^3$, two surfaces intersect in a curve, a surface and a curve intersect in a point and two curves are disjoint. In $\mathbb{R}^n$, if smooth manifolds E and F intersect at all, then 'in general' they intersect in a submanifold of dimension $\max\{0, \dim E + \dim F - n\}$. More precisely, if $\dim E + \dim F - n > 0$ then $\dim(E \cap \sigma(F)) = \dim E + \dim F - n$ for a set of rigid motions σ of positive measure, and is 0 for almost all other σ. (Of course, σ is now measured using the $\frac{1}{2}n(n+1)$ parameters required to specify a rigid transformation of $\mathbb{R}^n$.)

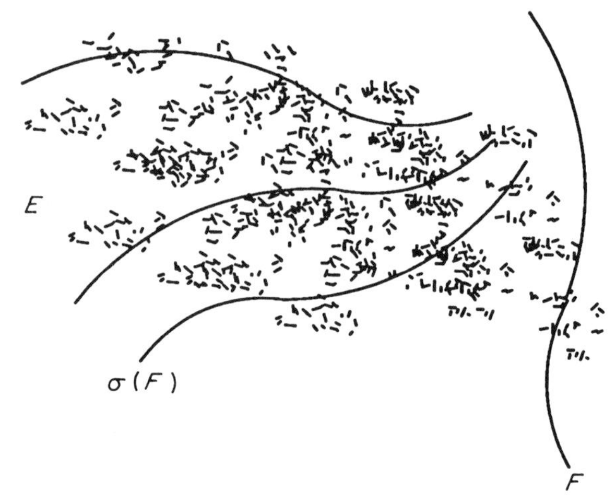

Figure 8.1 The intersection of a 'dust-like' set E with various congruent copies $\sigma(F)$ of a curve F. We are interested in the dimension of $E \cap \sigma(F)$ for 'typical' σ

8.1 Intersection formulae for fractals

Are there analogues of these formulae if E and F are fractals and we use Hausdorff dimension? In particular, is it true that 'in general'

$$\dim_H(E \cap \sigma(F)) \leqslant \max\{0, \dim_H E + \dim_H F - n\} \tag{8.1}$$

and 'often'

$$\dim_H(E \cap \sigma(F)) \geqslant \dim_H E + \dim_H F - n \tag{8.2}$$

as σ ranges over a group G of transformations, such as the group of translations, congruences or similarities (see figure 8.1)? Of course 'in general' means 'for almost all σ' and 'often' means 'for a set of σ of positive measure' with respect to a natural measure on the transformations in G. Generally, G can be parametrized by m coordinates in a straightforward way for some integer m and we can use Lebesgue measure on the parameter space $\mathbb{R}^m$.

We obtain upper bounds for $\dim_H(E \cap \sigma(F))$ when F is the group of translations; these hold automatically for the larger groups of congruences and similarities. We have already proved (8.1) in the special case in the plane where one of the sets is a straight line; this is essentially Corollary 7.10. The general result is easily deduced from this special case. Recall that $F + x = \{x + y : y \in F\}$ denotes the translation of F by the vector x.

Theorem 8.1

If E, F are Borel subsets of $\mathbb{R}^n$ then

$$\dim_H(E \cap (F + x)) \leqslant \max\{0, \dim_H(E \times F) - n\} \tag{8.3}$$

for almost all $x \in \mathbb{R}^n$.

Proof. We prove this when $n=1$; the proof for $n>1$ is similar, using a higher-dimensional analogue of Corollary 7.10. Let L_c be the line in the (x, y)-plane with equation $x = y + c$. Assuming that $\dim_H(E \times F) > 1$, it follows from Corollary 7.10 (rotating the lines through 45° and changing notation slightly) that

$$\dim_H((E \times F) \cap L_c) \leqslant \dim_H(E \times F) - 1 \tag{8.4}$$

for almost all $c \in \mathbb{R}$. But a point $(x, x-c) \in (E \times F) \cap L_c$ if and only if $x \in E \cap (F+c)$. Thus, for each c, the projection onto the x-axis of $(E \times F) \cap L_c$ is the set $E \cap (F+c)$. In particular, $\dim_H(E \cap (F+c)) = \dim_H((E \times F) \cap L_c)$, so the result follows from (8.4). □

Theorem 8.1 is some way from (8.1), but examples show that it is the best that we can hope to achieve, even if the group of translations is replaced by the group of all rigid motions. Unfortunately, inequality (7.6) is the opposite to what would be needed to deduce (8.1) from (8.3). Nevertheless, in many instances, we do have $\dim_H(E \times F) = \dim_H E + \dim_H F$; for example, if $\dim_H F = \overline{\dim}_B F$; see Corollary 7.4. Under such circumstances we recover (8.1), with $\sigma(F)$ as the translate $F + x$.

Lower bounds for $\dim_H(E \cap \sigma(F))$ of the form (8.2) are rather harder to obtain. The main known results are contained in the following theorem.

Theorem 8.2

Let $E, F \subset \mathbb{R}^n$ be Borel sets, and let G be a group of transformations on $\mathbb{R}^n$. Then

$$\dim_H(E \cap \sigma(F)) \geqslant \dim_H E + \dim_H F - n \tag{8.5}$$

for a set of motions $\sigma \in G$ of positive measure in the following cases:

(*a*) *G is the group of similarities and E and F are arbitrary sets*
(*b*) *G is the group of rigid motions, E is arbitrary and F is a rectifiable curve, surface, or manifold.*
(*c*) *G is the group of rigid motions and E and F are arbitrary, with either $\dim_H E > \frac{1}{2}(n+1)$ or $\dim_H F > \frac{1}{2}(n+1)$.*

* ***Outline of proof.*** The proof uses the potential theoretic methods of Section 4.3. In many ways, the argument resembles that of Projection theorem 6.1, but various technical difficulties make it much more complicated.

Briefly, if $s < \dim_H E$ and $t < \dim_H F$, there are mass distributions μ on E and ν on F with the energies $I_s(\mu)$ and $I_t(\nu)$ both finite. If ν happened to be absolutely continuous with respect to n-dimensional Lebesgue measure, i.e. if there were a function f such that $\nu(A) = \int_A f(x)\,\mathrm{d}x$ for each set A, then it would be natural to define a mass distribution η_σ on $E \cap \sigma(F)$ by $\eta_\sigma(A) = \int_A f(\sigma^{-1}((x))\,\mathrm{d}\mu(x)$. If we could show that $I_{s+t-n}(\eta_\sigma) < \infty$ for almost all σ, Theorem 4.13(*a*) would imply that $\dim(E \cap \sigma(F)) \geqslant s + t - n$ if $\eta_\sigma(\mathbb{R}^n) > 0$. Unfortunately, when F is a fractal, ν is supported by a set of zero n-dimensional volume, so is anything but absolutely continuous. To get around this difficulty, we can approximate ν by absolutely continuous mass distributions ν_δ supported by the δ-parallel body to F. Then, if $\nu_\delta(A) = \int_A f_\delta(x)\,dx$ and $\eta_{\sigma,\delta} = \int_A f_\delta(\sigma^{-1}(x))\,d\mu(x)$, we can estimate $I_{s+t-n}(\eta_{\sigma,\delta})$ and take the limit as $\delta \to 0$. Simplifying the integral $\int I_{s+t-n}(\eta_{\sigma,\delta})\,d\sigma$ isolates a term

$$\varphi_\delta(w) = \int_{G_0} \int_{\mathbb{R}^n} \nu_\delta(y)\nu_\delta(y + \sigma(w))\,dy\,dr$$

where integration with respect to σ is now over the subgroup G_0 of F which fixes the origin. Provided that

$$\varphi_\delta(w) \leqslant \text{constant}\,|w|^{t-n} \tag{8.6}$$

for all w and δ, it may be shown that $\int I_{s+t-n}(\nu_{\sigma,\delta})\,d\sigma < c < \infty$, where c is independent of δ. Letting $\delta \to 0$, the measures $\eta_{\sigma,\delta}$ 'converge' to measures η_σ on $E \cap \sigma(F)$, where $\int I_{s+t-n}(\eta_\sigma)\,d\sigma < c$. Thus $I_{s+t-n}(\eta_\sigma) < \infty$ for almost all σ, so, by Theorem 4.13(*a*), $\dim_H(E \cap \sigma(F)) \geqslant s + t - n$ whenever $\eta_\sigma(E \cap \sigma(F)) > 0$, which happens on a set of positive measure.

It may be shown that (8.6) holds if $I_t(\nu) < \infty$ in the cases (*a*), (*b*) and (*c*) listed. This is relatively easy to show for (*a*) and (*b*). Case (*c*) is more awkward; the only known method uses Fourier transform theory. □

The condition that $\dim_H E$ or $\dim_H F > \frac{1}{2}(n+1)$ in case (*c*) is a curious consequence of the use of Fourier transforms. It is not known whether the theorem remains valid for the group of congruences if $n \geqslant 2$ and $\frac{1}{2}n < \dim_H E, \dim_H F \leqslant \frac{1}{2}(n+1)$.

Example 8.3

Let $F \subset \mathbb{R}$ be the middle third Cantor set. For $\lambda, x \in \mathbb{R}$ write $\lambda F + x = \{\lambda y + x : x \in F\}$. Then $\dim_H(F \cap (F + x)) \leqslant 2(\log 2/\log 3) - 1$ for almost all $x \in \mathbb{R}$, and $\dim_H(F \cap (\lambda F + x)) = 2(\log 2/\log 3) - 1$ for a set of $(x, \lambda) \in \mathbb{R}^2$ of positive plane Lebesgue measure.

Calculation. We showed in Example 7.6 that $\dim_H(F \times F) = 2(\log 2/\log 3)$, so the stated dimensions follow from Theorems 8.1 and 8.2(*a*). □

*8.2 Sets with large intersection

We have seen that (8.1) need not always hold; in this section we examine a class of sets for which it fails dramatically. We construct a large class $\mathscr{C}^s$ of subsets of $\mathbb{R}$ of Hausdorff dimension at least s with the property that the intersection of any countable collection of sets in $\mathscr{C}^s$ still has dimension at least s. Sets of this type occur naturally in number theory; see Section 10.3.

The class $\mathscr{C}^s$ is defined in terms of the sums (2.1) used in the definition of Hausdorff measures. For any subset F of $\mathbb{R}$ we define

$$\mathscr{H}^s_\infty(F) = \inf\left\{ \sum_{i=1}^{\infty} |U_i|^s : \bigcup_{i=1}^{\infty} U_i \text{ is any cover of } F \right\}.$$

Thus $\mathscr{H}^s_\infty(F)$ is defined using covers of F without any diameter restriction. This ensures that $\mathscr{H}^s_\infty(I)$ is finite if I is a bounded interval, which would not be the case if we used $\mathscr{H}^s$. It is easy to see that $\mathscr{H}^s_\infty(F_1 \cup F_2) \leqslant \mathscr{H}^s_\infty(F_1) + \mathscr{H}^s_\infty(F_2)$ and that $\mathscr{H}^s_\infty(F_1) \leqslant \mathscr{H}^s_\infty(F_2)$ if $F_1 \subset F_2$.

Recall that $\overline{\lim}_{k\to\infty} E_k = \bigcap_{i=1}^{\infty} \bigcup_{k=i}^{\infty} E_k$ is the set of points that belong to infinitely many E_k. Let $0 < s < 1$ and let $[a, b] \subset \mathbb{R}$ be a proper closed interval. We say that a subset F of $[a, b]$ is a *member of the class* $\mathscr{C}^s[a, b]$ if

$$F \supset \overline{\lim_{k\to\infty}} E_k \tag{8.7}$$

where $\{E_k\}$ is a sequence of subsets of $[a, b]$, such that

(i) Each E_k is a finite union of disjoint closed intervals, and

(ii)
$$\lim_{k\to\infty} \mathscr{H}^s_\infty(I \cap E_k) = |I|^s \tag{8.8}$$

for every bounded closed interval I.

(Of course, we always have $\mathscr{H}^s_\infty(I \cap E_k) \leqslant |I|^s$.) We define $\mathscr{C}^s(-\infty, \infty)$ by saying that F is in $\mathscr{C}^s(-\infty, \infty)$ if $F \cap I \in \mathscr{C}^s[a, b]$ for every bounded interval $[a, b]$. The results below extend easily from $\mathscr{C}^s[a, b]$ to $\mathscr{C}^s(-\infty, \infty)$.

As an example of the sets we have in mind, we might take $E_k = \{x : |x - p/k| < k^{-3} \text{ for some integer } p\}$, so that $F = \overline{\lim}_{k\to\infty} E_k$ consists of the numbers which satisfy the inequality $|x - p/k| < k^{-3}$ for infinitely many positive integers k. As we shall see, $F \in \mathscr{C}^{1/3}(-\infty, \infty)$.

Any set in $\mathscr{C}^s[a, b]$ must be dense in $[a, b]$. For if F is in $\mathscr{C}^s[a, b]$ and I is a closed interval, then $I \cap E_{k_1}$ contains a closed interval I_1 if k_1 is large enough, by (8.8). Similarly, $I_1 \cap E_{k_2}$ contains a closed interval I_2 for some $k_2 > k_1$. Proceeding in this way, we get a sequence of closed intervals $I \supset I_1 \supset I_2 \supset \cdots$ with $I_r \subset E_{k_r}$ for each r. Thus the non-empty set $\bigcap_{r=1}^{\infty} I_r$ is contained in infinitely many E_k, so is contained in $F \cap I$.

By Proposition 3.4 any set in $\mathscr{C}^s[a, b]$ has box-counting dimension 1. We now show that these sets have Hausdorff dimension at least s. Moreover the intersection of any countable collection of sets in $\mathscr{C}^s[a, b]$ is also in $\mathscr{C}^s[a, b]$ and so has dimension at least s. Furthermore $f(F)$ is in $\mathscr{C}^s[f(a), f(b)]$ if F is in $\mathscr{C}^s[a, b]$, for a large class of functions f. The proofs below might well be omitted on a first reading. We require the following lemma; which extends (8.8) to unions of closed intervals.

Lemma 8.4

Let $\{E_k\}$ be a sequence of subsets of $\mathbb{R}$ such that

$$\lim_{k\to\infty} \mathscr{H}^s_\infty(I \cap E_k) = |I|^s \tag{8.9}$$

for every bounded closed interval I. Then, if A is a bounded set made up of a finite union of closed intervals,

$$\lim_{k\to\infty} \mathcal{H}^s_\infty(A\cap E_k) = \mathcal{H}^s_\infty(A). \tag{8.10}$$

Proof. Suppose that A consists of m disjoint intervals with minimum separation $d>0$. Given $\varepsilon>0$ we may, using (8.9), choose k_ε such that if $k\geqslant k_\varepsilon$

$$\mathcal{H}^s_\infty(I\cap E_k) > (1-\varepsilon)|I|^s \tag{8.11}$$

whenever $|I|\geqslant \varepsilon d$ and $I\subset A$. (Since $\mathcal{H}^s_\infty(E_k\cap I)$ varies continuously with I in the obvious sense, we may find a k_ε such that (8.11) holds simultaneously for all such I.) To estimate $\mathcal{H}^s_\infty(A\cap E_k)$ let $\{U_i\}$ be a cover of $A\cap E_k$. We may assume that this cover is finite, since $A\cap E_k$ is compact (see Section 2.4) and also that the U_i are closed intervals with endpoints in A, which are disjoint except possibly at endpoints. We divide the sets U_i into two batches according to whether $|U_i|\geqslant d$ or $|U_i|<d$. The set $A\backslash\bigcup_{|U_i|\geqslant d}U_i$ consists of disjoint intervals $V_1,\ldots,V_r$ where $r\leqslant m$, and

$$A\subset \bigcup_{|U_i|\geqslant d} U_i \cup \bigcup_j \bar{V}_j. \tag{8.12}$$

Observe that any U_i with $|U_i|<d$ is contained in an interval of A, and so in one of the $\bar{V}_j$. For each j the sets U_i contained in $\bar{V}_j$ cover $\bar{V}_j\cap E_k$, so

$$\sum_{\{i:U_i\subset \bar{V}_j\}} |U_i|^s \geqslant \mathcal{H}^s_\infty(\bar{V}_j\cap E_k) > (1-\varepsilon)|V_j|^s$$

if $|V_j|\geqslant \varepsilon d$, by (8.11). Hence

$$\sum_i |U_i|^s \geqslant \sum_{|U_i|\geqslant d} |U_i|^s + \sum_{|V_j|\geqslant \varepsilon d}\ \sum_{U_i\subset \bar{V}_j} |U_i|^s \geqslant \sum_{|U_i|\geqslant d} |U_i|^s + \sum_{|V_j|\geqslant \varepsilon d} (1-\varepsilon)|V_j|^s. \tag{8.13}$$

From (8.12)

$$\mathcal{H}^s_\infty(A) \leqslant \sum_{|U_i|\geqslant d}|U_i|^s + \sum_{|V_j|\geqslant \varepsilon d}|V_j|^s + \sum_{|V_j|<\varepsilon d}|V_j|^s \leqslant \sum_{|U_i|\geqslant d}|U_i|^s + \sum_{|V_j|\geqslant \varepsilon d}|V_j|^s + (\varepsilon d)^s r.$$

Combining with (8.13) we see that

$$\mathcal{H}^s_\infty(A) \leqslant (1-\varepsilon)^{-1}\sum_i |U_i|^s + (\varepsilon d)^s m$$

for any cover $\{U_i\}$ of $A\cap E_k$. Thus

$$\mathcal{H}^s_\infty(A) \leqslant (1-\varepsilon)^{-1}\mathcal{H}^s_\infty(A\cap E_k) + (\varepsilon d)^s m$$

if $k\geqslant k_\varepsilon$, which implies (8.10). □

Proposition 8.5

If $F\in\mathscr{C}^s[a,b]$ *then* $\mathcal{H}^s(F)>0$, *and in particular* $\dim_H F\geqslant s$.

Proof. For simplicity of notation assume that $[a,b]=[0,1]$. Suppose

$\overline{\lim}_{k\to\infty} E_k \subset F \subset \bigcup_i U_i$ where the U_i are open sets. Taking $I=[0,1]$ in (8.8) we may find a number k_1 such that $\mathscr{H}^s_\infty(E_{k_1}) > \frac{1}{2}$. Since E_{k_1} is a finite union of closed intervals, Lemma 8.4 implies that there is a number $k_2 > k_1$ such that $\mathscr{H}^s_\infty(E_{k_1} \cap E_{k_2}) > \frac{1}{2}$. Proceeding in this way, we get a sequence $k_1 < k_2 < \cdots$ such that $\mathscr{H}^s_\infty(E_{k_1} \cap \cdots \cap E_{k_r}) > \frac{1}{2}$ for all r. We have $\bigcap_{i=1}^\infty E_{k_i} \subset F \subset \bigcup_i U_i$; since $E_{k_1} \cap \cdots \cap E_{k_r}$ is a decreasing sequence of compact (i.e. closed and bounded) sets and $\bigcup_i U_i$ is open, there is an integer r such that $E_{k_1} \cap \cdots \cap E_{k_r} \subset \bigcup_i U_i$. It follows that $\sum_i |U_i|^s \geq \mathscr{H}^s_\infty(E_{k_1} \cap \cdots \cap E_{k_r}) > \frac{1}{2}$ for any cover of F by open sets, so $\mathscr{H}^s(F) \geq \frac{1}{2}$. □

Proposition 8.6

Let $F_j \in \mathscr{C}^s[a,b]$ for $j=1,2,\ldots$. Then $\bigcap_{j=1}^\infty F_j \in \mathscr{C}^s[a,b]$.

Proof. For each j there is a sequence of sets $E_{j,k}$, each a finite union of closed intervals, such that $F_j \supset \overline{\lim}_{k\to\infty} E_{j,k}$, where $\lim_{k\to\infty} \mathscr{H}^s_\infty(I \cap E_{j,k}) = \mathscr{H}^s_\infty(I)$ for every interval I. By Lemma 8.4

$$\lim_{k\to\infty} \mathscr{H}^s_\infty(A \cap E_{j,k}) = \mathscr{H}^s_\infty(A) \tag{8.14}$$

for any finite union of closed intervals A. There are countably many intervals $[c,d] \subset [a,b]$ with c and d both rational: let $I_1, I_2, \ldots$ be an enumeration of all such intervals.

For each r we define a set G_r as follows. Using (8.14) we may choose $k_1 \geq r$ large enough to make

$$\mathscr{H}^s_\infty(I_m \cap E_{1,k_1}) > \mathscr{H}^s_\infty(I_m) - 1/r$$

simultaneously for $m=1,\ldots,r$. Using (8.14) again, taking $A = I_m \cap E_{1,k_1}$, we may find $k_2 \geq r$ such that

$$\mathscr{H}^s_\infty(I_m \cap E_{1,k_1} \cap E_{2,k_2}) > \mathscr{H}^s_\infty(I_m) - 1/r$$

for $m=1,\ldots,r$. Continuing in this way, we get $k_1,\ldots,k_r \geq r$ such that

$$\mathscr{H}^s_\infty\left(I_m \cap \bigcap_{j=1}^r E_{j,k_j}\right) > \mathscr{H}^s_\infty(I_m) - 1/r \tag{8.15}$$

for all $m=1,\ldots,r$. For each r, let G_r be the finite union of closed intervals

$$G_r = \bigcap_{j=1}^r E_{j,k_j}. \tag{8.16}$$

Let $I \subset [a,b]$ be any closed interval. Given $\varepsilon > 0$, there is an interval $I_m \subset I$ such that $\mathscr{H}^s_\infty(I_m) > \mathscr{H}^s_\infty(I) - \varepsilon/2$. If $r \geq m$ and $r > 2/\varepsilon$, (8.15) gives that $\mathscr{H}^s_\infty(I \cap G_r) \geq \mathscr{H}^s_\infty(I_m \cap G_r) > \mathscr{H}^s_\infty(I_m) - 1/r > \mathscr{H}^s_\infty(I) - \varepsilon$, so

$$\lim_{r\to\infty} \mathscr{H}^s_\infty(I \cap G_r) = \mathscr{H}^s_\infty(I).$$

Let j be any positive integer. If $r > j$ and $x \in G_r$ then $x \in E_{j,k_j}$, by (8.16). Thus if

$x \in \overline{\lim}_{r\to\infty} G_r$, then $x \in E_{j,k_j}$ for infinitely many k_j, so $x \in \overline{\lim}_{i\to\infty} E_{j,i} \subset F_j$. Hence $\overline{\lim}_{r\to\infty} G_r \subset F_j$ for each j, so $\bigcap_{i=1}^{\infty} F_i \in \mathscr{C}^s[a,b]$. □

Corollary 8.7

Let $F_j \in \mathscr{C}^s[a,b]$ for $j = 1, 2, \ldots$. Then $\dim_{\mathrm{H}} \bigcap_{j=1}^{\infty} F_j \geqslant s$.

Proof. This is immediate from Propositions 8.5 and 8.6. □

Clearly, if F is in $\mathscr{C}^s(-\infty, \infty)$ then so is the translate $F + x$. Hence, given a set F in $\mathscr{C}^s(-\infty, \infty)$ and a sequence of numbers $x_1, x_2, \ldots$, we have $\bigcap_{i=1}^{\infty}(F + x_i)$ a member of $\mathscr{C}^s(-\infty, \infty)$, so that this intersection has dimension at least s. The same idea may be applied using more general transformations of F.

Proposition 8.8

Let $f:[a,b] \to \mathbb{R}$ be a mapping with a continuous derivative such that $|f'(x)| > c$ for some constant $c > 0$. If $F \in \mathscr{C}^s[a,b]$, then $f(F) \in \mathscr{C}^s[f(a), f(b)]$.

Proof. This may be proved in the same sort of way as Proposition 8.4. We omit the (rather tedious) details. □

In a typical $\mathscr{C}^s$ set the E_k are made up of intervals which have lengths and spacings tending to 0 as $k \to \infty$.

Example 8.9

Fix $\alpha > 2$. Let $E_k = \{x : |x - p/k| \leqslant k^{-\alpha}$ for some integer $p\}$, so that E_k is a union of equally spaced intervals of length $2k^{-\alpha}$. Then $F = \overline{\lim}_{k\to\infty} E_k$ is a member of $\mathscr{C}^s(-\infty, \infty)$ for all $s < 1/\alpha$.

Proof. Take $0 < s < 1/\alpha$ and a bounded closed interval I. We must show that

$$\lim_{k\to\infty} \mathscr{H}^s_\infty(I \cap E_k) = |I|^s. \tag{8.17}$$

The interval I contains m complete intervals of E_k, each of length $2k^{-\alpha}$, where $m \geqslant k|I| - 2$. Let μ be the mass distribution on $I \cap E_k$ obtained by distributing a mass $1/m$ uniformly across each complete interval of E_k contained in I. To estimate $\mathscr{H}^s_\infty(I \cap E_k)$, let U be a set in a covering of $I \cap E_k$; we may assume that U is a closed interval and that the ends of U are points of $I \cap E_k$. Then U intersects at most $k|U| + 2$ intervals of $I \cap E_k$. If $1/2k \leqslant |U| \leqslant |I|$ then

$$\begin{aligned} \mu(U) &\leqslant (k|U| + 2)/m \leqslant (k|U| + 2)/(k|I| - 2) \leqslant (|U| + 2k^{-1})/(|I| - 2k^{-1}) \\ &\leqslant |U|^s(|U|^{1-s} + 2k^{-1}|U|^{-s})/(|I| - 2k^{-1}) \end{aligned}$$

$$\leqslant |U|^s(|U|^{1-s} + 2^{s+1}k^{s-1})/(|I| - 2k^{-1})$$

$$\leqslant \frac{|U|^s}{|I|^s}\frac{(|U|^{1-s}|I|^{s-1} + 2^{s+1}k^{s-1}|I|^{s-1})}{(1 - 2k^{-1}|I|^{-1})}$$

$$\leqslant \frac{|U|^s}{|I|^s}\frac{(1 + 2^{s+1}k^{s-1}|I|^{s-1})}{(1 - 2k^{-1}|I|^{-1})}. \tag{8.18}$$

If k is large enough and $|U| < 2k$, then U can intersect just one interval of E_k so $|U| \leqslant 2k^{-\alpha}$, since the endpoints of U are in E_k. A mass of $1/m$ is distributed evenly across this interval of length $2k^{-\alpha}$, so

$$\mu(U) \leqslant |U|/2k^{-\alpha}m \leqslant |U|^s|U|^{1-s}/2k^{-\alpha}m \leqslant |U|^s(2k^{-\alpha})^{1-s}/2k^{-\alpha}(k|I| - 2)$$
$$\leqslant |U|^s 2^{-s}k^{s\alpha-1}/(|I| - 2k^{-1}). \tag{8.19}$$

With I and $\varepsilon > 0$ given, then, provided k is sufficiently large,

$$\mu(U) \leqslant (1 + \varepsilon)|U|^s/|I|^s$$

for all covering intervals U, using (8.18) and (8.19). Hence if $I \cap E_k \subset \bigcup_i U_i$ then

$$1 = \mu(I \cap E_k) \leqslant \sum_i \mu(U_i) \leqslant (1 + \varepsilon)|I|^{-s}\sum_i |U_i|^s$$

so $\mathscr{H}^s_\infty(I \cap E_k) \geqslant |I|^s/(1 + \varepsilon)$, from which (8.17) follows. □

In this example, F belongs to $\mathscr{C}^s(-\infty, \infty)$ if $s < 1/\alpha$, so $\dim_H F \geqslant 1/\alpha$, by Proposition 8.5. Moreover, it is clear that the translate $F + x$ is in $\mathscr{C}^s(-\infty, \infty)$ for any real number x, so by Proposition 8.6 $\bigcap_{i=1}^\infty (F + x_i)$ belongs to $\mathscr{C}^s(-\infty, \infty)$ for any countable set $x_1, x_2, \ldots$, implying that $\dim_H \bigcap_{i=1}^\infty (F + x_i) \geqslant 1/\alpha$. More generally, $f(F)$ is in $\mathscr{C}^s(-\infty, \infty)$ for all 'reasonable' functions f by Proposition 8.8, and this generates a large stock of $\mathscr{C}^s$ sets, countable intersections of which also have dimension at least $1/\alpha$.

In Section 10.3 we shall indicate how Example 8.9 may be improved to give F in $\mathscr{C}^s(-\infty, \infty)$ for all $s < 2/\alpha$, with corresponding consequences for dimensions.

8.3 Notes and references

The study of intersections of sets as they are moved relative to one another is part of the subject known as integral geometry. An full account in the classical setting is given by Santaló (1976). The main references for the fractal intersection formulae of Section 8.1 are Kahane (1986) and Mattila (1984, 1985). There are several definitions of classes of sets with large intersections such as those given by Baker and Schmidt (1970), Falconer (1958b) and Dodson, Rynne and Vickers (to appear).

Exercises

8.1 Let E and F be rectifiable curves in $\mathbb{R}^2$ and let σ be a rigid motion. Prove Poincaré's formula of classical integral geometry

$$4 \times \text{length}(E)\text{length}(F) = \int (\text{number of points in } (E \cap \sigma(F)))\,\mathrm{d}\sigma$$

where integration is with respect to the natural measure on the set of rigid motions. (Hint: show this first when E and F are line segments, then for polygons, and obtain the general result by approximation.)

8.2 Show that if a curve C bounds a (compact) convex set in the plane, then the length of C is given by

$$\frac{1}{2}\int_{\theta=0}^{2\pi} \text{length}(\text{proj}_\theta C)\,\mathrm{d}\theta.$$

(Hint: take E as C and F as a long line segment in the result of Exercise 8.1.)

8.3 In the plane, let E be the product of two middle third Cantor sets and let F be (i) a circle and (ii) the von Koch curve. In each case, what can be said about the Hausdorff dimension of $E \cap \sigma(F)$ for congruence transformations σ?

8.4 Show that the conclusion of Theorem 8.1 may be strengthened to give that $E \cap (F + x)$ is empty for almost all x if $\dim_{\mathrm{H}}(E \times F) < n$.

8.5 By taking E as a suitable set dense in a region of $\mathbb{R}^2$ and F as a unit line segment, show that (8.5) fails if Hausdorff dimension is replaced by box dimensions, even for the group of similarities.

8.6 Let $1 < s < 2$. Construct a plane s-set F in the unit disc B such that if E is any straight line segment of length 2 that intersects the interior of B then $E \cap F$ is an $(s-1)$-set.

8.7 Let E_k be the set of real numbers with base-3 expansion $m \cdot a_1 a_2 \cdots$ such that $a_k = 0$ or 2. Show that $F = \overline{\lim}_{k\to\infty} E_k$ is in class $\mathscr{C}^s(-\infty, \infty)$ for all $0 < s < 1$. (Note that F is the set of numbers with infinitely many base-3 digits different from 1.) Deduce that $\dim_{\mathrm{H}} F = 1$ and that $\dim_{\mathrm{H}}(\bigcap_{i=1}^{\infty}(F + x_i)) = 1$ for any countable sequence $x_1, x_2, \ldots$.

Part II
APPLICATIONS AND EXAMPLES

Chapter 9 Fractals defined by transformations—self-similar and self-affine sets

9.1 Iterated function schemes

We begin this chapter by describing a general construction for fractals, of which the Cantor set, von Koch curve and other standard examples are special cases.

Many fractals are made up of parts which are, in some way, similar to the whole. For example, the middle third Cantor set is the union of two similar copies of itself, and the von Koch curve is made up of four similar copies. These self-similarities are not only properties of the fractals, they may actually be used to define them—an approach which is often extremely useful.

Let D be a closed subset of $\mathbb{R}^n$. A mapping $S: D \to D$ is called a *contraction* on D if there is a number c with $0 < c < 1$ such that $|S(x) - S(y)| \leqslant c|x - y|$ for all x, y in D. Clearly any contraction is a continuous mapping. If equality holds, i.e. if $|S(x) - S(y)| = c|x - y|$, then S transforms sets into geometrically similar ones, and we call S a *similarity*.

Let $S_1, \ldots, S_m$ be contractions. We call a subset F of D *invariant* for the transformations S_i if

$$F = \bigcup_{i=1}^{m} S_i(F). \tag{9.1}$$

As we shall see, such invariant sets are often fractals.

This is most easily illustrated when F is the middle third Cantor set. Let $S_1, S_2: \mathbb{R} \to \mathbb{R}$ be given by $S_1(x) = \frac{1}{3}x$; $S_2(x) = \frac{1}{3}x + \frac{2}{3}$. Then $S_1(F)$ and $S_2(F)$ are just the left and right 'halves' of F, so that $F = S_1(F) \cup S_2(F)$. Thus F is invariant for the mappings S_1 and S_2, the two mappings which represent the fundamental self-similarities of the Cantor set.

We show that families of contractions, or *iterated function schemes* as they are sometimes known, define unique (non-empty) compact invariant sets. This means, for example, that the middle third Cantor set is completely specified as the compact invariant set of the mappings S_1 and S_2 given above.

We define a metric or distance between subsets of D. Let $\mathscr{S}$ denote the class of all non-empty compact subsets of D. Recall that the δ-*parallel body* of $A \in \mathscr{S}$ is the set of points within distance δ of A, i.e. $A_\delta = \{x \in D: |x - a| \leqslant \delta$ for some

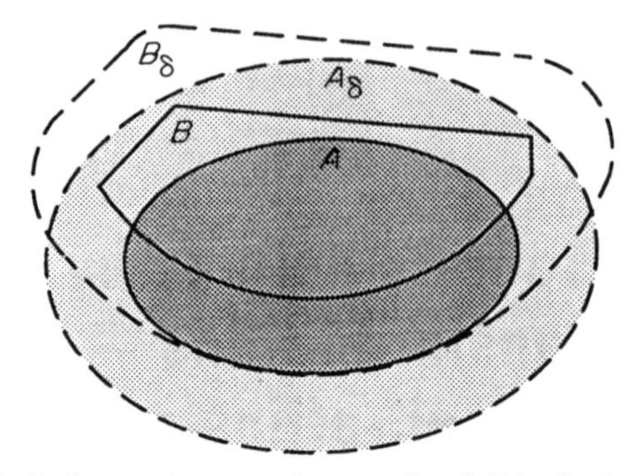

Figure 9.1 The Hausdorff distance between the sets A and B is the least $\delta > 0$ such that the δ-parallel body A_δ of A contains B and the δ-parallel body B_δ of B contains A

$a \in A\}$. We make $\mathcal{S}$ into a metric space by defining the distance $d(A, B)$ between two sets A, B to be the least δ such that the δ-parallel body of A contains B and vice-versa:

$$d(A, B) = \inf\{\delta : A \subset B_\delta \text{ and } B \subset A_\delta\} \tag{9.2}$$

(see figure 9.1). A simple check shows that d is a metric or distance function, known as the *Hausdorff metric* on $\mathcal{S}$. Thus (i) $d(A, B) \geqslant 0$, with equality if and only if $A = B$, (ii) $d(A, B) = d(B, A)$, and (iii) $d(A, B) \leqslant d(A, C) + d(C, B)$ for any A, B and C in $\mathcal{S}$. In particular, if $d(A, B)$ is small then, in one sense, A and B are close to each other.

Theorem 9.1

Let $S_1, \ldots, S_m$ be contractions on $D \subset \mathbb{R}^n$ so that

$$|S_i(x) - S_i(y)| \leqslant c_i |x - y| \qquad (x, y \in D)$$

with $c_i < 1$ for each i. Then there exists a unique non-empty compact set F that is invariant for the S_i, i.e. which satisfies

$$F = \bigcup_{i=1}^{m} S_i(F).$$

Moreover, if we define a transformation S on the class $\mathcal{S}$ of non-empty compact sets by

$$S(E) = \bigcup_{i=1}^{m} S_i(E) \tag{9.3}$$

and write S^k for the kth iterate of S given by $S^0(E) = E$, $S^k(E) = S(S^{k-1}(E))$ for $k \geqslant 1$, then

$$F = \bigcap_{k=1}^{\infty} S^k(E) \tag{9.4}$$

for any set E in $\mathcal{S}$ such that $S_i(E) \subset E$ for each i.

Proof. Note that sets in $\mathscr{S}$ are transformed by S into other sets in $\mathscr{S}$. Let E be any set in $\mathscr{S}$ such that $S_i(E) \subset E$ for all i; for example $D \cap B_r(0)$ will do, provided that r is sufficiently large. Then $S^k(E) \subset S^{k-1}(E)$ so that $S^k(E)$ is a decreasing sequence of non-empty compact sets, which necessarily have non-empty compact intersection $F = \bigcap_{k=1}^{\infty} S^k(E)$. Since $S^k(E)$ is a decreasing sequence, it follows that $S(F) = F$, so F is invariant.

To show that the invariant set is unique, note that if $A, B \in \mathscr{S}$ then

$$d(S(A), S(B)) = d\left(\bigcup_{i=1}^{m} S_i(A), \bigcup_{i=1}^{m} S_i(B) \right) \leqslant \max_{1 \leqslant i \leqslant m} d(S_i(A), S_i(B))$$

(since if δ is such that the δ-parallel body $(S_i(A))_\delta$ contains $S_i(B)$ for each i, then $(\bigcup_{i=1}^{m} S_i(A))_\delta$ contains $\bigcup_{i=1}^{m} S_i(B)$). Thus

$$d(S(A), S(B)) \leqslant \left(\max_{1 \leqslant i \leqslant m} c_i \right) d(A, B). \tag{9.5}$$

It follows that if $S(A) = A$ and $S(B) = B$ are both invariant sets, then $d(A, B) = 0$, which implies that $A = B$. □

In fact, the sequence of iterates $S^k(E)$ converges to F for any initial set E in $\mathscr{S}$, in the sense that $d(S^k(E), F) \to 0$. This follows since (9.5) implies that $d(S(E), F) = d(S(E), S(F)) \leqslant c d(E, F)$, so that $d(S^k(E), F) \leqslant c^k d(E, F)$ where $c = \max_{1 \leqslant i \leqslant m} c_i < 1$. Thus the $S^k(E)$ provide increasingly good approximations to F. (If F is a fractal these approximations are sometimes called *pre-fractals* for F.)

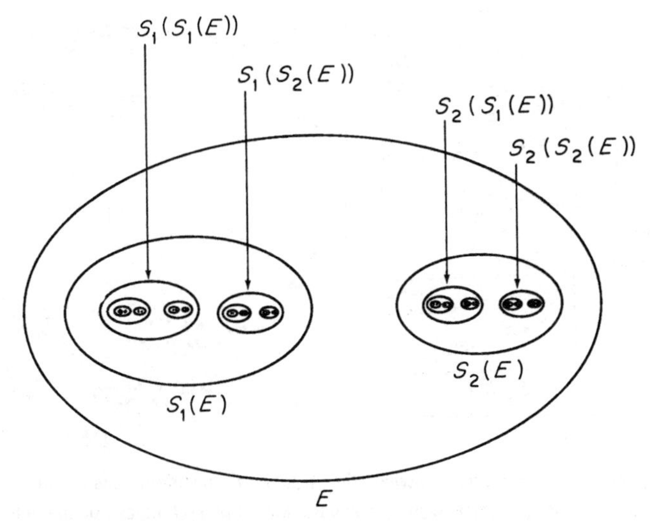

Figure 9.2 Construction of the invariant set F for transformations S_1 and S_2 which map the large ellipse E onto the ellipses $S_1(E)$ and $S_2(E)$. The sets $S^k(E) = \bigcup_{i_j=1,2} S_{i_1} \circ \cdots \circ S_{i_k}(E)$ give increasingly good approximations to F

For each k

$$S^k(E) = \bigcup_{J_k} S_{i_1} \circ \cdots \circ S_{i_k}(E) = \bigcup_{J_k} S_{i_1}(S_{i_2}(\cdots(S_{i_k}(E)))) \tag{9.6}$$

where the union is over the set J_k of all k-term sequences $(i_1, \ldots, i_k)$ with $1 \leqslant i_j \leqslant m$; see figure 9.2. (Recall that $S_{i_1} \circ \cdots \circ S_{i_k}$ denotes the composition of mappings, so that $(S_{i_1} \circ \cdots \circ S_{i_k})(x) = S_{i_1}(S_{i_2}(\cdots(S_{i_k}(x))\cdots)).$) If $S_i(E)$ is contained in E for each i and x is a point of F, it follows from (9.4) that there is a (not necessarily unique) sequence $(i_1, i_2, \ldots)$ such that $x \in S_{i_1} \circ \cdots \circ S_{i_k}(E)$ for all k. Thus

$$F = \bigcup \{x_{i_1, i_2, \ldots}\}$$

where

$$x_{i_1, i_2, \ldots} = \bigcap_{k=1}^{\infty} S_{i_1} \circ \cdots \circ S_{i_k}(E). \tag{9.7}$$

This expression for $x_{i_1, i_2, \ldots}$ is independent of E provided that $S_i(E)$ is contained in E for all i.

Notice that if the union in (9.1) is disjoint then F must be totally disconnected, since if $x_{i_1, i_2, \ldots} \neq x_{i'_1, i'_2, \ldots}$ we may find k such that $(i_1, \ldots, i_k) \neq (i'_1, \ldots, i'_k)$ so that the disjoint closed sets $S_{i_1} \circ \cdots \circ S_{i_k}(F)$ and $S_{i'_1} \circ \cdots \circ S_{i'_k}(F)$ disconnect the two points.

Again this may be illustrated by $S_1(x) = \frac{1}{3}x$, $S_2(x) = \frac{1}{3}x + \frac{2}{3}$ and F the Cantor set. If $E = [0, 1]$ then $S^k(E) = E_k$, the set of 2^k basic intervals of length 3^{-k} obtained at the kth stage of the usual Cantor set construction; see figure 0.1. Moreover, $x_{i_1, i_2, \ldots}$ is the point of the Cantor set with base-3 expansion $0{\cdot}a_1 a_2 \ldots$, where $a_k = 0$ if $i_k = 1$ and $a_k = 2$ if $i_k = 2$. The pre-fractals $S^k(E)$ provide the usual construction of many fractals for a suitably chosen initial set E; the $S_{i_1} \circ \cdots \circ S_{i_k}(E)$ are called the *basic sets* of the construction.

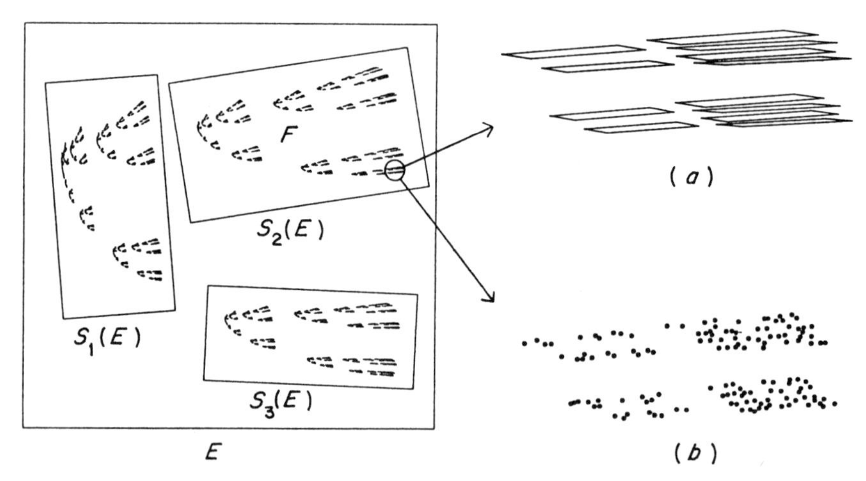

Figure 9.3 Two ways of computer drawing the fractal F, invariant under the three affine transformations S_1, S_2 and S_3 which map the square onto the rectangles. In method (*a*) the 3^k parallelograms $S_{i_1}(S_{i_2}(\cdots(S_{i_k}(E))\cdots))$ for $i_j = 1, 2, 3$ are drawn ($k = 6$ here). In method (*b*), the sequence of points x_k is plotted by choosing S_{i_k} at random from S_1, S_2 and S_3 for successive k and letting $x_k = S_{i_k}(x_{k-1})$

This theory provides us with two methods for computer drawing of invariant sets in the plane, as indicated in figure 9.3. For the first method, take any initial set E (such as a square) and draw the kth approximation $S^k(E)$ to F given by (9.6) for a suitable value of k. The set $S^k(E)$ is made up of m^k small sets—either these can be drawn in full, or a representative point of each can be plotted. If E can be chosen as a line segment in such a way that $S_1(E),\ldots,S_m(E)$ join up to form a polygonal curve with endpoints the same as those of E, then the sequence of polygonal curves $S^k(E)$ provides increasingly good approximations to the fractal curve F. Taking E as the initial interval in the von Koch curve construction is an example of this, with $S^k(E)$ just the kth step of the construction (E_k in figure 0.2). Careful recursive programming is helpful when using this method.

For the second method, take x_0 as any initial point, select a contraction S_{i_1} from $S_1,\ldots,S_m$ at random, and let $x_1 = S_{i_1}(x_0)$. Continue in this way, choosing S_{i_k} from $S_1,\ldots,S_m$ at random (with equal probability) and letting $x_k = S_{i_k}(x_{k-1})$ for $k = 1, 2, \ldots$ For large enough k, the points x_k will be indistinguishably close to F, with x_k close to $S_{i_k}\circ\cdots\circ S_{i_1}(F)$, so the sequence $\{x_k\}$ will appear randomly distributed across F. A plot of the sequence $\{x_k\}$ from, say, the hundredth term onwards may give a good impression of F.

9.2 Dimensions of self-similar sets

One of the advantages of using an iterated function scheme is that the dimension of the invariant set is often relatively easy to calculate or estimate in terms of the defining contractions. In this section we discuss the case where $S_1,\ldots,S_k:\mathbb{R}^n \to \mathbb{R}^n$ are *similarities*, i.e. with

$$|S_i(x) - S_i(y)| = c_i|x - y| \qquad (x, y \in \mathbb{R}^n) \tag{9.8}$$

where $0 < c_i < 1$ (c_i is called the *ratio* of S_i). Thus each S_i transforms subsets of $\mathbb{R}^n$ into geometrically similar sets. A set that is invariant under such a collection of similarities is called a (*strictly*)-*self-similar set*, being a union of a number of smaller similar copies of itself. Standard examples include the middle third Cantor set, the Sierpiński gasket and the von Koch curve, see figures 0.1–0.5. We show that, under certain conditions, a self-similar set F has Hausdorff and box dimensions equal to the value of s satisfying

$$\sum_{i=1}^{m} c_i^s = 1 \tag{9.9}$$

and further that F has positive and finite $\mathcal{H}^s$-measure. A calculation similar to the 'heuristic calculation' of Example 2.7 indicates that the value given by (9.9) is at least plausible. If $F = \bigcup_{i=1}^m S_i(F)$ with the union 'nearly disjoint', we have that

$$\mathcal{H}^s(F) = \sum_{i=1}^{m} \mathcal{H}^s(S_i(F)) = \sum_{i=1}^{m} c_i^s \mathcal{H}^s(F) \tag{9.10}$$

using (9.8) and Scaling property 2.1. On the assumption that $0 < \mathscr{H}^s(F) < \infty$ at the 'jump' value, $s = \dim_H F$, we get that s satisfies (9.9).

For this argument to give the right answer, we require a condition that ensures that the components $S_i(F)$ of F do no overlap 'too much'. We say that the S_i satisfy the *open set condition* if there exists a non-empty bounded open set V such that

$$V \supset \bigcup_{i=1}^{m} S_i(V) \tag{9.11}$$

with the union disjoint. (In the middle third Cantor set example, the open set condition holds for S_1 and S_2 with V as the open interval $(0, 1)$.) We show that, provided that the S_i satisfy the open set condition, the Hausdorff dimension of the invariant set is given by (9.9).

We require the following geometrical result.

Lemma 9.2

Let $\{V_i\}$ be a collection of disjoint open subsets of $\mathbb{R}^n$ such that each V_i contains a ball of radius $a_1 r$ and is contained in a ball of radius $a_2 r$. Then any ball B of radius r intersects at most $(1 + 2a_2)^n a_1^{-n}$ of the closures $\bar{V}_i$.

Proof. If $\bar{V}_i$ meets B, then $\bar{V}_i$ is contained in the ball concentric with B of radius $(1 + 2a_2)r$. Suppose that q of the sets $\bar{V}_i$ intersect B. Then, summing the volumes of the corresponding interior balls of radii $a_1 r$, it follows that $q(a_1 r)^n \leqslant (1 + 2a_2)^n r^n$, giving the stated bound for q. □

The derivation of the lower bound in the following theorem is a little awkward. The reader may find it helpful to follow through the proof with the middle third Cantor set in mind, or by referring to the 'general example' of figure 9.2. Alternatively, the proof of Proposition 9.7 covers the case when the sets $S_1(F), \ldots, S_m(F)$ are disjoint, and is rather simpler.

Theorem 9.3

Suppose that the open set condition (9.11) holds for the similarities S_i on $\mathbb{R}^n$ with ratios $c_i (1 \leqslant i \leqslant m)$. If F is the invariant set satisfying

$$F = \bigcup_{i=1}^{m} S_i(F) \tag{9.12}$$

then $\dim_H F = \dim_B F = s$, where s is given by

$$\sum_{i=1}^{m} c_i^s = 1. \tag{9.13}$$

Moreover, for this value of s, $0 < \mathscr{H}^s(F) < \infty$.

Proof. Let s satisfy (9.13). For any set A we write $A_{i_1,\ldots,i_k} = S_{i_1} \circ \cdots \circ S_{i_k}(A)$. Let J_k denote the set of all k-term sequences $(i_1, \ldots, i_k)$ with $1 \leqslant i_j \leqslant m$. It follows, by using (9.12) repeatedly, that

$$F = \bigcup_{J_k} F_{i_1,\ldots,i_k}.$$

We check that these covers of F provide a suitable upper estimate for the Hausdorff measure. Since the mapping $S_{i_1} \circ \cdots \circ S_{i_k}$ is a similarity of ratio $c_{i_1} \cdots c_{i_k}$, then

$$\sum_{J_k} |F_{i_1,\ldots,i_k}|^s = \sum_{J_k} (c_{i_1} \cdots c_{i_k})^s |F|^s = \left(\sum_{i_1} c_{i_1}^s\right) \cdots \left(\sum_{i_k} c_{i_k}^s\right) |F|^s = |F|^s \tag{9.14}$$

by (9.13). For any $\delta > 0$, we may choose k such that $|F_{i_1,\ldots,i_k}| \leqslant (\max_i c_i)^k \leqslant \delta$, so $\mathscr{H}^s_\delta(F) \leqslant |F|^s$ and hence $\mathscr{H}^s(F) \leqslant |F|^s$.

The lower bound is more awkward. Let I be the set of all infinite sequences $I = \{(i_1, i_2, \ldots) : 1 \leqslant i_j \leqslant m\}$, and let $I_{i_1,\ldots,i_k} = \{(i_1, \ldots, i_k, q_{k+1}, \ldots) : 1 \leqslant q_j \leqslant m\}$ be the 'cylinder' consisting of those sequences in I with initial terms $(i_1, \ldots, i_k)$. We may put a mass distribution μ on I such that $\mu(I_{i_1,\ldots,i_k}) = (c_{i_1} \cdots c_{i_k})^s$. Since $(c_{i_1} \cdots c_{i_k})^s = \sum_{i=1}^m (c_{i_1} \cdots c_{i_k} c_i)^s$, i.e. $\mu(I_{i_1,\ldots,i_k}) = \sum_{i=1}^m \mu(I_{i_1,\ldots,i_k,i})$, it follows that μ is indeed a mass distribution on subsets of I with $\mu(I) = 1$. We may transfer μ to a mass distribution $\tilde{\mu}$ on F in a natural way by defining $\tilde{\mu}(A) = \mu\{(i_1, i_2, \ldots) : x_{i_1,i_2,\ldots} \in A\}$ for subsets A of F. (Recall that $x_{i_1,i_2,\ldots} = \bigcap_{k=1}^\infty F_{i_1,\ldots,i_k}$.) It is easily checked that $\tilde{\mu}(F) = 1$.

We show that $\tilde{\mu}$ satisfies the conditions of the Mass distribution principle 4.2. Let V be the open set of (9.11). Since $\bar{V} \supset S(\bar{V}) = \bigcup_{i=1}^m S_i(\bar{V})$, the decreasing sequence of iterates $S^k(\bar{V})$ converges to F; see (9.4). In particular $\bar{V} \supset F$ and $\bar{V}_{i_1,\ldots,i_k} \supset F_{i_1,\ldots,i_k}$ for each finite sequence $(i_1, \ldots, i_k)$. Let B be any ball of radius $r < 1$. We estimate $\tilde{\mu}(B)$ by considering the sets $V_{i_1,\ldots,i_k}$ with diameters comparable with that of B and with closures intersecting $F \cap B$.

We curtail each infinite sequence $(i_1, i_2, \ldots) \in I$ after the first term i_k for which

$$\left(\min_i c_i\right) r \leqslant c_{i_1} c_{i_2} \cdots c_{i_k} \leqslant r \tag{9.15}$$

and let Q denote the finite set of all (finite) sequences obtained in this way. Then for every infinite sequence $(i_1, i_2, \ldots) \in I$ there is exactly one value of k with $(i_1, \ldots, i_k) \in Q$. Since $V_1, \ldots, V_m$ are disjoint, so are $V_{i_1,\ldots,i_k,1}, \ldots, V_{i_1,\ldots,i_k,m}$ for each $(i_1, \ldots, i_k)$. Using this in a nested way, it follows that the collection of open sets $\{V_{i_1,\ldots,i_k} : (i_1, \ldots, i_k) \in Q\}$ is disjoint. Similarly $F \subset \bigcup_Q F_{i_1,\ldots,i_k} \subset \bigcup_Q \bar{V}_{i_1,\ldots,i_k}$.

We choose a_1 and a_2 so that V contains a ball of radius a_1 and is contained in a ball of radius a_2. Then, for $(i_1, \ldots, i_k) \in Q$, the set $V_{i_1,\ldots,i_k}$ contains a ball of radius $c_{i_1} \cdots c_{i_k} a_1$ and therefore one of radius $(\min_i c_i) a_1 r$, and is contained in a ball of radius $c_{i_1} \cdots c_{i_k} a_2$ and hence in a ball of radius $a_2 r$. Let Q_1 denote those sequences $(i_1,\ldots,i_k)$ in Q such that B intersects $\bar{V}_{i_1,\ldots,i_k}$. By Lemma 9.2 there are at most $q=(1+2a_2)^n a_1^{-n}(\min_i c_i)^{-n}$ sequences in Q_1. Then

$$\tilde{\mu}(B)=\tilde{\mu}(F\cap B)\leqslant \mu\{(i_1,i_2,\ldots):x_{i_1,i_2,\ldots}\in F\cap B\}$$
$$\leqslant \mu\left\{\bigcup_{Q_1} I_{i_1,\ldots,i_k}\right\}$$

since, if $x_{i_1,i_2,\ldots}\in F\cap B\subset \bigcup_{Q_1}\bar{V}_{i_1,\ldots,i_k}$, then there is an integer k such that $(i_1,\ldots,i_k)\in Q_1$. Thus

$$\tilde{\mu}(B)\leqslant \sum_{Q_1}\mu(I_{i_1,\ldots,i_k})$$
$$=\sum_{Q_1}(c_{i_1}\cdots c_{i_k})^s\leqslant \sum_{Q_1} r^s\leqslant r^s q$$

using (9.15). Since any set U is contained in a ball of radius $|U|$, we have $\tilde{\mu}(U)\leqslant |U|^s q$, so the Mass distribution principle 4.2 gives $\mathscr{H}^s(F)\geqslant q^{-1}>0$, and $\dim_{\mathrm{H}} F=s$.

If Q is any set of infinite sequences such that for every $(i_1,i_2,\ldots)\in I$ there is exactly one integer k with $(i_1,\ldots,i_k)\in Q$, it follows inductively from (9.13) that $\sum_Q(c_{i_1}c_{i_2}\cdots c_{i_k})^s=1$. Thus, if Q is chosen as in (9.15), Q contains at most $(\min_i c_i)^{-s}r^{-s}$ sequences. For each sequence $(i_1,\ldots,i_k)$ in Q we have $|\bar{V}_{i_1,\ldots,i_k}|=c_{i_1}\cdots c_{i_q}|\bar{V}|\leqslant r|\bar{V}|$, so F may be covered by $(\min_i c_i)^{-s}r^{-s}$ sets of diameter $r|\bar{V}|$ for each $r<1$. It follows from Equivalent definition 3.1(iv) that $\overline{\dim}_{\mathrm{B}} F\leqslant s$; since the Hausdorff dimension is also s, this completes the proof.

□

If the open set condition is not assumed in Theorem 9.3, it may be shown that we still have $\dim_{\mathrm{H}} F=\dim_{\mathrm{B}} F$ though this value may be less than s.

Theorem 9.3 enables us to find the dimension of many self-similar fractals.

Example 9.4 Sierpiński gasket

The Sierpiński gasket F is constructed from an equilateral triangle by repeatedly removing inverted equilateral triangles; see figure 0.3. *Then* $\dim_{\mathrm{H}} F=\dim_{\mathrm{B}} F=\log 3/\log 2$.

Calculation. The gasket F is the invariant set under the three obvious similarities of ratios $\frac{1}{2}$ which map the triangle E_0 onto the triangles of E_1. The open set condition holds, taking V as the interior of E_0. Thus, by Theorem 9.3, $\dim_{\mathrm{H}} F=\dim_{\mathrm{B}} F=\log 3/\log 2$, which is the solution of $\sum_1^3(\frac{1}{2})^s=1$. □

The next example involves similarity transformations of more than one ratio.

Example 9.5 Modified von Koch curve.

Fix $0<a\leqslant\frac{1}{3}$ and construct a curve F by repeatedly replacing the middle proportion a of each interval by the other two sides of an equilateral triangle; see figure 9.4. *Then $\dim_{\mathrm{H}} F=\dim_{\mathrm{B}} F$ is the solution of $2a^s+2(\frac{1}{2}(1-a))^s=1$.*

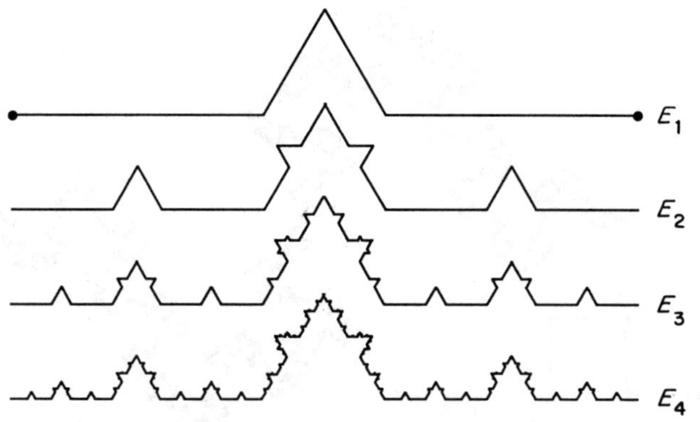

Figure 9.4 Construction of a modified von Koch curve—see Example 9.5 (E_1 is a generator for the curve)

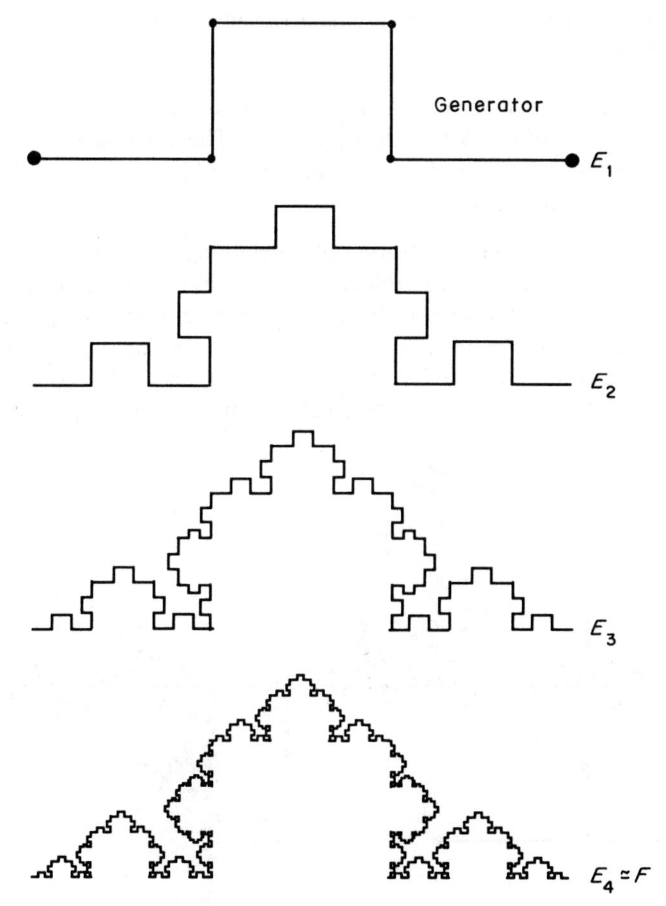

Figure 9.5 Stages in the construction of a fractal curve from a generator. The lengths of the segments in the generator are $\frac{1}{3}, \frac{1}{4}, \frac{1}{3}, \frac{1}{4}, \frac{1}{3}$, and the Hausdorff and box dimensions of F are given by $3(\frac{1}{3})^s + 2(\frac{1}{4})^s = 1$ or $s = 1.34$

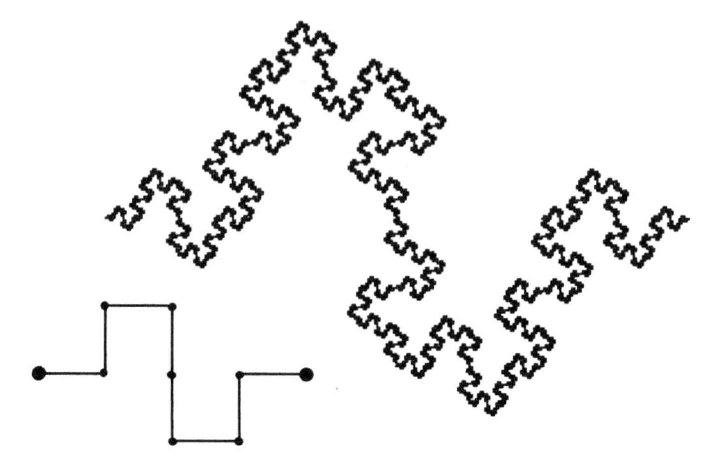

Figure 9.6 A fractal curve and its generator. The Hausdorff and box dimensions of the curve are equal to $\log 8/\log 4 = 1\frac{1}{2}$

Calculation. The curve F is invariant under the similarities that map the unit interval onto each of the four intervals in E_1. The open set condition holds, taking V as an isosceles triangle of base length 1 and height $\frac{1}{2}\sqrt{3}$, so Theorem 9.3 gives the dimension stated. □

There is a convenient method of specifying certain self-similar sets diagramatically, in particular self-similar curves such as Example 9.5. A *generator* consists of a number of straight line segments and two points specially identified. We associate with each line segment the similarity that maps the two special points onto the endpoints of the segment. A sequence of sets approximating to the self-similar invariant set may be built up by iterating the

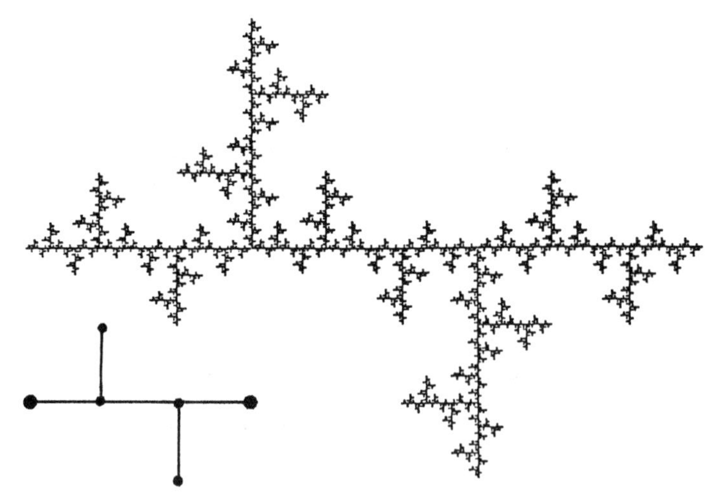

Figure 9.7 A tree-like fractal and its generator. The Hausdorff and box dimensions are equal to $\log 5/\log 3 = 1.465$

process of replacing each line segment by a similar copy of the generator; see figures 9.5–9.7 for some examples. Note that the similarities are defined by the generator only to within a reflection, but the orientation may be specified by displaying the first step of the construction.

9.3 Some variations

The calculations underlying Theorem 9.3 may be adapted to estimate the dimension of the invariant set F of a collection of contractions that are not similarities.

Proposition 9.6

Let $S_1, \ldots, S_m$ be contractions on a closed subset D of $\mathbb{R}^n$ such that

$$|S_i(x) - S_i(y)| \leqslant c_i |x - y| \qquad (x, y \in D)$$

with $c_i < 1$ for each i. Then $\dim_H F \leqslant s$ and $\overline{\dim}_B F \leqslant s$, where $\sum_{i=1}^{m} c_i^s = 1$.

Proof. These estimates are essentially those of the first and last paragraphs of the proof of Theorem 9.3, noting that we have the inequality $|A_{i_1,\ldots,i_k}| \leqslant c_{i_1} \cdots c_{i_k} |A|$ for any set A, rather than equality. □

We next obtain a lower bound for dimension in the case where the components $S_i(F)$ of F are disjoint, in which case F must be totally disconnected. Note that this will certainly be the case if there is *some* compact set E with $S_i(E) \subset E$ for all i and with the $S_i(E)$ disjoint.

Proposition 9.7

Let $S_1, \ldots, S_m$ be contractions on a closed subset D of $\mathbb{R}^n$ such that

$$b_i |x - y| \leqslant |S_i(x) - S_i(y)| \qquad (x, y \in D) \tag{9.16}$$

with $0 < b_i < 1$ for each i. Suppose that F is invariant for the S_i,

$$F = \bigcup_{i=1}^{m} S_i(F), \tag{9.17}$$

with this union disjoint. Then $\dim_H F \geqslant s$ where

$$\sum_{i=1}^{m} b_i^s = 1. \tag{9.18}$$

Proof. Let $d > 0$ be the minimum distance between any pair of the disjoint compact sets $S_1(F), \ldots, S_m(F)$, i.e. $d = \min_{i \neq j} \inf\{|x - y| : x \in S_i(F), y \in S_j(F)\}$.

Let $F_{i_1,\ldots,i_k} = S_{i_1}\circ\cdots\circ S_{i_k}(F)$ and define μ by $\mu(F_{i_1,\ldots i_k}) = (b_{i_1}\cdots b_{i_k})^s$. Since

$$\sum_{i=1}^{m} \mu(F_{i_1,\ldots,i_k,i}) = \sum_{i=1}^{m} (b_{i_1}\cdots b_{i_k} b_i)^s$$
$$= (b_{i_1}\cdots b_{i_k})^s = \mu(F_{i_1,\ldots,i_k})$$
$$= \mu\left(\bigcup_{i=1}^{k} F_{i_1,\ldots,i_k,i}\right)$$

it follows that μ is a mass distribution on F with $\mu(F) = 1$.

If $x\in F$, there is a unique infinite sequence $i_1, i_2, \ldots$ such that $x \in F_{i_1,\ldots,i_k}$ for each k. For $0 < r < d$ let k be the least integer such that

$$b_{i_1}\cdots b_{i_k} d \leqslant r < b_{i_1}\cdots b_{i_{k-1}} d.$$

If $i'_1,\ldots,i'_k$ is distinct from $i_1,\ldots,i_k$, the sets $F_{i_1,\ldots,i_k}$ and $F_{i'_1,\ldots,i'_k}$ are disjoint and separated by a gap of at least $b_{i_1}\cdots b_{i_{k-1}} d > r$. (To see this, note that if j is the least integer such that $i_j \neq i'_j$ then $F_{i_j,\ldots,i_k} \subset F_{i_j}$ and $F_{i'_j,\ldots,i'_k} \subset F_{i'_j}$ are separated by d, so $F_{i_1,\ldots,i_k}$ and $F_{i'_1,\ldots,i'_k}$ are separated by at least $b_{i_1}\cdots b_{i_{j-1}} d$.) It follows that $F \cap B_r(x) \subset F_{i_1,\ldots,i_k}$ so

$$\mu(F\cap B_r(x)) \leqslant \mu(F_{i_1,\ldots,i_k}) = (b_{i_1}\cdots b_{i_k})^s \leqslant d^{-s} r^s.$$

If U intersects F, then $U \subset B_r(x)$ for some $x\in F$ with $r = |U|$. Thus $\mu(U) \leqslant d^{-s}|U|^s$, so by the Mass distribution principle 4.2 $\mathscr{H}^s(F) > 0$ and $\dim_H F \geqslant s$. □

Example 9.8 'Non-linear' Cantor set

Let $D = [\frac{1}{2}(1+\sqrt{3}), (1+\sqrt{3})]$ and let $S_1, S_2 : D \to D$ be given by $S_1(x) = 1 + 1/x$, $S_2(x) = 2 + 1/x$. Then $0.44 < \dim_H F \leqslant \underline{\dim}_B F \leqslant \overline{\dim}_B F < 0.66$ where F is the invariant set for S_1 and S_2. (This example arises in connection with number theory; see Section 10.2.)

Calculation. We note that $S_1(D) = [\frac{1}{2}(1+\sqrt{3}), \sqrt{3}]$ and $S_2(D) = [\frac{1}{2}(3+\sqrt{3}), 1+\sqrt{3}]$ so we can use Propositions 9.6 and 9.7 to estimate $\dim_H F$. By the mean-value theorem, if $x \neq y \in D$, then $(S_i(x) - S_i(y))/(x-y) = S'_i(c_i)$ for some $c_i \in D$, for $i = 1, 2$. Thus for $i = 1, 2$.

$$\inf_{x\in D} |S'_i(x)| \leqslant \frac{|S_i(x) - S_i(y)|}{|x-y|} \leqslant \sup_{x\in D} |S'_i(x)|.$$

Since $S'_1(x) = S'_2(x) = -1/x^2$ it follows that

$$\tfrac{1}{2}(2-\sqrt{3})|x-y| \leqslant |S_i(x) - S_i(y)| \leqslant 2(2-\sqrt{3})|x-y|$$

for both $i = 1$ and $i = 2$. According to Propositions 9.6 and 9.7 lower and upper bounds for the dimensions are given by the solutions of $2(\frac{1}{2}(2-\sqrt{3}))^s = 1$ and $2(2(2-\sqrt{3}))^s = 1$ which are $s = \log 2/\log 2(2+\sqrt{3}) = 0.34$ and $\log 2/\log\frac{1}{2}(2+\sqrt{3}) = 1.11$ respectively.

For a subset of the real line, an upper bound greater than 1 is not of much use. One way of getting better estimates is to note that F is also the invariant set of the *four* mappings on $[0,1]$

$$S_i \circ S_j = i + 1/(j + 1/x) = i + x/(jx + 1) \qquad (i,j = 1,2).$$

By calculating derivatives and using the mean-value theorem as before, we get that

$$(S_i \circ S_j)'(x) = (jx + 1)^{-2}$$

so

$$(j(1+\sqrt{3})+1)^{-2}|x-y| \leqslant |S_i \circ S_j(x) - S_i \circ S_j(y)| \leqslant (\tfrac{1}{2}j(1+\sqrt{3})+1)^{-2}|x-y|.$$

Lower and upper bounds for the dimensions are now given by the solutions of $2(2+\sqrt{3})^{-2s} + 2(3+2\sqrt{3})^{-2s} = 1$ and $2(\frac{1}{2}(3+\sqrt{3}))^{-2s} + 2(2+\sqrt{3})^{-2s} = 1$, giving $0.44 < \dim_{\mathrm{H}} F < 0.66$, a considerable improvement on the previous estimates. In fact $\dim_{\mathrm{H}} F = 0.531$, a value that may be obtained by looking at yet higher-order iterates of the S_i. □

*[The rest of this subsection may be omitted.]

The technique used in Example 9.8 to improve the dimension estimates is often useful for invariant sets of transformations that are not strict similarities. If F is invariant for the contractions $S_1, \ldots, S_m$ on D then F is also invariant for the collection of m^k transformations $\{S_{i_1} \circ \cdots \circ S_{i_k}\}$ for each k. If the S_i are, say, twice differentiable on an open set containing F, it may be shown that when k is large, the contractions $S_{i_1} \circ \cdots \circ S_{i_k}$ are in a sense, close to similarities on D. In particular, for transformations on a subset D of $\mathbb{R}$, if $b = \inf_{x \in D} |(S_{i_1} \circ \cdots \circ S_{i_k})'(x)|$ and $c = \sup_{x \in D} |(S_{i_1} \circ \cdots \circ S_{i_k})'(x)|$, then

$$b|x-y| \leqslant |S_{i_1} \circ \cdots \circ S_{i_k}(x) - S_{i_1} \circ \cdots \circ S_{i_k}(y)| \leqslant c|x-y| \quad (x, y \in D)$$

If k is large then b/c will be close to 1, and applying Propositions 9.6 and 9.7 to the m^k transformations $S_{i_1} \circ \cdots \circ S_{i_k}$, gives good upper and lower estimates for the dimensions.

We may take this further. If the S_i are twice differentiable on a subset D of $\mathbb{R}$,

$$\frac{|S_{i_1} \circ \cdots \circ S_{i_k}(x) - S_{i_1} \circ \cdots \circ S_{i_k}(y)|}{|x-y|} \sim |(S_{i_1} \circ \cdots \circ S_{i_k})'(w)|$$

for large k, where x, y and w are any points of D. The composition of mappings $S_{i_1} \circ \cdots \circ S_{i_k}$ is close to a similarity on D, so by comparison with Theorem 9.3 we would expect the dimension of the invariant set F to be close to the value of s for which

$$\sum_{J_k} |(S_{i_1} \circ \cdots \circ S_{i_k})'(w)|^s = 1 \tag{9.19}$$

where the sum is over the set J_k of all k-term sequences. This expectation motivates the following theorem.

Theorem 9.9

Let $V \subset \mathbb{R}$ be an open interval. Let $S_1, \ldots, S_m$ be contractions on $\bar{V}$ that are twice differentiable on V with $a \leqslant |S_i'(w)| \leqslant c$ for all i and $w \in V$, where $0 < a$ and $c < 1$ are constants. Suppose that the S_i satisfy the open set condition (9.11) *with open set V. Then the limit*

$$\lim_{k \to \infty} \left[\sum_{J_k} |(S_{i_1} \circ \cdots \circ S_{i_k})'(w)|^s \right]^{1/k} = \varphi(s) \tag{9.20}$$

exists for each $s > 0$, is independent of $w \in V$, and is decreasing in s. If F is the invariant set for the S_i then $\dim_H F = \dim_B F$ is the solution of $\varphi(s) = 1$, and, F is an s-set, i.e. *$0 < \mathscr{H}^s(F) < \infty$ for this value of s.*

Note on Proof. The main difficulty is to show that the limit (9.20) exists—this depends on the differentiability condition on the S_i. Given this, the argument outlined above may be used to show that the value of s satisfying (9.19) is a good approximation to the dimension when k is large; letting $k \to \infty$ then gives the result.

Similar ideas, but involving the rate of convergence to the limit in (9.20), are needed to show that $0 < \mathscr{H}^s(F) < \infty$. □

There are higher-dimensional analogues of Theorem 9.9. Suppose that the contractions $S_1, \ldots, S_m$ on a domain D in the complex plane are complex analytic mappings. Then the S_i are conformal, or in other words are locally like similarity transformations, contracting at the same rate in every direction. We have

$$S_i(z) = S_i(z_0) + S_i'(z_0)(z - z_0) + \text{terms in } (z - z_0)^2 \text{ and higher powers}$$

so that if $z - z_0$ is small

$$S_i(z) \simeq S_i(z_0) + S_i'(z_0)(z - z_0) \tag{9.21}$$

where $S_i'(z_0)$ is a complex number with $|S_i'(z_0)| < 1$. But the right-hand side of (9.21) is just a similarity expressed in complex notation. In this setting, Theorem 9.9 holds, by the same sort of argument as in the 1-dimensional case.

9.4 Self-affine sets

Self-affine sets form an important class of sets, which include self-similar sets as a particular case. An *affine transformation* $S: \mathbb{R}^n \to \mathbb{R}^n$ is a transformation of the form

$$S(x) = T(x) + b$$

where T is a linear transformation on $\mathbb{R}^n$ (representable by an $n \times n$ matrix) and b is a vector in $\mathbb{R}^n$. Thus an affine transformation S is a combination of a translation, rotation, dilation and, perhaps, a reflection. In particular, S maps spheres to ellipsoids, squares to parallelograms, etc. Unlike similarities, affine transformations contract with differing ratios in different directions.

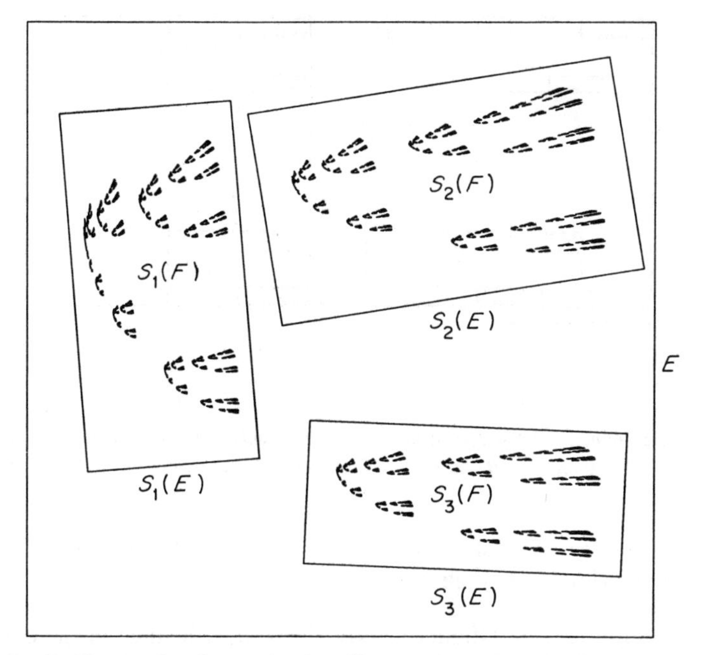

Figure 9.8 A self-affine set invariant under the affine transformations S_1, S_2 and S_3 that map the square E onto the rectangles shown

If $S_1, \ldots, S_m$ are affine contractions on $\mathbb{R}^n$, the unique compact invariant set F for the S_i guaranteed by Theorem 9.1 is termed a *self-affine set*. An example is given in figure 9.8: S_1, S_2 and S_3 are defined as the transformations that map the square E onto the three rectangles in the obvious way. (In the figure the invariant set F is represented as the aggregate of $S_{i_1} \circ \cdots \circ S_{i_k}(E)$ over all sequences $(i_1, \ldots, i_k)$ with $i_j = 1, 2, 3$ for suitably large k. Clearly F is made up of the three affine copies of itself: $S_1(F), S_2(F)$ and $S_3(F)$.)

It is natural to look for a formula for the dimension of self-affine sets that generalizes formula (9.13) for self-similar sets. We would hope that the dimension depends on the affine transformations in a reasonably simple way, easily expressible in terms of the matrices and vectors that represent the affine transformation. Unfortunately, the situation is much more complicated than this—the following example shows that if the affine transformations are varied in a continuous way, the dimension of the self-affine set need not change continuously.

Example 9.10

Let S_1, S_2 be the affine contractions on $\mathbb{R}^2$ that map the unit square onto the rectangles R_1 and R_2 of sides $\frac{1}{2}$ and ε where $0 < \varepsilon < \frac{1}{2}$, as in figure 9.9. The rectangle R_1 abuts the y-axis, but the end of R_2 is distance λ from the y-axis. If F is the

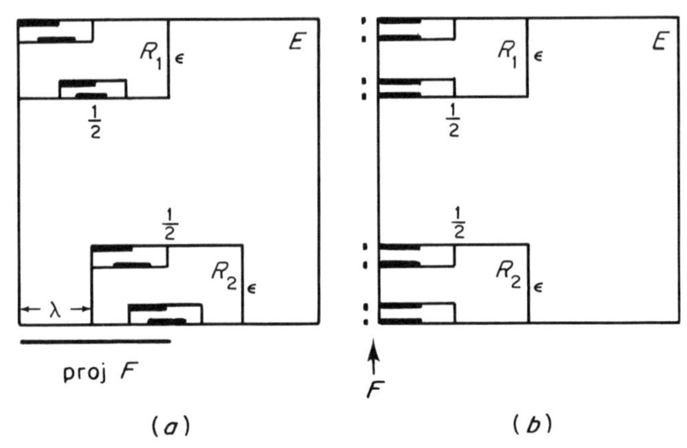

Figure 9.9 Discontinuity of the dimension of self-affine sets. The affine mappings S_1 and S_2 map the unit square E onto R_1 and R_2. In (*a*) $\lambda > 0$ and $\dim_H F \geqslant \dim_H \text{proj}\, F = 1$, but in (*b*) $\lambda = 0$, and $\dim_H F = \log 2/-\log \epsilon < 1$

invariant set for S_1 *and* S_2, *we have* $\dim_H F \geqslant 1$ *when* $\lambda > 0$, *but* $\dim_H F = \log 2/\text{-}\log \varepsilon < 1$ *when* $\lambda = 0$.

Calculation. Suppose $\lambda > 0$ (figure 9.9(a)). Then the kth stage of the construction $E_k = \cup S_{i_1} \circ \cdots \circ S_{i_k}(E)$ consists of 2^k rectangles of sides 2^{-k} and ε^k with the projection of E_k onto the x-axis, $\text{proj}\, E_k$, containing the interval $[0, 2\lambda]$. Since $F = \bigcap_{i=1}^{\infty} E_k$ it follows that $\text{proj}\, F$ contains the interval $[0, 2\lambda]$. (Another way of seeing this is by noting that $\text{proj}\, F$ must be invariant under the transformations $\tilde{S}_1, \tilde{S}_2 : \mathbb{R} \to \mathbb{R}$ given by $\tilde{S}_1(x) = \frac{1}{2}x$, $\tilde{S}_2(x) = \frac{1}{2}x + \lambda$, for which the unique invariant set is the interval $[0, 2\lambda]$.) Thus $\dim_H F \geqslant \dim_H \text{proj}\, F = \dim_H [0, 2\lambda] = 1$.

If $\lambda = 0$, the situation changes (figure 9.9(b)). E_k consists of 2^k rectangles of sides 2^{-k} and ε^k which all have their left-hand ends abutting the y-axis, with E_k contained in the strip $\{(x, y) : 0 \leqslant x \leqslant 2^{-k}\}$. Letting $k \to \infty$ we see that F is a uniform Cantor set contained in the y-axis, which may be obtained by repeatedly removing a proportion $1 - 2\varepsilon$ from the centre of each interval. Thus $\dim_H F = \log 2/-\log \varepsilon < 1$ (see Example 4.5). □

With such discontinuous behaviour, which becomes even worse for more involved sets of affine transformations, it is likely to be difficult to obtain a general expression for the dimension of self-affine sets. However, one situation which has been completely analysed is the self-affine set obtained by the following recursive construction; a specific case is illustrated in figures 9.10 and 9.11.

Example 9.11

Let the unit square E_0 *be divided into a* $p \times q$ *array of rectangles of sides* $1/p$ *and* $1/q$ *where* p *and* q *are positive integers with* $p < q$. *Select a subcollection of these rectangles to form* E_1, *and let* N_j *denote the number of rectangles selected from the* j*th column for* $1 \leqslant j \leqslant p$; *see figure* 9.9. *Iterate this construction in the usual*

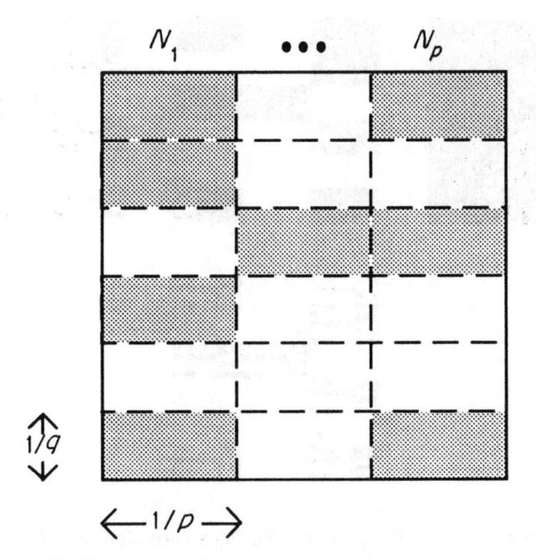

Figure 9.10 Data for the self-affine set of Example 9.11. The affine transformations map the square onto selected $1/p \times 1/q$ rectangles from the $p \times q$ array

way, with each rectangle replaced by an affine copy of E_1, and let F be the limiting set obtained. Then

$$\dim_{\mathrm{H}} F = \log\left(\sum_{j=1}^{p} N_j^{\log p/\log q} \right) \frac{1}{\log p}$$

and

$$\dim_{\mathrm{B}} F = \frac{\log p_1}{\log p} + \log\left(\frac{1}{p_1} \sum_{j=1}^{p} N_j \right) \frac{1}{\log q}$$

where p_1 is the number of columns containing at least one rectangle of E_1.

Calcuation. Omitted. □

Notice in this example that the dimension depends not only on the number of rectangles selected at each stage, but also on their relative positions. Moreover $\dim_{\mathrm{H}} F$ and $\dim_{\mathrm{B}} F$ are not, in general, equal.

∗[Rest of this subsection may be omitted.]

The above example is rather specific in that the affine transformations are all translates of each other. Obtaining a dimension formula for general self-affine sets is an intractable problem. We briefly outline an approach which leads to an expression for the dimension of the invariant set for the affine contractions $S_i(x) = T_i(x) + b_i$ $(1 \leqslant i \leqslant m)$ for *almost all* sequences of vectors $b_1, \ldots, b_m$.

Let $T: \mathbb{R}^n \to \mathbb{R}^n$ be a linear mapping that is contracting and non-singular. The *singular values* $1 > \alpha_1 \geqslant \alpha_2 \geqslant \cdots \geqslant \alpha_n > 0$ of T may be defined in two ways: they are the lengths of the principle semi-axes of the ellipsoid $T(B)$ where B is the unit

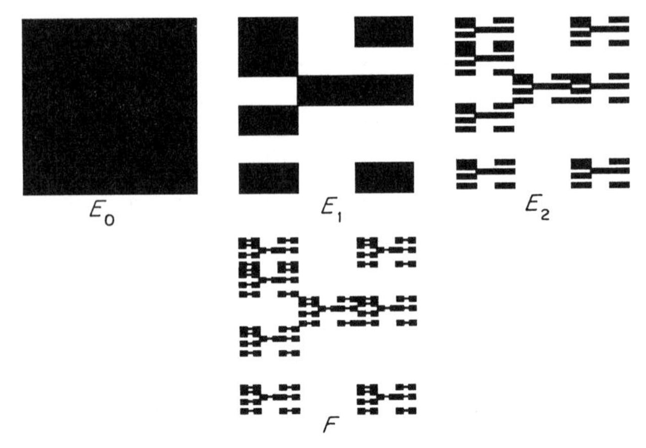

Figure 9.11 Construction of a self-affine set of the type considered in Example 9.11. Such sets may have different Hausdorff and box dimensions

ball in $\mathbb{R}^n$, and they are the positive square roots of the eigenvalues of T^*T, where T^* is the adjoint of T. Thus the singular values reflect the contractive effect of T in different directions. For $0 \leqslant s \leqslant n$ we define the *singular value function*

$$\varphi^s(T) = \alpha_1 \alpha_2 \cdots \alpha_{r-1} \alpha_r^{s-r+1} \tag{9.22}$$

where r is the integer for which $r-1 < s \leqslant r$. Then $\varphi^s(T)$ is continuous and strictly decreasing in s. Moreover, for fixed s, φ^s may be shown to be submultiplicative, i.e.

$$\varphi^s(TU) \leqslant \varphi^s(T)\varphi^s(U)$$

for any linear mappings T and U. We introduce the kth level sums $\Sigma_k^s \equiv \sum_{J_k} \varphi^s(T_{i_1} \circ \cdots \circ T_{i_k})$ where J_k denotes the set of all k-term sequences $(i_1, \ldots, i_k)$ with $1 \leqslant i_j \leqslant m$. For fixed s

$$\begin{aligned}
\Sigma_{k+q}^s &= \sum_{J_{k+q}} \varphi^s(T_{i_1} \circ \cdots \circ T_{i_{k+q}}) \\
&\leqslant \sum_{J_{k+q}} \varphi^s(T_{i_1} \circ \cdots \circ T_{i_k}) \varphi^s(T_{i_{k+1}} \circ \cdots \circ T_{i_{k+q}}) \\
&= \left(\sum_{J_k} \varphi^s(T_{i_1} \circ \cdots \circ T_{i_k}) \right) \left(\sum_{J_q} \varphi^s(T_{i_1} \circ \cdots \circ T_{i_q}) \right) = \Sigma_k^s \Sigma_q^s
\end{aligned}$$

i.e. the sequence Σ_k^s is submultiplicative in k. By a standard property of submultiplicative sequences, $(\Sigma_k^s)^{1/k}$ converges to a number Σ_∞^s as $k \to \infty$. Since φ^s is decreasing in s, so is Σ_∞^s. Provided that $\Sigma_\infty^n \leqslant 1$, there is a unique s, which we denote by $d(T_1, \ldots, T_m)$, such that $1 = \Sigma_\infty^s = \lim_{k \to \infty} (\sum_{J_k} \varphi^s(T_{i_1} \circ \cdots \circ T_{i_k}))^{1/k}$. Equivalently

$$d(T_1, \ldots, T_m) = \inf\left\{ s: \sum_{k=1}^{\infty} \sum_{J_k} \varphi^s(T_{i_1} \circ \cdots \circ T_{i_k}) < \infty \right\}. \tag{9.23}$$

Theorem 9.12

Let $T_1, \ldots, T_m$ be linear contractions and let $b_1, \ldots, b_m \in \mathbb{R}^n$ be vectors. If F is the affine invariant set satisfying

$$F = \bigcup_{i=1}^{m} (T_i(F) + b_i)$$

then $\dim_H F = \dim_B F = d(T_1, \ldots, T_m)$ for almost all $(b_1, \ldots, b_m) \in \mathbb{R}^{nm}$ in the sense of nm-dimensional Lebesgue measure.

Partial Proof. We show that $\dim_H F \leqslant d(T_1, \ldots, T_m)$ for any $b_1, \ldots, b_m$. Write S_i for the contracting affine transformation $S_i(x) = T_i(x) + b_i$. Let B be a large ball so that $S_i(B) \subset B$ for all i. Given $\delta > 0$ we may choose k large enough to get $|S_{i_1} \circ \cdots \circ S_{i_k}(B)| < \delta$ for every k-term sequence $(i_1, \ldots, i_k) \in J_k$. By (9.6)

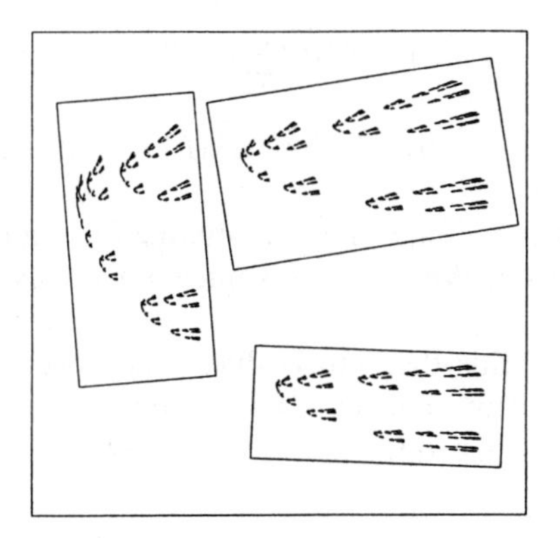

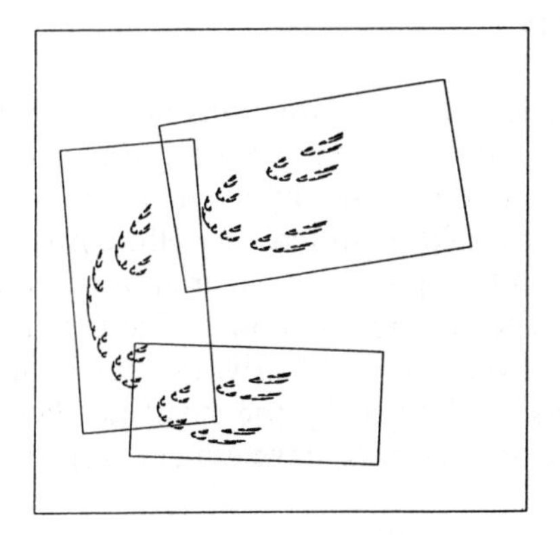

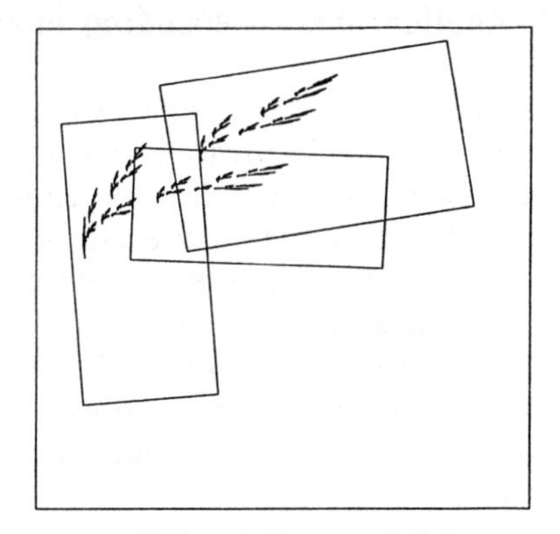

Figure 9.12 Each of the fractals depicted above is invariant under the set of transformations that map the square onto the three rectangles. The affine transformations for each fractal differ only by translations, so by Theorem 9.12 the three fractals all have the same dimension (unless we have been very unlucky in our positioning!). A computation gives this common value of Hausdorff and box dimension as about 1.42

$F \subset \bigcup_{J_k} S_{i_1} \circ \cdots \circ S_{i_k}(B)$. But $S_{i_1} \circ \cdots \circ S_{i_k}(B)$ is a translate of the ellipsoid $T_{i_1} \circ \cdots \circ T_{i_k}(B)$ which has principal axes of lengths $\alpha_1 |B|, \ldots, \alpha_n |B|$, where $\alpha_1, \ldots, \alpha_n$ are the singular values of $T_{i_1} \circ \cdots \circ T_{i_k}$. Thus $S_{i_1} \circ \cdots \circ S_{i_k}(B)$ is contained in a rectangular parallelepiped P of side lengths $\alpha_1 |B|, \ldots, \alpha_n |B|$. If $0 \leqslant s \leqslant n$ and r is the least integer greater than or equal to s, we may divide P into at most

$$\left(\frac{2\alpha_1}{\alpha_r}\right)\left(\frac{2\alpha_2}{\alpha_r}\right)\cdots\left(\frac{2\alpha_{r-1}}{\alpha_r}\right) \leqslant 2^n \alpha_1 \cdots \alpha_{r-1} \alpha_r^{1-r}$$

cubes of side $\alpha_r |B| < \delta$. Hence $S_{i_1} \circ \cdots \circ S_{i_k}(B)$ may be covered by a collection of cubes U_i with $|U_i| < \delta\sqrt{n}$ such that

$$\begin{aligned} \sum_i |U_i|^s &\leqslant 2^n \alpha_1 \cdots \alpha_{r-1} \alpha_r^{1-r} \alpha_r^s |B|^s \\ &\leqslant 2^n |B|^s \varphi^s(T_{i_1} \circ \cdots \circ T_{i_k}). \end{aligned}$$

Taking such a cover of $S_{i_1} \circ \cdots \circ S_{i_k}(B)$ for each $(i_1, \ldots, i_k) \in J_k$ it follows that

$$\mathscr{H}^s_{\delta\sqrt{n}}(F) \leqslant 2^n |B|^s \sum_{J_k} \varphi^s(T_{i_1} \circ \cdots \circ T_{i_k}).$$

But $k \to \infty$ as $\delta \to 0$, so by (9.23), $\mathscr{H}^s(F) = 0$ if $s > d(T_1, \ldots, T_m)$. Thus $\dim_H F \leqslant d(T_1, \ldots, T_m)$.

The lower estimate for $\dim_H F$ may be obtained using the potential theoretic techniques of Section 4.3. We omit the somewhat involved details. □

One consequence of this theorem is that, unless we have been unfortunate enough to hit on an exceptional set of parameters, the fractals in figure 9.12 all have the same dimension, estimated at about 1.42.

9.5 Applications to encoding images

In this chapter, we have seen how a small number of contractions can determine objects of a highly intricate fractal structure. This has applications to data compression—if a complicated picture can be encoded by a small amount of information, then the picture can be transmitted or stored very efficiently.

It is desirable to know which objects can be represented as, or approximated by, invariant sets of an iterated function scheme, and also how to find functions that provide a good representation of a given object. Clearly, the possibilities using, say, three or four transformations are limited by the small number of parameters at our disposal. Such sets are also likely to have a highly repetitive structure.

However, a little experimentation drawing self-affine sets on a computer (see end of Section 9.1) can produce surprisingly good pictures of naturally occurring objects such as ferns, grasses, trees, clouds, etc. The fern and grass in figure 9.13 are invariant for just four and six affine transformations, respectively. Self-similarity and self-affinity are indeed present in nature.

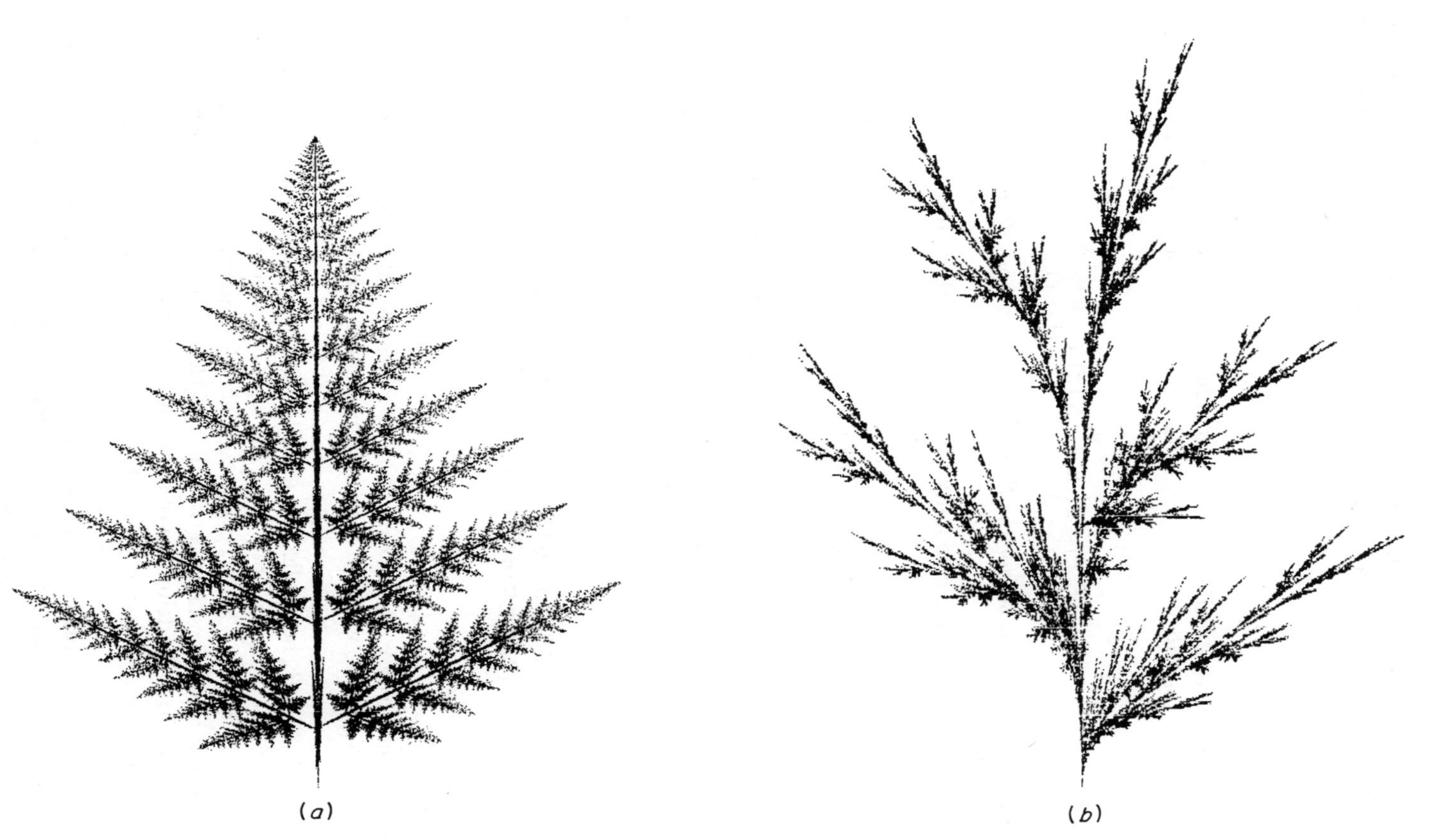

Figure 9.13 The fern (*a*) and grass (*b*) are the invariant sets of just four and six affine transformations, respectively

The following theorem, sometimes known as the collage theorem, gives an idea of how good an approximation a set is to the invariant set of a collection of contractions.

Theorem 9.13

Let $S_1, \ldots, S_m$ be contractions on $\mathbb{R}^n$ and suppose that $|S_i(x) - S_i(y)| \leqslant c|x-y|$ for all $x, y \in \mathbb{R}^n$ and all i, where $c < 1$. Let $E \subset \mathbb{R}^n$ be any non-empty compact set. Then

$$d(E, F) \leqslant d\left(E, \bigcup_{i=1}^{m} S_i(E)\right)\frac{1}{(1-c)} \tag{9.24}$$

where F is the invariant set for the S_i, and d is the Hausdorff metric (9.2).

Proof. Using the triangle inequality for the Hausdorff metric followed by the invariance of F

$$\begin{aligned} d(E, F) &\leqslant d\left(E, \bigcup_{i=1}^{m} S_i(E)\right) + d\left(\bigcup_{i=1}^{m} S_i(E), F\right) \\ &= d\left(E, \bigcup_{i=1}^{m} S_i(E)\right) + d\left(\bigcup_{i=1}^{m} S_i(E), \bigcup_{i=1}^{m} S_i(F)\right) \\ &\leqslant d\left(E, \bigcup_{i=1}^{m} S_i(E)\right) + c\,d(E, F) \end{aligned}$$

by (9.5), as required. □

A consequence of Theorem 9.13 is that any compact subset of $\mathbb{R}^n$ can be approximated arbitrarily closely by a self-similar set.

Corollary 9.14

Let E be a non-empty compact subset of $\mathbb{R}^n$. Given $\delta > 0$ there exist contracting similarities $S_1, \ldots, S_m$ with invariant set F satisfying $d(E, F) < \delta$.

Proof. Let $B_1, \ldots, B_m$ be a collection of balls that cover E and which have centres in E and radii at most $\frac{1}{4}\delta$. Then $E \subset \bigcup_{i=1}^{m} B_i \subset E_{\frac{1}{4}\delta}$, where $E_{\frac{1}{4}\delta}$ is the $\frac{1}{4}\delta$-parallel body to E. For each i, let S_i be any contracting similarity of ratio less than $\frac{1}{2}$ that maps E into B_i. Then $S_i(E) \subset B_i \subset (S_i(E))_{\frac{1}{2}\delta}$, so $(\bigcup_{i=1}^{m} S_i(E)) \subset E_{\frac{1}{4}\delta}$ and $E \subset \bigcup_{i=1}^{m} (S_i(E))_{\frac{1}{2}\delta}$. By definition of the Hausdorff metric, $d(E, \bigcup_{i=1}^{m} S_i(E)) < \frac{1}{2}\delta$. It follows from (9.24) that $d(E, F) < \delta$ where F is invariant for the S_i. □

The method of approximation by invariant sets used in the above proof is rather coarse—it is likely to yield a very large number of transformations that take little account of the fine structure of F. A rather more subtle approach is required to obtain convincing images with a small number of transformations. One method which often gives good results is to draw a rough outline of the object and then cover it, as closely as possible, by a number of smaller similar or affine copies. The similarities thus determined may be used to compute an invariant set which may be compared with the object being modelled. Theorem 9.13 guarantees that the invariant set will be a good approximation if the union of the smaller copies is close to the object. A trial and error process allows modification and improvements to the picture.

More complex objects may be built up by superposition of the invariant sets of several different sets of transformations.

Ideally, it would be desirable to have a 'camera' which could be pointed at an object to produce a 'photograph' consisting of a specified number of affine transformations whose invariant set is a good approximation to the object. Obviously, the technical problems involved are considerable, and at the moment such a device is a long way from reality. One approach is to scan the object to estimate various geometric parameters, and use these to impose restrictions on the transformations.

For example, for a 'natural fractal' such as a fern, we might estimate the dimension by a box-counting method. The assumption that the similarities or affinities sought must provide an invariant set of this dimension gives, at least theoretical, restrictions on the possible set of transformations, using results such as Theorem 9.3 or 9.12. In practice, however, such information is rather hard to utilise, and we certainly need many further parameters for it to be of much use.

Very often, invariant sets in the plane that provide good pictures of physical objects will have positive area, so will not be fractals in the usual sense. Nevertheless, such sets may well be bounded by fractal curves, a feature that adds realism to pictures of natural objects. However, fractal properties of boundaries of invariant sets seem hard to analyse.

These ideas may be extended to provide shaded or coloured images, by assigning a probability p_i to each of the transformations S_i, where $0 \leqslant p_i \leqslant 1$ and $\sum_{i=1}^{m} p_i = 1$. Without going into details, these data define a mass distribution μ on the invariant set F such that $\mu(A) = \sum_{i=1}^{m} p_i \mu(S_i^{-1}(A))$, and the set may be shaded, or even coloured, according to the local density of μ.

This leads to the following modification of the second method of drawing invariant sets mentioned at the end of Section 9.1. Let x_0 be any initial point. We choose S_{j_1} from $S_1, \ldots, S_m$ at random in such a way that the probability of choosing S_i is p_i, and let $x_1 = S_{j_1}(x_0)$. We continue in this way, so that $x_k = S_{j_k}(x_{k-1})$ where S_{j_k} equals S_i with probability p_i, Plotting the sequence $\{x_k\}$ (after omitting the first 100 terms, say) gives a rendering of the invariant set F, but in such a way that a proportion $p_{i_1} \cdots p_{i_q}$ of the points tends to lie in the part $S_{i_1} \circ \cdots \circ S_{i_q}(F)$ for each $i_1, \ldots, i_q$. This variable point density provides a direct shading of F. Alternatively, the colour of a point of F can be determined by some rule, which depends on the number of $\{x_k\}$ falling close to each point. The computer artist may care to experiment with the endless possibilities that this method provides—certainly, some very impressive colour pictures have been produced using relatively few transformations.

It is perhaps appropriate to end this section with some of the 'pros and cons' of representing images using iterated function schemes. By utilizing the self-similarity and repetition in nature, and, indeed, in man-made objects, the method often enables scenes to be described by a small number (perhaps fewer than 100) of transformations and probabilities in an effective manner. This represents an enormous compression of information compared, for example, with that required to detail the colour in each square of a fine mesh. The corresponding disadvantage is that there is a high correlation between different parts of the picture—the method is excellent for giving an overall picture of a tree, but is no use if the exact arrangement of the leaves on different branches is important. Given a set of affine transformations, reproduction of the image is computationally straightforward, is well suited to parallel computation, and is stable—small changes in the transformations lead to small changes in the invariant set. The transformations define the image at arbitrarily small scales, and it is easy to produce a close-up of a small region. At present, the main disadvantage of the method is the difficulty of obtaining a set of transformations to represent a given object or picture.

9.6 Notes and references

The first systematic account of iterated function schemes seems to be that of Hutchinson (1981), though similar ideas were certainly around earlier. The derivation of the formula for the dimension of self-similar sets is essentially contained in Moran (1946). Computer pictures of self-similar and other invariant sets are widespread, the works of Mandelbrot (1982), Dekking (1982), Barnsley and Demko (1985) and Barnsley (1988) containing many interesting and beautiful examples. For material relating to Theorem 9.9, see Ruelle (1983). Full details of Example 9.11 are given in McMullen (1984) and of Theorem 9.12 in Falconer (1988). A discussion of self-affine sets is given by Mandelbrot (1986). Applications to image processing are described in Barnsley (1988) and Barnsley and Sloan (1988).

Exercises

9.1 Verify that the Hausdorff metric (9.2) satisfies the conditions for a metric.

9.2 Find a pair of similarity transformations on $\mathbb{R}$ for which the interval $[0, 1]$ is the (unique non-empty compact) invariant set. Now find infinitely many such pairs of transformations.

9.3 Find sets of (i) four and (ii) three similarity transformations on $\mathbb{R}$ for which the middle third Cantor set is the invariant set. Check that (9.13) has solution $\log 2/\log 3$ in each case.

9.4 Find an open set satisfying the open set condition for the four basic similarity transformations that define the von Koch curve (figure 0.2). Deduce from Theorem 9.3 that the von Koch curve does indeed have box and Hausdorff dimension of $\log 4/\log 3$.

9.5 Verify that the set depicted in figure 0.5 has Hausdorff and box dimensions given by $4(\frac{1}{4})^s + (\frac{1}{2})^s = 1$.

9.6 Sketch the first few steps in the construction of a self-similar set with generator •—┬—•. What are the Hausdorff and box dimensions of this fractal? (The stem of the T is one quarter of the total length of the top.)

9.7 Let F be the set obtained by a Cantor-type construction in which each interval is replaced by two intervals, one of a quarter of the length at the left-hand end, and one of half the length at the right-hand end. Thus E_0 is the interval $[0, 1]$, E_1 consists of the intervals $[0, \frac{1}{4}]$ and $[\frac{1}{2}, 1]$, etc. Find the Hausdorff and box dimensions of F.

9.8 Show that any self-similar set F satisfying the conditions of Theorem 9.3 has $c_1 \leqslant \underline{D}(F, x) \leqslant \bar{D}(F, x) \leqslant c_2$ for all $x \in F$, where c_1 and c_2 are positive constants. (See equations (5.2) and (5.3) for the definition of the densities.)

9.9 Let S_1, $S_2 : [0, 1] \to [0, 1]$ be given by $S_1(x) = x/(2 + x)$ and $S_2(x) = 2/(2 + x)$. Show that the invariant set F associated with these transformations satisfies $0.53 < \dim_H F < 0.80$.

9.10 Let $S_1, \ldots, S_m$ be bi-Lipschitz contractions on a subset D of $\mathbb{R}^n$ and let F be the compact invariant set satisfying (9.1). Show that, if V is any open set intersecting F, then F and $F \cap V$ have equal Hausdorff, equal upper box and equal lower box dimensions. Deduce from Corollary 3.9 that $\dim_P F = \overline{\dim}_B F$.

9.11 Verify the Hausdorff dimension formula in Example 9.11 in the cases (*a*) where $N_j = N$ for $1 \leqslant j \leqslant p$ and (*b*) where $N_j = N$ or 0 for $1 \leqslant j \leqslant p$, where N is an integer with $1 < N < q$. (Hint: see Example 7.13.)

9.12 Calculate the Hausdorff and box dimensions of the set in figure 9.11.

9.13 Write a computer program to draw self-similar sets in the plane, given a generator of the set.

9.14 Write a computer program to draw the invariant set of a given collection of transformations of a plane region (see the end of Section 9.1). Examine the invariant sets of similarities, affinities and try some non-linear transformations. If you are feeling really enterpising, you might write a program to estimate the dimension of these sets using a box-counting method.

Chapter 10 Examples from number theory

Fractals can often be defined in number theoretic terms; for instance, the middle third Cantor set consists of the numbers between 0 and 1 which have a base-3 expansion containing only the digits 0 and 2. We examine three classes of fractal that occur in classical number theory—these examples serve to illustrate some of the ideas encountered earlier in the book.

10.1 Distribution of digits of numbers

In this section we consider base-m expansions of real numbers, where $m \geqslant 2$ is a fixed integer. Let $p_0, p_1, \ldots, p_{m-1}$ be 'proportions' summing to 1, so that $0 \leqslant p_i \leqslant 1$ and $\sum_{i=0}^{m-1} p_i = 1$. Let $F(p_0, \ldots, p_{m-1})$ be the set of numbers x in $[0, 1)$ with base-m expansions containing the digits $0, 1, \ldots, m-1$ in proportions $p_0, \ldots, p_{m-1}$ respectively. More precisely, if $n_j(x|_k)$ denotes the number of times the digit j occurs in the first k places of the base-m expansion of x, then

$$F(p_0, \ldots, p_{m-1}) = \{x \in [0, 1): \lim_{k \to \infty} n_j(x|_k)/k = p_j \text{ for all } j = 1, \ldots, m\}. \quad (10.1)$$

Thus we think of $F(\frac{1}{3}, \frac{2}{3})$ as the numbers with 'two thirds' of their base-2 digits being 1 and the rest being 0.

It is well-known that almost all numbers (in the sense of Lebesgue measure) are *normal* to all bases; that is, they have base-m expansions containing equal proportions of the digits $0, 1, \ldots, m-1$ for all m. In our notation, $F(m^{-1}, \ldots, m^{-1})$ has Lebesgue measure 1, and therefore dimension 1, for all m. Paradoxically, no specific example of a number that is normal to all bases has ever been exhibited. We may use Hausdorff dimension to describe the size of the sets $F(p_0, \ldots, p_{m-1})$ when the p_j are not all equal. (Such sets are dense in $[0, 1)$ so have box dimension 1.)

A mass distribution technique is used in the following proof—the mass distribution occurs naturally as a probability measure. Note that we adopt the usual convention that $0 \times \log 0 = 0$.

Proposition 10.1

With $F = F(p_0, \ldots, p_{m-1})$ *as above,*

$$\dim_{\mathrm{H}} F = -\frac{1}{\log m} \sum_{i=0}^{m-1} p_i \log p_i.$$

Proof. The proof is best thought of probabilistically. We imagine that base-m numbers $x = 0.i_1 i_2\ldots$ are selected at random in such a way that the kth digit i_k takes the value j with probability p_j, independently for each k. Thus we take $[0, 1)$ as our sample space and define a probability measure P on subsets of $[0, 1)$ such that if $I_{i_1,\ldots,i_k}$ is the 'basic interval' containing the numbers with base-m expansion beginning $0.i_1 \cdots i_k$ then the probability of a number being in this interval is

$$\mathsf{P}(I_{i_1,\ldots,i_k}) = p_{i_1} \cdots p_{i_k}. \tag{10.2}$$

Given j, the events 'the kth digit of x is a j' are independent for $k = 1, 2, \ldots$. A consequence of the strong law of large numbers (see Exercise 1.15) is that, with probability 1, the proportion of occurrences of an event in a number of repeated independent trials tends to the probability of the event occurring. Thus, with probability 1,

$$n_j(x|_k)/k = (\text{number of occurrences of } j \text{ in the first } k \text{ digits})/k \to p_j$$

as $k \to \infty$ for all j. Hence $\mathsf{P}(F) = 1$. We write $I_k(x) = I_{i_1,\ldots,i_k}$ for the basic interval of length m^{-k} to which x belongs. For a fixed y, the probability that $x \in I_k(y)$ is given by

$$\log \mathsf{P}(I_k(y)) = n_0(y|_k) \log p_0 + \cdots + n_{m-1}(y|_k) \log p_{m-1}$$

by taking logarithms of (10.2). If $y \in F$ then $n_j(y|_k)/k \to p_j$ as $k \to \infty$ for each j, so

$$\begin{aligned} \frac{1}{k} \log \frac{\mathsf{P}(I_k(y))}{|I_k(y)|^s} &= \frac{1}{k} \log \mathsf{P}(I_k(y)) - \frac{1}{k} \log m^{-ks} \\ &\to \sum_{i=0}^{m-1} p_i \log p_i + s \log m. \end{aligned}$$

Hence, for all y in F, the 'interval density'

$$\lim_{k \to \infty} \frac{\mathsf{P}(I_k(y))}{|I_k(y)|^s} = \begin{cases} 0 & \text{if } s < \theta \\ \infty & \text{if } s > \theta \end{cases}$$

where

$$\theta = -\frac{1}{\log m} \sum_{i=0}^{m-1} p_i \log p_i.$$

We are now virtually in the situation of Proposition 4.9. The same results hold and may be proved in the same way, if the 'spherical densities' $\lim_{r \to 0} \mu(B_r(x))/r^s$ are replaced by these interval densities. Thus $\mathcal{H}^s(F) = \infty$ if $s < \theta$ and $\mathcal{H}^s(F) = 0$ if $s > \theta$, as required. □

10.2 Continued fractions

Instead of defining sets of numbers in terms of base-m expansions, we may use the continued fraction expansions. Any number x that is not an integer may be written as

$$x = a_0 + 1/x_1$$

where a_0 is an integer and $x_1 > 1$. Similarly, if x_1 is not an integer, then

$$x_1 = a_1 + 1/x_2$$

with $x_2 > 1$, so

$$x = a_0 + 1/(a_1 + 1/x_2).$$

Proceeding in this way,

$$x = a_0 + 1/(a_1 + 1/(a_2 + 1/(\cdots + 1/(a_{k-1} + 1/x_k))))$$

for each k, provided that at no stage is x_k an integer. We call the sequence of integers $a_0, a_1, a_2, \ldots$ the *partial quotients* of x, and write

$$x = a_0 + \frac{1}{a_1 +}\,\frac{1}{a_2 +}\,\frac{1}{a_3 + \cdots}$$

for the *continued fraction expansion* of x. This expansion terminates if and only if x is rational, otherwise taking a finite number of terms,

$$a_0 + 1/(a_1 + 1/(a_2 + 1/(\cdots + 1/a_k)))$$

provides a sequence of rational approximations to x which converge to x as $k \to \infty$. (Continued fractions are, in fact, closely allied to the theory of Diophantine approximation; see Section 10.3.)

Examples of continued fractions include

$$\sqrt{2} = 1 + \frac{1}{2+}\,\frac{1}{2+}\,\frac{1}{2+\cdots}$$

$$\sqrt{3} = 1 + \frac{1}{1+}\,\frac{1}{2+}\,\frac{1}{1+}\,\frac{1}{2+\cdots}.$$

More generally, any quadratic surd (i.e. root of a quadratic equation with integer coefficients) has eventually periodic partial quotients.

Sets of numbers defined by conditions on their partial quotients are often fractals, as the following example illustrates.

Example 10.2

Let F be the set of positive real numbers x with infinite partial fraction expressions which have all partial quotients equal to 1 or 2. Then F is a fractal with $0.44 < \dim_H F < 0.66$.

Proof. It is easy to see that F is closed and bounded. Moreover, $x \in F$ precisely when $x = 1 + 1/y$ or $x = 2 + 1/y$ with $y \in F$. Letting $S_1(x) = 1 + 1/x$ and $S_2(x) = 2 + 1/x$, it follows that $F = S_1(F) \cup S_2(F)$; in other words, F is invariant for the S_i in the sense of (9.1). In fact F is just the set analysed in Example 9.8 which we showed to have Hausdorff dimension between 0.44 and 0.66. □

Obviously, varying the conditions on the partial quotients will lead to other fractals that are the invariant sets of certain transformations.

10.3 Diophantine approximation

How closely can a given irrational number x be approximated by a rational p/q with denominator q no larger than q_0? Diophantine approximation is the study of such problems, which can crop up in quite practical situations (see Section 13.6). A classical theorem of Dirichlet (see Exercise 10.7) states that for any real number x, there are infinitely many positive integers q such that

$$\left|x-\frac{p}{q}\right| \leqslant \frac{1}{q^2}$$

for some integer p; such p/q are 'good' rational approximations to x. Equivalently,

$$\|qx\| \leqslant q^{-1}$$

for infinitely many q, where $\|y\| = \min_{m\in\mathbb{Z}}|m-y|$ denotes the distance from y to the nearest integer.

There are variations on Dirichlet's result that apply to *almost all* numbers x. It may be shown that if $\psi(q)$ is a decreasing function of q with $0 \leqslant \psi(q) \leqslant \frac{1}{2}$ then

$$\|qx\| \leqslant \psi(q) \tag{10.3}$$

is satisfied by infinitely many q for almost all x or almost no x (in the sense of 1-dimensional Lebesgue measure) according to whether $\sum_1^\infty \psi(q)$ diverges or converges. In the latter case, the set of x for which (10.3) does have infinitely many solutions is often a fractal.

We speak of a number x such that

$$\|qx\| \leqslant q^{1-\alpha} \tag{10.4}$$

for infinitely many positive integers q as being *α-well approximable*. It is natural to ask how large this set of numbers is when $\alpha > 2$, and, indeed, whether such irrational numbers exist at all. We prove Jarník's Theorem, that the set of α-well approximable numbers has Hausdorff dimension $2/\alpha$.

It is almost immediate from Example 4.7 (check!) that the set of α-well approximable numbers has dimension at least $1/\alpha$. An extra subtlety is required to obtain a value of $2/\alpha$. The idea is as follows. Let G_q be the set of $x\in[0,1]$ satisfying (10.4). A factorization argument shows that, if n is large and p_1, p_2 are primes with $n < p_1, p_2 \leqslant 2n$, then G_{p_1} and G_{p_2} are disjoint (except for points very close to 0 or 1). Thus the set

$$H_n = \bigcup_{\substack{p \text{ prime} \\ n<p\leqslant 2n}} G_p$$

consists of, roughly, $\sum_{n<p\leqslant 2n} 1/p \simeq n^2/\log n$ reasonably regularly spaced intervals of lengths at least $2(2n)^{-\alpha}$. We then show that if n_k is a rapidly increasing sequence, the intersection $\bigcap_{k=1}^{\infty} H_{n_k}$ has dimension at least $2/\alpha$, and note that any number in this intersection lies in infinitely many G_p and so is α-well-approximable.

Jarník's theorem 10.3

Suppose $\alpha > 2$. *Let* F *be the set of real numbers* $x \in [0, 1]$ *for which the inequality*

$$\| qx \| \leqslant q^{1-\alpha} \tag{10.5}$$

is satisfied by infinitely many positive integers q. *Then* $\dim_{\mathrm{H}} F = 2/\alpha$.

$*$ ***Proof.*** For each q let G_q denote the set of $x \in [0, 1]$ satisfying (10.5). Then G_q consists of $q-1$ intervals of length $2q^{-\alpha}$ and two 'end' intervals of length $q^{-\alpha}$. Clearly, $F \subset \bigcup_{q=k}^{\infty} G_q$ for each k, so taking the intervals of G_q for $q \geqslant k$ as a cover of F gives that $\mathscr{H}_{\delta}^{s}(F) \leqslant \sum_{q=k}^{\infty} (q+1)(2q^{-\alpha})^s$ if $2k^{-\alpha} \leqslant \delta$. If $s > 2/\alpha$ the series $\sum_{1}^{\infty} (q+1)(2q^{-\alpha})^s$ converges, so $\mathscr{H}^s(F) = 0$. Hence $\dim_{\mathrm{H}} F \leqslant 2/\alpha$.

Let G'_q be the set of $x \in (q^{-\alpha}, 1 - q^{-\alpha})$ satisfying (10.5), so that G'_q is just G_q with the end intervals removed. Let n be a positive integer, and suppose p_1 and p_2 are prime numbers satisfying $n < p_1 < p_2 \leqslant 2n$. We show that G'_{p_1} and G'_{p_2} are disjoint and reasonably well separated. For if $1 \leqslant r_1 < p_1$ and $1 \leqslant r_2 < p_2$, then $p_1 r_2 \neq p_2 r_1$ since p_1 and p_2 are prime. Thus

$$\left| \frac{r_1}{p_1} - \frac{r_2}{p_2} \right| = \frac{1}{p_1 p_2} | p_2 r_1 - p_1 r_2 | \geqslant \frac{1}{p_1 p_2} \geqslant \frac{1}{4n^2}$$

that is, the distance between the centres of any pairs of intervals from G'_{p_1} and G'_{p_2} is at least $1/4n^2$. Since these intervals have lengths at most $2n^{-\alpha}$, the distance between any point of G'_{p_1} and any point of G'_{p_2} is at least $\frac{1}{4}n^{-2} - 2n^{-\alpha} \geqslant \frac{1}{8}n^{-2}$ provided that $n \geqslant n_0$ for some sufficiently large n_0. For such n the set

$$H_n = \bigcup_{\substack{p \text{ prime} \\ n < p \leqslant 2n}} G'_p$$

is a disjoint union of the intervals in the G'_p, so H_n is made up of intervals of lengths at least $(2n)^{-\alpha}$ which are separated by gaps of lengths at least $\frac{1}{8}n^{-2}$. If $I \subset [0, 1]$ is any interval with $3/|I| < n < p \leqslant 2n$ then at least $p|I|/3 \geqslant n|I|/3$ of the intervals of G'_p are completely contained in I. A version of the prime number theorem states that the number of primes between 2 and n is asymptotically $n/\log n$, so there are at least $n/(2 \log n)$ primes in the range $(n, 2n]$, if $n \geqslant n_1$ for some large $n_1 \geqslant n_0$. Thus at least

$$\frac{n^2 |I|}{6 \log n} \tag{10.6}$$

intervals of H_n are contained in I provided that $n \geqslant n_1$ and $|I| \geqslant 3/n$.

To complete the proof, we use Example 4.6. With n_1 as above, let $n_k =$

$\max\{n_{k-1}^k, 3.2^a n_{k-1}^a\}$, for $k=2,3,\ldots$, where $a>\alpha$ is an integer. Let $E_0=[0,1]$ and for $k=1,2,\ldots$ let E_k consist of those intervals of H_{n_k} that are completely contained in E_{k-1}. The intervals of E_k are of lengths at least $(2n_k)^{-\alpha}$ and are separated by gaps of at least $\varepsilon_k=\frac{1}{8}n_k^{-2}$. Using (10.6), each interval of E_{k-1} contains at least m_k intervals of E_k, where

$$m_k = \frac{n_k^2(2n_{k-1})^{-\alpha}}{6\log n_k} = \frac{cn_k^2 n_{k-1}^{-\alpha}}{\log n_k}$$

if $k\geqslant 2$, where $c=2^{-\alpha}/6$. (We take $m_1=1$.) By (4.7)

$$\begin{aligned}
&\dim_{\mathrm{H}}\left(\bigcap_{k=1}^{\infty} E_k\right)\\
&\geqslant \varliminf_{k\to\infty} \frac{\log[c^{k-2}n_1^{-\alpha}(n_2\cdots n_{k-2})^{2-\alpha}n_{k-1}^2(\log n_2)^{-1}\cdots(\log n_{k-1})^{-1}]}{-\log[cn_{k-1}^{-\alpha}(8\log n_k)^{-1}]}\\
&= \varliminf_{k\to\infty} \frac{\log[c^{k-2}n_1^{-\alpha}(n_2\cdots n_{k-2})^{2-\alpha}(\log n_2)^{-1}\cdots(\log n_{k-1})^{-1}]+2\log n_{k-1}}{-\log(c/8)+\log k(\log n_{k-1})+\alpha\log n_{k-1}}\\
&= 2/\alpha
\end{aligned}$$

since the dominant terms in numerator and denominator are those in $\log n_{k-1}$. (Note that $\log n_k = k\log n_{k-1}$ so $\log n_k = ck!$ for k sufficiently large.) If $x\in E_k\subset H_{n_k}$ for all k, then x lies in infinitely many of the G'_p and so $x\in F$. Hence $\dim_{\mathrm{H}} F\geqslant 2/\alpha$. □

*[The rest of this section may be omitted.]

Obviously, the set F of Jarník's theorem is dense in $[0,1]$, with $\dim_{\mathrm{H}}(F\cap I)=2/\alpha$ for any interval I. However, considerably more than this is true, F is a 'set of large intersection' of the type discussed in Section 8.2, and this has some surprising consequences. For the definition of $\mathscr{C}^s$, in the following proposition, see (8.7) to (8.8).

Proposition 10.4

Suppose $\alpha>2$. If F is the set of positive numbers such that $\|qx\|\leqslant q^{1-\alpha}$ for infinitely many q, then $F\in\mathscr{C}^s[0,\infty)$ for all $s<2/\alpha$.

Note on proof. This follows the proof of Jarník's Theorem 10.3 up to the definition of H_n. Then a combination of the method of Example 8.9 and prime number theorem estimates are used to show that $\lim_{n\to\infty}\mathscr{H}^s_\infty(I\cap H_n)=\mathscr{H}^s_\infty(I)$. Slightly different methods are required to estimate the number of intervals of H_n that can intersect a covering interval U, depending on whether $|I|<1/n$ or $|I|\geqslant 1/n$.

The first deduction from Proposition 10.4 is that $\dim_{\mathrm{H}} F=2/\alpha$, which we know already from Jarník's Theorem. However, Proposition 8.8 tells us that smooth bijective images of F are also in $\mathscr{C}^s$. Thus if $s<2/\alpha$ then $f(F\cap[a,b])$ is in $\mathscr{C}^s[f(a),f(b)]$ for any continuously differentiable function $f:[a,b]\to\mathbb{R}$ with $|f'(x)|>c>0$. Taking the functions given by $f_m(x)=x^{1/m}$ we have that $f_m(F)\cap[1,2]$ is in $\mathscr{C}^s[1,2]$ for $s<2/\alpha$. It follows from Proposition 8.6 that $\bigcap_{m=1}^{\infty} f_m(F)\cap[1,2]$ is in $\mathscr{C}^s[1,2]$, so

$$\dim_{\mathrm{H}} \bigcap_{m=1}^{\infty} f_m(F)=2/\alpha.$$

But

$$f_m(F)=\{x:\|qx^m\|\leqslant q^{1-\alpha}\text{ for infinitely many } q\}$$

so we have shown that the set of x for which all positive integral powers are α-well-approximable has Hausdorff dimension $2/\alpha$.

Clearly, many variations are possible using different sequences of functions f_m.

10.4 Notes and references

There are a wide range of introductory books on number theory, of which the classic by Hardy and Wright (1960) is hard to beat. The dimensional analysis of the distribution of base-m digits may be found in Billingsley (1965). Continued fractions are discussed in most basic number theory texts, and Rogers (1970) and Bumby (1985) discuss dimensional aspects. Full accounts of Diophantine approximation are to be found in the books of Schmidt (1980) and Cassels (1957). Proofs of Jarník's Theorem have been given by Jarník (1931), Besicovitch (1934), Eggleston (1952) and Kaufman (1981). See Dodson, Rynne and Vickers (to appear) for relations to sets of large intersection.

Exercises

10.1 Show that the set $F(p_0,\ldots,p_{m-1})$ in (10.1) is invariant for a set of m similarity transformations in the sense of (9.1). (It is not, of course, compact.)

10.2 Find the Hausdorff dimension of the set of numbers whose base-3 expansions have 'twice as many 2s as 1s' (i.e. those x such that $2\lim_{k\to\infty} n_1(x|_k)/k=\lim_{k\to\infty} n_2(x|_k)/k$ with both these limits existing).

10.3 Find the continued fraction representations of (i) 41/9 and (ii) $\sqrt{5}$.

10.4 What number has continued fraction representation

$$1+\frac{1}{1+}\frac{1}{1+}\frac{1}{1+\cdots}?$$

10.5 Use the continued fraction representation of $\sqrt{2}$ (with partial quotients $1,2,2,2,\ldots$) to obtain some good rational approximations to $\sqrt{2}$. (In fact, the number obtained by curtailing a partial fraction at the kth partial quotient gives the best rational approximation by any number with equal or smaller denominator.)

10.6 Obtain estimates for the Hausdorff and box dimensions of the set of positive numbers whose continued fraction expansions have partial quotients containing only the digits 2 and 3.

10.7 Let x be a real number and Q a positive integer. By considering the set of numbers $\{rx(\text{mod } 1): r = 0, 1, \ldots, Q\}$, prove Dirichlet's theorem: i.e. that there is an integer q with $0 \leqslant q \leqslant Q$ such that $\| qx \| \leqslant Q^{-1}$. Deduce that there are infinitely many positive integers q such that $\| qx \| \leqslant q^{-1}$.

10.8 Let n and d be positive integers. Show that if the Diophantine equation $x^n - dy^n = 1$ has infinitely many solutions (x, y) with x and y positive integers, then $d^{1/n}$ must be n-well-approximable.

10.9 Fix $\alpha > 3$ let F be the set of (x, y) in $\mathbb{R}^2$ such that $\| qx \| \leqslant q^{1-\alpha}$ and $\| qy \| \leqslant q^{1-\alpha}$ are satisfied simultaneously for infinitely many positive integers q. Show, in a similar way to the first part of the proof of Theorem 10.3, that $\dim_H F \leqslant 3/\alpha$. (In fact, it may be shown, using a generalization of the remainder of the proof, that $\dim_H F = 3/\alpha$.)

10.10 Show that the set of real numbers x, such that $(x + m)^2$ is α-well-approximable for all integers m, has Hausdorff dimension $2/\alpha$.

Chapter 11 Graphs of functions

A variety of interesting fractals, both of theoretical and practical importance, occur as graphs of functions. Indeed, many phenomena display fractal features when plotted as functions of time. Examples include atmospheric pressure, levels of reservoirs and prices on the stock market, at least when recorded over fairly long time spans.

11.1 Dimensions of graphs

We consider functions $f:[a,b]\to\mathbb{R}$. Under certain circumstances the graph

$$\text{graph}\, f = \{(t, f(t)): a \leqslant t \leqslant b\}$$

regarded as a subset of the (t, x)-coordinate plane may be a fractal. (Note that we work with coordinates (t, x) rather than (x, y) for consistency with the rest of the book, and because the independent variable is frequently time.) If f has a continuous derivative, then it is not difficult to see that graph f has dimension 1 and, indeed, is a regular 1-set. The same is true if f is of bound variation; that is, if $\sum_{i=0}^{m-1}|f(t_i)-f(t_{i+1})| \leqslant$ constant for all dissections $0 = t_0 < t_1 < \cdots < t_m = 1$. However, it is possible for a continuous function to be sufficiently irregular to have a graph of dimension strictly greater than 1. Perhaps the best known example is

$$f(t) = \sum_{k=1}^{\infty} \lambda^{(s-2)k} \sin(\lambda^k t)$$

where $1 < s < 2$ and $\lambda > 1$. This function, essentially Weierstrass's example of a continuous function that is nowhere differentiable, has box dimension s, and is believed to have Hausdorff dimension s.

We first derive some simple but widely applicable estimates for the box dimension of graphs. Given a function f and an interval $[t_1, t_2]$, we write R_f for the *maximum range* of f over an interval,

$$R_f[t_1, t_2] = \sup_{t_1 < t, u < t_2} |f(t) - f(u)|$$

Proposition 11.1

Let $f:[0,1]\to\mathbb{R}$ be continuous. Suppose that $0 < \delta < 1$, and m is the least integer greater than or equal to $1/\delta$. Then, if N_δ is the number of squares of the δ-mesh

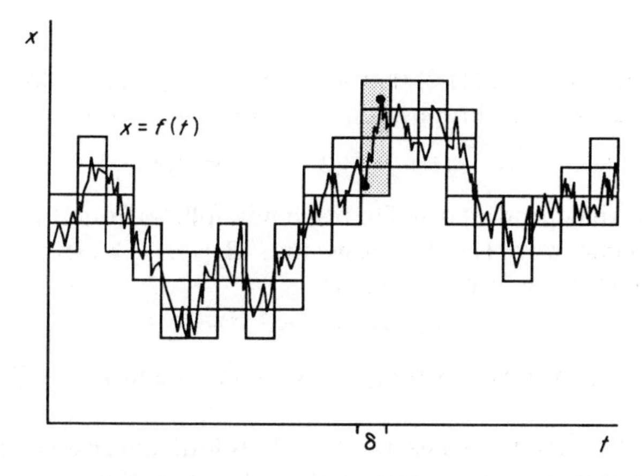

Figure 11.1 The number of δ-mesh squares in a column above an interval of width δ that intersect graph f is approximately the range of f over that interval divided by δ. Summing these numbers gives estimates for the box dimension of graph f

that intersect graph f,

$$\delta^{-1}\sum_{i=0}^{m-1} R_f[i\delta,(i+1)\delta] \leqslant N_\delta \leqslant 2m + \delta^{-1}\sum_{i=0}^{m-1} R_f[i\delta,(i+1)\delta]. \qquad (11.1)$$

Proof. The number of mesh squares of side δ in the column above the interval $[i\delta,(i+1)\delta]$ that intersect graph f is at least $R_f[i\delta,(i+1)\delta]/\delta$ and at most $2 + R_f[i\delta,(i+1)\delta]/\delta$, using that f is continuous. Summing over all such intervals gives (11.1). This is illustrated in figure 11.1. □

This proposition may be applied immediately to functions satisfying a Hölder condition.

Corollary 11.2

Let $f:[0,1]\to\mathbb{R}$ *be a continuous function.*

(*a*) *Suppose*

$$|f(t)-f(u)| \leqslant c|t-u|^{2-s} \qquad (0 \leqslant t,u \leqslant 1) \qquad (11.2)$$

where $c>0$ *and* $1\leqslant s\leqslant 2$. *Then* $\mathscr{H}^s(F)<\infty$ *and* $\dim_{\mathrm{H}}\text{graph } f \leqslant \dim_{\mathrm{B}}\text{graph } f \leqslant s$. *This remains true if* (11.2) *holds when* $|t-u|<\delta$ *for some* $\delta>0$.

(*b*) *Suppose that there are numbers* $c>0, \delta_0>0$ *and* $1\leqslant s<2$ *with the following property: for each* $t\in[0,1]$ *and* $0<\delta\leqslant\delta_0$ *there exists* u *such that* $|t-u|\leqslant\delta$ *and*

$$|f(t)-f(u)| \geqslant c\delta^{2-s}. \qquad (11.3)$$

Then $s \leqslant \underline{\dim}_{\mathrm{B}}\text{graph } f$.

Proof.

(*a*) It is immediate from (11.2) that $R_f[t_1, t_2] \leqslant c|t_1 - t_2|^{2-s}$ for $0 \leqslant t_1, t_2 \leqslant 1$. With notation as in Proposition 11.1, $m < (1 + \delta^{-1})$ so

$$N_\delta \leqslant (1 + \delta^{-1})(2 + c\delta^{-1}\delta^{2-s}) \leqslant c_1 \delta^{-s}$$

where c_1 is independent of δ. The result now follows from Proposition 4.1.

(*b*) In the same way, (11.3) implies that $R_f[t_1, t_2] \geqslant c|t_1 - t_2|^{2-s}$. Since $\delta^{-1} \leqslant m$, we have from (11.1) that

$$N_\delta \geqslant \delta^{-1}\delta^{-1}c\delta^{2-s} = c\delta^{-s}$$

so Equivalent definition 3.1(iii) gives $s \leqslant \underline{\dim}_B$ graph f. □

Unfortunately, lower bounds for the Hausdorff dimension of graphs are generally very much more awkward to find than box dimensions.

Example 11.3. The Weierstrass function

Suppose $\lambda > 1$ *and* $1 < s < 2$. *Define* $f: [0, 1] \to \mathbb{R}$ *by*

$$f(t) = \sum_{k=1}^{\infty} \lambda^{(s-2)k} \sin(\lambda^k t). \tag{11.4}$$

Then, provided λ *is large enough,* $\dim_B$ graph $f = s$.

Calculation. Given $0 < h < 1$, let N be the integer such that

$$\lambda^{-(N+1)} \leqslant h < \lambda^{-N}. \tag{11.5}$$

Then

$$|f(t+h) - f(t)| \leqslant \sum_{k=1}^{N} \lambda^{(s-2)k} |\sin(\lambda^k(t+h)) - \sin(\lambda^k t)|$$

$$+ \sum_{k=N+1}^{\infty} \lambda^{(s-2)k} |\sin(\lambda^k(t+h)) - \sin(\lambda^k t)|$$

$$\leqslant \sum_{k=1}^{N} \lambda^{(s-2)k} \lambda^k h + \sum_{k=N+1}^{\infty} 2\lambda^{(s-2)k}$$

using the mean-value theorem on the first N terms of the sum, and an obvious estimate on the remainder. Summing these geometric series,

$$|f(t+h) - f(t)| \leqslant \frac{h\lambda^{(s-1)N}}{1 - \lambda^{1-s}} + \frac{2\lambda^{(s-2)(N+1)}}{1 - \lambda^{s-2}}$$

$$\leqslant ch^{2-s}$$

where c is independent of h, using (11.5). Corollary 11.2(*a*) now gives that $\overline{\dim}_B$ graph $f \leqslant s$.

In the same way, but splitting the sum into three parts—the first $N - 1$ terms,

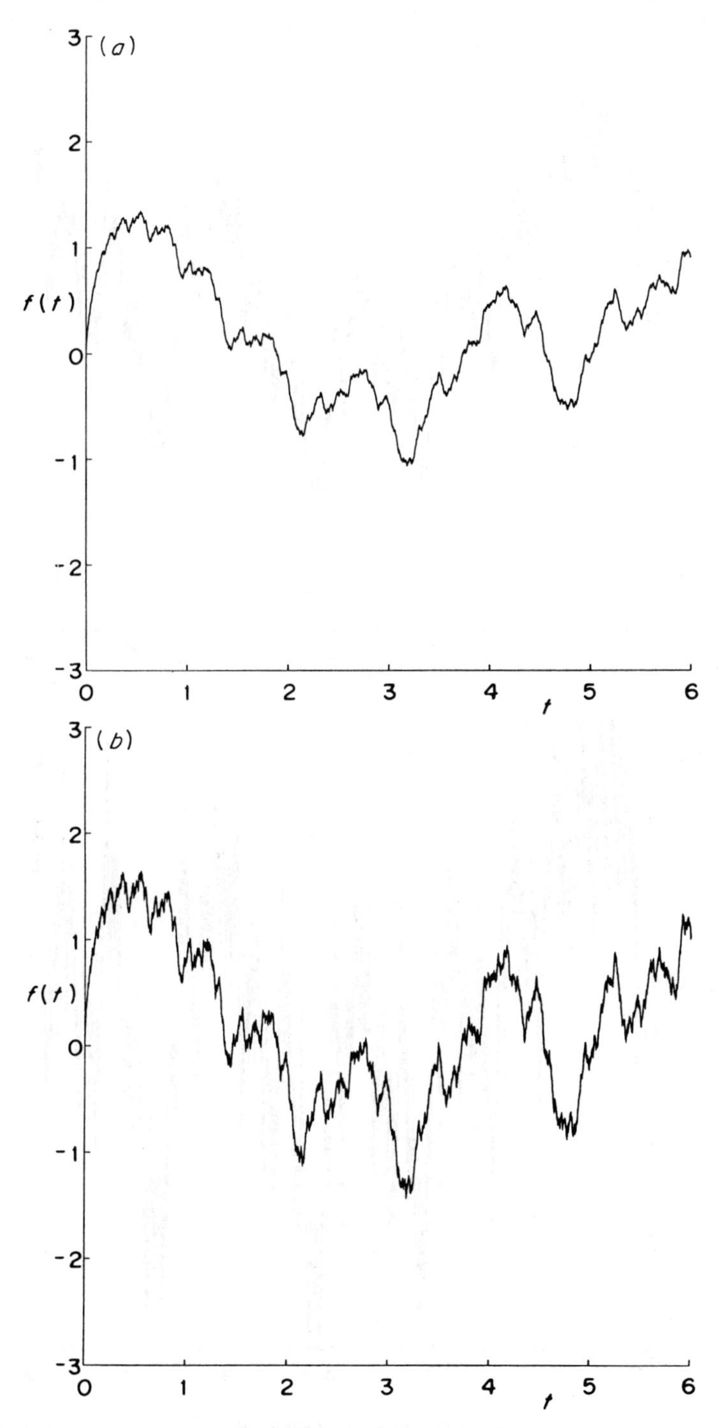

Figure 11.2 The Weierstrass function $f(t)=\sum_{k=0}^{\infty}\lambda^{(s-2)k}\sin(\lambda^k t)$ with $\lambda=1.5$ and (*a*) $s=1.1$, (*b*) $s=1.3$, (*c*) $s=1.5$, (*d*) $s=1.7$

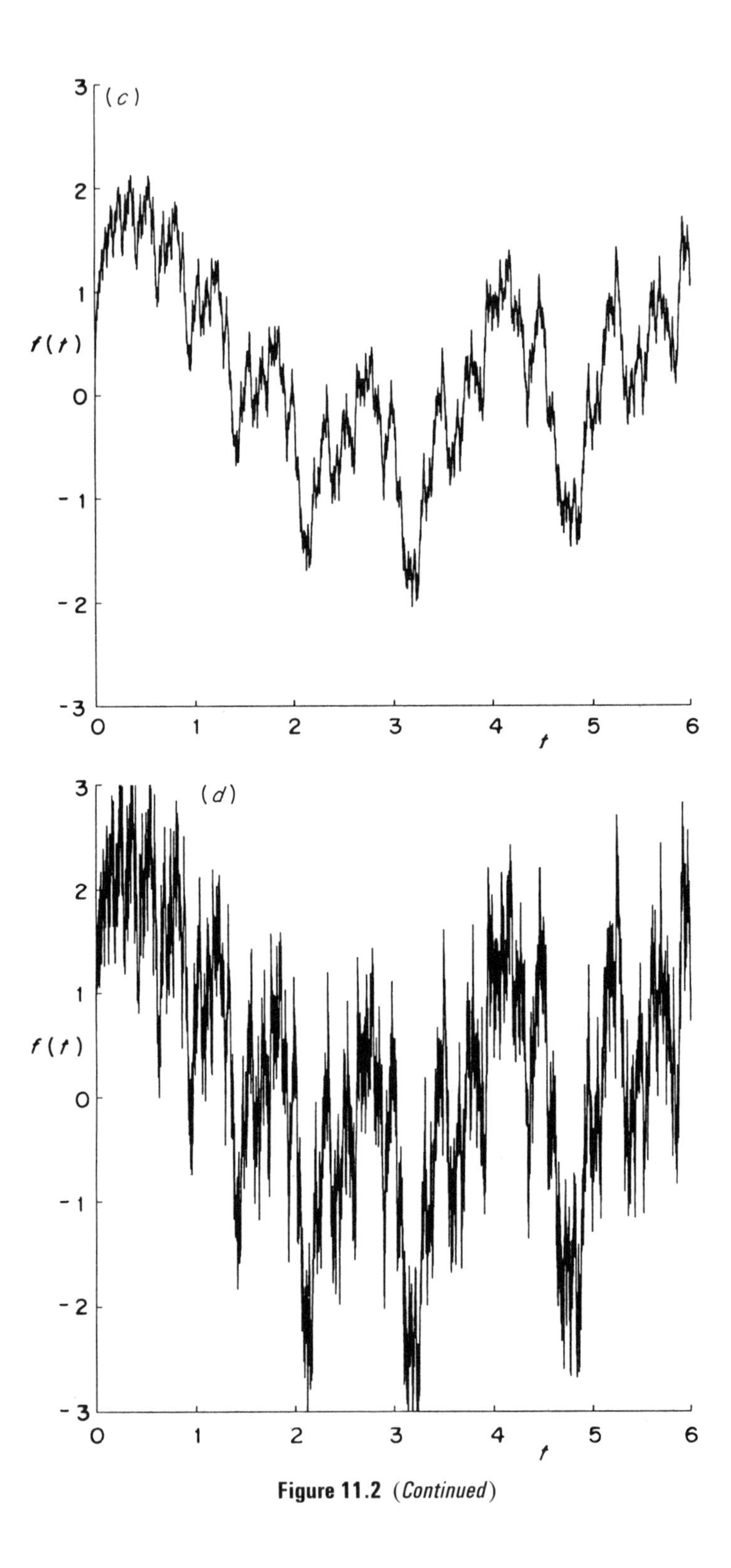

Figure 11.2 (*Continued*)

the Nth term, and the rest—we get that

$$|f(t+h)-f(t)-\lambda^{(s-2)N}(\sin \lambda^N(t+h)-\sin \lambda^N t)| \leqslant \frac{\lambda^{(s-2)N-s+1}}{1-\lambda^{1-s}}+\frac{2\lambda^{(s-2)(N+1)}}{1-\lambda^{s-2}} \tag{11.6}$$

if $\lambda^{-(N+1)} \leqslant h < \lambda^N$.

Suppose $\lambda > 2$ is large enough for the right-hand side of (11.6) to be less than $\frac{1}{20}\lambda^{(s-2)N}$ for all N. For $\delta < \lambda^{-1}$, take N such that $\lambda^{-N} \leqslant \delta < \lambda^{-(N-1)}$. For each t, we may choose h with $\lambda^{-(N+1)} \leqslant h < \lambda^{-N}$ such that $|\sin \lambda^N(t+h) - \sin \lambda^N t| > \frac{1}{10}$, so by (11.6)

$$|f(t+h)-f(t)| \geqslant \tfrac{1}{20}\lambda^{(s-2)N} \geqslant \tfrac{1}{20}\lambda^{s-2}\delta^{2-s}.$$

It follows from Corollary 11.2(b) that $\underline{\dim}_B$ graph $f \geqslant s$. □

Various cases of the Weierstrass function are shown in figure 11.2.

It is immediate from the above estimate that the Hausdorff dimension of the graph of the Weierstrass function (11.4) is at most s. It is widely believed that it equals s, at least for 'most' values of λ. This has not yet been proved rigorously—it could be that there are coverings of the graph of the function by sets of widely varying sizes that give a smaller value. Even to show that $\dim_H$ graph $f > 1$ is not trivial. The known lower bounds come from mass distribution methods depending on estimates for $\mathscr{L}\{t:(t, f(t)) \in B\}$ where B is a disc and $\mathscr{L}$ is Lebesgue measure. The rapid small-scale oscillation of f ensures that the graph is inside B relatively rarely, so that this measure is small. In this way it is possible to show that there is a constant c such that

$$s \geqslant \dim_H \text{graph } f \geqslant s - c/\log \lambda$$

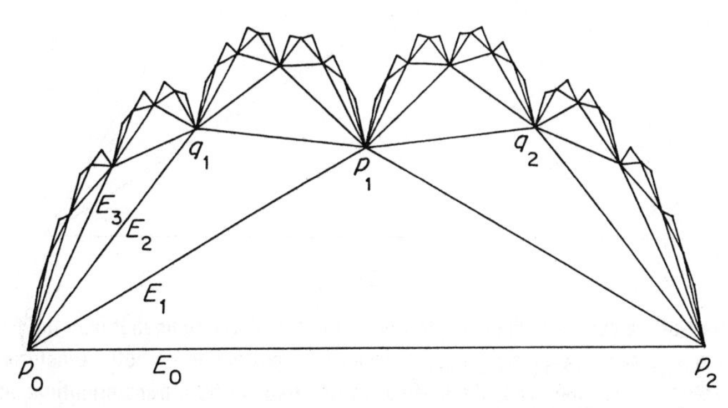

Figure 11.3 Stages in the construction of a self-affine curve F. The affine transformations S_1 and S_2 map the triangle $p_0p_1p_2$ onto the triangles $p_0q_1p_1$ and $p_1q_2p_2$, respectively, and transform vertical lines to vertical lines. The rising sequence of polygonal curves E_0, $E_1, \ldots$ are given by $E_{k+1} = S_1(E_k) \cup S_2(E_k)$ and provide increasingly good approximations to F (shown in figure 11.4(a) for this case)

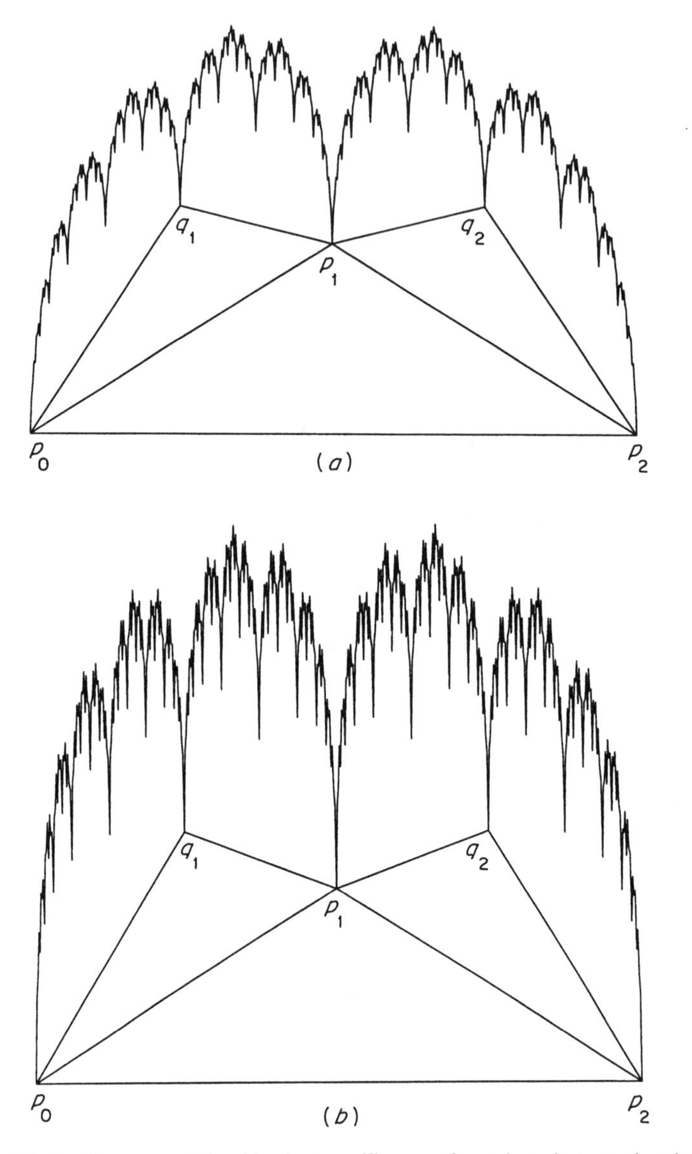

Figure 11.4 Self-affine curves defined by the two affine transformations that map the triangle $p_0p_1p_2$ onto $p_0q_1p_1$ and $p_1q_2p_2$ respectively. In (*a*) the vertical contraction of both transformations is 0.7 giving $\dim_B \text{graph } f = 1.49$, and in (*b*) the vertical contractions of both transformations are 0.8, giving $\dim_B \text{graph } f = 1.68$

so when λ is large the Hausdorff dimension cannot be much less than the conjectured value.

The Weierstrass function (11.4) is representative of a much wider class of functions to which these methods apply. If g is a suitable periodic function, a similar method can often show that

$$f(t) = \sum_{k=1}^{\infty} \lambda^{(s-2)k} g(\lambda^k t) \tag{11.7}$$

has $\dim_B \text{graph } f = s$. At first such functions seem rather contrived, but their occurrence as repellers in certain dynamical systems (see Exercise 13.7) gives them a new importance.

In Section 9.4 we saw that self-affine sets are often fractals; by suitable choice of affine transformations, they can also be graphs of functions. Let $S_i (1 \leqslant i \leqslant m)$ be affine transformations represented in matrix notation with respect to (t, x) coordinates by

$$S_i \begin{bmatrix} t \\ x \end{bmatrix} = \begin{bmatrix} 1/m & 0 \\ a_i & c_i \end{bmatrix} \begin{bmatrix} t \\ x \end{bmatrix} + \begin{bmatrix} (i-1)/m \\ b_i \end{bmatrix}. \tag{11.8}$$

Thus the S_i transform vertical lines to vertical lines, with the vertical strip $0 \leqslant t \leqslant 1$ mapped onto the strip $(i-1)/m \leqslant t \leqslant i/m$. We suppose that

$$1/m < c_i \leqslant 1 \tag{11.9}$$

so that contraction in the t direction is greater than in the x direction.

Let $p_1 = (0, b_1/(1-c_1))$ and $p_m = (1, (a_m + b_m)/(1-c_m))$ be the fixed points of S_1 and S_m. We assume that the matrix entries have been chosen so that

$$S_i(p_m) = S_{i+1}(p_1) \qquad (1 \leqslant i \leqslant m-1) \tag{11.10}$$

so that the segments $[S_i(p_1), S_i(p_m)]$ join up to form a polygonal curve E_1. To avoid trivial cases, we assume that the points $S_1(p_1), \ldots, S_m(p_1), p_m$ are not all collinear. The invariant set F of the S_i (see (9.1)) may be constructed by repeatedly replacing line segments by affine images of E_1; see figures 11.3 and 11.4. Condition (11.10) ensures that the segments join up with the result that F is the graph of some continuous function $f:[0,1] \to \mathbb{R}$.

Example 11.4 Self-affine curves

Let $F = \text{graph } f$ be the self-affine curve described above. Then $\dim_B F = 1 + \log(c_1 + \cdots + c_m)/\log m$.

Calculation. Let T_i be the 'linear part' of S_i, represented by the matrix

$$\begin{bmatrix} 1/m & 0 \\ a_i & c_i \end{bmatrix}.$$

Let $I_{i_1,\ldots,i_k}$ be the interval of the t-axis consisting of those t with base-m expansion beginning $0.i'_1 \cdots i'_k$ where $i'_j = i_j - 1$. Then the part of F above $I_{i_1,\dots,i_k}$ is the affine image $S_{i_1} \circ \cdots \circ S_{i_k}(F)$, which is a translate of $T_{i_1} \circ \cdots \circ T_{i_k}(F)$. The matrix representing $T_{i_1} \circ \cdots \circ T_{i_k}$ is easily seen by induction to be

$$\begin{bmatrix} m^{-k} & 0 \\ m^{1-k}a_{i_1} + m^{2-k}c_{i_1}a_{i_2} + \cdots + c_{i_1}c_{i_2}\cdots c_{i_{k-1}}a_{i_k} & c_{i_1}c_{i_2}\cdots c_{i_k} \end{bmatrix}.$$

This is a shear transformation, contracting vertical lines by a factor $c_{i_1}c_{i_2}\cdots c_{i_k}$. Observe that the bottom left-hand entry is bounded by

$$\begin{aligned} &|m^{1-k}a + m^{2-k}c_{i_1}a + \cdots + c_{i_1}\cdots c_{i_{k-1}}a| \\ &\leqslant ((mc)^{1-k} + (mc)^{2-k} + \cdots + 1)c_{i_1}\cdots c_{i_{k-1}}a \\ &\leqslant rc_{i_1}\cdots c_{i_{k-1}} \end{aligned}$$

where $a = \max|a_i|$, $c = \min\{c_i\} > 1/m$ and $r = a/(1-(mc)^{-1})$. Thus the image $T_{1_i} \circ \cdots \circ T_{i_k}(F)$ is contained in a rectangle of height $(r+h)c_{i_1}\cdots c_{i_k}$ where h is the height of F. On the other hand, if q_1, q_2, q_3 are three non-collinear points chosen from $S_1(p_1), \dots, S_m(p_1), p_m$, then $T_{i_1} \circ \cdots \circ T_{i_k}(F)$ contains the points $T_{i_1} \circ \cdots \circ T_{i_k}(q_j)$ $(j = 1, 2, 3)$. The height of the triangle with these vertices is at least $c_{i_1}\cdots c_{i_k}d$ where d is the vertical distance from q_2 to the segment $[q_1, q_3]$. Thus the range of the function f, over $I_{i_1,\dots,i_k}$ satisfies

$$dc_{i_1}\cdots c_{i_k} \leqslant R_f[I_{i_1,\dots,i_k}] \leqslant r_1 c_{i_1}\cdots c_{i_k}$$

where $r_1 = r + h$.

For fixed k, we sum this over the m^k intervals $I_{i_1,\dots,i_k}$ of lengths m^{-k} to get, using Proposition 11.1,

$$m^k d \Sigma c_{i_1}\cdots c_{i_k} \leqslant N_{m^{-k}}(F) \leqslant 2m^k + m^k r_1 \Sigma c_{i_1}\cdots c_{i_k}$$

where $N_{m^{-k}}(F)$ is the number of mesh squares of side m^{-k} that intersect F. For each j the number c_{i_j} ranges through the values $c_1, \dots, c_m$, so that $\Sigma c_{i_1}\cdots c_{i_k} = (c_1 + \cdots + c_m)^k$. Thus

$$dm^k(c_1 + \cdots + c_m)^k \leqslant N_{m^{-k}}(F) \leqslant 2m^k + r_1 m^k (c_1 + \cdots + c_m)^k.$$

Taking logarithms and using Definition 3.1(iii) of box dimension gives the value stated. □

Self-affine functions are useful for *fractal interpolation*. Suppose we wish to find a fractal curve of a given dimension passing through the points $(i/m, x_i)$ for $i = 0, 1, \dots, m$. By choosing transformations (11.8) in such a way that S_i maps the segment $[p_1, p_m]$ onto the segment $[((i-1)/m, x_{i-1}), (i/m, x_i)]$ for each i, the construction described above gives a self-affine function with graph passing through the given points. By adjusting the values of the matrix entries we can ensure that the curve has the required box dimension; there is also some freedom to vary the apperance of the curve in other ways. Fractal interpolation has been used very effectively to picture mountain skylines.

Of course, self-affine functions can be generalized so that the S_i do not all

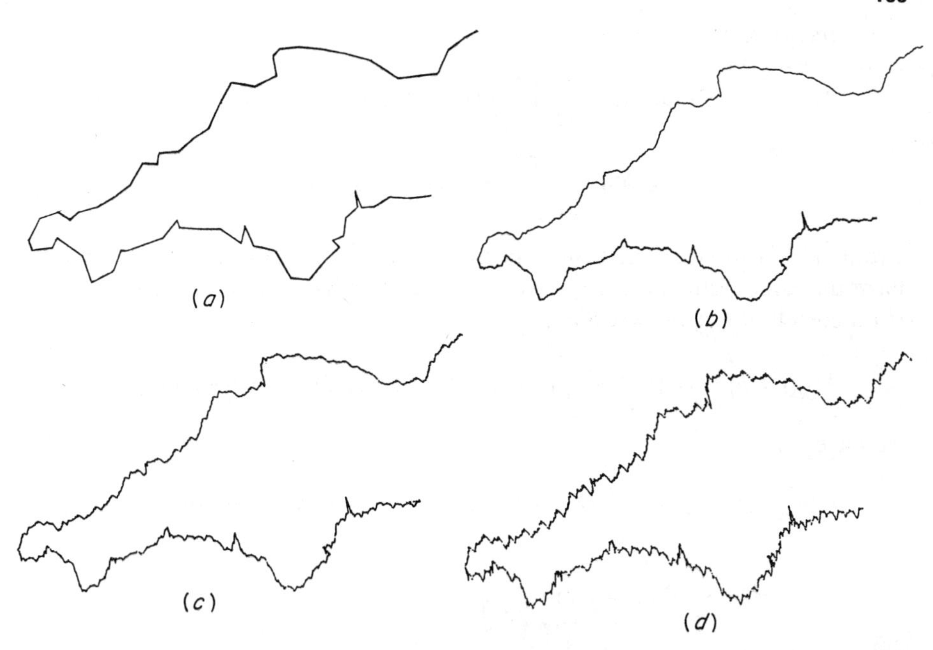

Figure 11.5 Fractal interpolation on the northern and southern halves of a map of South-West England, using the vertices of the polygon in figure (*a*) as data points. The dimensions of the self-affine curves fitted to these data points are (*b*) 1.1, (*c*) 1.2 and (*d*) 1.3

have the same contraction ratio in the t direction. This leads to fractal interpolation of points that are not equally spaced. With additional effort, the box dimension of such curves may be found.

One example of fractal interpolation is illustrated in figure 11.5.

*11.2 Autocorrelation of fractal functions

As we have remarked, quantities varying with time often turn out to have fractal graphs. One way in which their fractal nature is often manifested is by a power-law behaviour of the correlation between measurements separated by time h. In this section we merely outline the ideas involved; we make no attempt to be rigorous. In particular, the limits used are all assumed to exist.

For convenience of analysis, we assume that $f:(-\infty,\infty)\to\mathbb{R}$ is a continuous bounded function and we consider the average behaviour of f over long periods $[-T, T]$. (Similar ideas hold if f is just defined on $[0,\infty)$, or on a finite interval, by extending f to $\mathbb{R}$ in a periodic manner). We write $\overline{f}$ for the average value of f, i.e.

$$\overline{f} = \lim_{T\to\infty}\frac{1}{2T}\int_{-T}^{T} f(t)\,\mathrm{d}t.$$

A measure of the correlation between f at times separated by h is provided by the *autocorrelation function*

$$C(h) = \lim_{T\to\infty} \frac{1}{2T}\int_{-T}^{T} (f(t+h)-\bar{f})(f(t)-\bar{f})\,\mathrm{d}t \tag{11.11}$$

$$= \lim_{T\to\infty} \frac{1}{2T}\int_{-T}^{T} f(t+h)f(t)\,\mathrm{d}t - (\bar{f})^2. \tag{11.12}$$

From (11.11) we see that $C(h)$ is positive if $f(t+h)-\bar{f}$ and $f(t)-\bar{f}$ tend to have the same sign, and is negative if they tend to have opposite signs. If there is no correlation, $C(h)=0$. Since

$$\int (f(t+h)-f(t))^2\,\mathrm{d}t = \int f(t+h)^2\,\mathrm{d}t + \int f(t)^2\,\mathrm{d}t - 2\int f(t+h)f(t)\,\mathrm{d}t$$

we have

$$C(h) = (\bar{f})^2 - \overline{f^2} - \tfrac{1}{2}\lim_{T\to\infty}\frac{1}{2T}\int_{-T}^{T}(f(t+h)-f(t))^2\,\mathrm{d}t$$

$$= C(0) - \tfrac{1}{2}\lim_{T\to\infty}\frac{1}{2T}\int_{-T}^{T}(f(t+h)-f(t))^2\,\mathrm{d}t \tag{11.13}$$

where

$$\overline{f^2} = \lim_{T\to\infty}\frac{1}{2T}\int_{-T}^{T} f(t)^2\,\mathrm{d}t$$

is the mean square of f, assumed to be positive and finite. With $C(h)$ in the form (11.13) we can infer a plausible relationship between the autocorrelation function of f and the dimension of graph f. The clue is in Corollary 11.2. Suppose that f is a function satisfying (11.2) and also satisfying (11.3) in a 'reasonably uniform way'. Then, there are constants c_1 and c_2 such that

$$c_1 h^{4-2s} \leqslant \frac{1}{2T}\int_{-T}^{T}(f(t+h)-f(t))^2\,\mathrm{d}t \leqslant c_2 h^{4-2s} \tag{11.14}$$

for small h. Obviously, this is not directly equivalent to (11.2) and (11.3), but in many reasonably 'time-homogeneous' situations, the conditions do correspond. Thus if the autocorrelation function of f satisfies

$$C(0) - C(h) \simeq ch^{4-2s}$$

it is not unreasonable to expect the box dimension of graph f to equal s.

The autocorrelation function is closely connected with the *power spectrum* of f, defined by

$$S(\omega) = \lim_{T\to\infty}\frac{1}{2T}\left|\int_{-T}^{T} f(t)\,\mathrm{e}^{it\omega}\,\mathrm{d}t\right|^2. \tag{11.15}$$

For functions with any degree of long-term regularity, $S(\omega)$ is likely to exist.

The power spectrum reflects the strength of the frequency ω in the harmonic decomposition of f.

We show that the power spectrum is the Fourier transform of the autocorrelation function. By working with $f(t)-\bar{f}$ we may assume that f has zero mean. Let $f_T(t)$ be given by $f(t)$ if $|t| \leqslant T$ and be 0 otherwise, and define

$$C_T(h) = \frac{1}{2T}\int_{-\infty}^{\infty} f_T(t+h) f_T(t)\,\mathrm{d}t$$

$$= \frac{1}{2T} f_T^{-} * f_T(-h)$$

where $f_T^{-}(t) = f_T(-t)$ and $*$ denotes convolution. By the convolution theorem for Fourier transforms (see Section 4.4) this equation transforms to

$$\hat{C}_T(\omega) = \frac{1}{2T}\hat{f}_T^{-}(\omega)\hat{f}_T(\omega)$$

$$= \frac{1}{2T}|\hat{f}_T(\omega)|^2$$

where $\hat{C}_T(\omega) = \int_{-\infty}^{\infty} C_T(t)\,\mathrm{e}^{it\omega}\,\mathrm{d}t$ and $\hat{f}_T(\omega) = \int_{-\infty}^{\infty} f_T(t)\,\mathrm{e}^{it\omega}\,\mathrm{d}t$ are the usual Fourier transforms. (Note that we cannot work with the transform of f itself, since the integral would diverge.) Letting $T \to \infty$ we see that $C_T(h) \to C(h)$ for each h and $\hat{C}_T(\omega) \to S(\omega)$ for each ω. It may be shown that this implies that

$$\hat{C}(\omega) = S(\omega).$$

Clearly S and C are both real and even functions, so the transforms are cosine transforms. Thus

$$S(\omega) = \int_{-\infty}^{\infty} C(t)\,\mathrm{e}^{it\omega}\,\mathrm{d}t = \int_{-\infty}^{\infty} C(t)\cos(\omega t)\,\mathrm{d}t \tag{11.16}$$

and, by the inversion formula for Fourier transforms,

$$C(h) = \frac{1}{2\pi}\int_{-\infty}^{\infty} S(\omega)\mathrm{e}^{-ih\omega}\mathrm{d}\omega = \frac{1}{2\pi}\int_{-\infty}^{\infty} S(\omega)\cos(\omega h)\mathrm{d}\omega. \tag{11.17}$$

In this analysis we have not gone into questions of convergence of the integrals too carefully, but in most practical situations the argument can be justified.

Autocorrelations provide us with several methods of estimating the dimension of the graph of a function or 'signal' f. We may compute the autocorrelation function $C(h)$ or equivalently, the mean-square change in signal in time h over a long period, so from (11.13)

$$2[C(0) - C(h)] \simeq \frac{1}{2T}\int_{-T}^{T} (f(t+h) - f(t))^2\,\mathrm{d}t. \tag{11.18}$$

If the power-law behaviour

$$C(0) - C(h) \simeq ch^{4-2s} \tag{11.19}$$

is observed for small h, we might expect the box dimension of graph f to be s. In other words,

$$\dim_B \text{graph } f = 2 - \lim_{h \to 0} \frac{\log(C(0) - C(h))}{2 \log h} \tag{11.20}$$

if this limit exists. We might then seek functions with graphs known to have this dimension, such as those of Examples 11.3 and 11.4 or the fractional Brownian functions of Section 16.2 to provide simulations of signals with similar characteristics.

Alternatively, we can work from the power spectrum $S(\omega)$ and use (11.17) to find the autocorrelation function. We need to know about $C(0) - C(h)$ for small h; typically this depends on the behaviour of its transform $S(\omega)$ when ω is large. The situation of greatest interest is when the power spectrum obeys a power law $S(\omega) \sim c/\omega^\alpha$ for large ω, in which case

$$C(0) - C(h) \sim bh^{\alpha-1} \tag{11.21}$$

for small h, for some constant b. To see this formally note that from (11.17)

$$\pi(C(0) - C(h)) = \int_0^\infty S(\omega)(1 - \cos(\omega h))\,d\omega = 2\int_0^\infty S(\omega)\sin^2(\tfrac{1}{2}\omega h)\,d\omega$$

and taking $S(\omega) = \omega^{-\alpha}$ gives

$$\tfrac{1}{2}\pi(C(0) - C(h)) = \int_0^\infty \omega^{-\alpha}\sin^2 \tfrac{1}{2}\omega h\,d\omega = h^{\alpha-1}\int_0^\infty u^{-\alpha}\sin^2 \tfrac{1}{2}u\,du$$

having substituted $u = \omega h$. It may be shown that (11.21) also holds if S is any sufficiently smooth function such that $S(\omega) \sim c\omega^{-\alpha}$ as $\omega \to \infty$. Comparing (11.19) and (11.21) suggests that graph f has box dimension s where $4 - 2s = \alpha - 1$, or $s = \frac{1}{2}(5 - \alpha)$. Thus it is reasonable to expect a signal with a $1/\omega^\alpha$ power spectrum to have a graph of dimension $\frac{1}{2}(5 - \alpha)$.

In practice, curves of dimension $\frac{1}{2}(5 - \alpha)$ often provide good simulations and display similar characteristics to signals observed to have $1/\omega^\alpha$ power spectra.

11.3 Notes and references

The dimension of fractal graphs was first studied by Besicovitch and Ursell (1937). For more recent work on Weierstrass-type curves see Berry and Lewis (1980) (containing many computer drawings) and Mauldin and Williams (1986b). Self-affine curves are discussed in Bedford (1989). The theory of autocorrelation functions is given in most books on time series analysis, for example, Papoulis (1962).

Exercises

11.1 Verify that if $f:[0,1]\to\mathbb{R}$ has a continuous derivative then graph f is a regular 1-set.

11.2 Let $f,g:[0,1]\to\mathbb{R}$ be continuous functions and define the sum function $f+g$ by $(f+g)(t)=f(t)+g(t)$ in the obvious way. Suppose that f is a Lipschitz function. By setting up a Lipschitz mapping between graph $(f+g)$ and graph g, show that $\dim_H \text{graph}\,(f+g)=\dim_H \text{graph}\, g$, with a similar result for box dimensions.

11.3 Let $f,g:[0,1]\to\mathbb{R}$ be continuous functions such that the box dimension of their graphs exist. Use Proposition 11.1 to show that $\dim_B \text{graph}\,(f+g)$ equals the greater of $\dim_B \text{graph}\, f$ and $\dim_B \text{graph}\, g$, provided that these dimensions are unequal. Give an example to show that this proviso is necessary.

11.4 Show that any function satisfying the conditions of Corollary 11.2(*b*) must be nowhere differentiable. Deduce that the Weierstrass function of Example 11.3 and the self-affine curves of Example 11.4, are nowhere differentiable.

11.5 For $\lambda>1$ and $1<s<2$ let $f:[0,1]\to\mathbb{R}$ be a Weierstrass function modified to include 'phases' θ_k:

$$f(t)=\sum_{k=1}^{\infty}\lambda^{(s-2)k}\sin(\lambda^k t+\theta_k).$$

Show that $\dim_B$ graph $f=s$, provided that λ is large enough.

11.6 Let $g:\mathbb{R}\to\mathbb{R}$ be the 'zig-zag' function of period 4 given by

$$g(4k+t)=\begin{cases} t & (0\leqslant t<1)\\ 2-t & (1\leqslant t<3)\\ t-4 & (3\leqslant t<4)\end{cases}$$

where k is an integer and $0\leqslant t<4$. Let $1<s<2$ and $\lambda>1$ and let $f:\mathbb{R}\to\mathbb{R}$ be given by

$$f(t)=\sum_{k=1}^{\infty}\lambda^{(s-2)k}g(\lambda^k t).$$

Show that $\dim_B$ graph $f=s$, provided that λ is sufficiently large.

11.7 Suppose that the function $f:[0,1]\to\mathbb{R}$ satisfies the Holder condition (11.2). Let F be a subset of $[0,1]$. Obtain an estimate for $\dim_H f(F)$ in terms of $\dim_H F$.

11.8 Let $f:[0,1]\to\mathbb{R}$ be a function. Suppose that

$$\int_0^1\int_0^1 [|f(t)-f(u)|^2+|t-u|^2]^{-s/2}\,dt\,du<\infty$$

for some s with $1<s<2$. Show, using Theorem 4.13, that $\dim_H \text{graph}\, f\geqslant s$.

11.9 Let D be the unit square $[0,1]\times[0,1]$ and let $f:D\to\mathbb{R}$ be a continuous function such that

$$|f(x)-f(y)|\leqslant c|x-y|^{3-s}\qquad (x,y\in D).$$

Show that the surface $\{(x, f(x)):x \in D\}$ has box dimension at most s. Similarly, find a surface analogue to part (*b*) of Corollary 11.2.

11.10 Investigate the graphs of Weierstrass-type functions (11.7) using a computer. Examine the effect of varying s and λ, and experiment with various functions g.

11.11 Write a computer program to draw self-affine curves given by (11.8). Investigate the effect of varying the values of the c_i.

Chapter 12 Examples from pure mathematics

Fractal constructions have provided counterexamples, and sometimes solutions, to a variety of problems where more regular constructions have failed. In this chapter we look at several instances from differing areas of pure mathematics.

12.1 Duality and the Kakeya problem

The method of duality converts sets of points in the plane to sets of lines and may be used to create new fractals from old. The techniques can be applied to construct sets with particular properties; for example, to construct a plane set of zero area that contains a line running in every direction.

For any point (a,b) of $\mathbb{R}^2$, we let $L(a,b)$ denote the set of points on the line $y=a+bx$, see figure 12.1. If F is any subset of $\mathbb{R}^2$ we define the *line set* $L(F)$ to be the union of the lines corresponding to the points of F, i.e. $L(F)=\cup\{L(a,b):(a,b)\in F\}$. Writing L_c for the vertical line $x=c$, we have

$$L(a,b)\cap L_c=(c,a+bc)=(c,(a,b)\cdot(1,c)),$$

where '$\cdot$' is the usual scalar product in $\mathbb{R}^2$; thus for a subset F of $\mathbb{R}^2$

$$L(F)\cap L_c=\{(c,(a,b)\cdot(1,c)):(a,b)\in F\}.$$

Taking a scalar product with the vector $(1,c)$ may be interpreted geometrically as projecting onto the line in the direction of $(1,c)$ and stretching by a factor $(1+c^2)^{1/2}$. Thus the set $L(F)\cap L_c$ is geometrically similar to $\text{proj}_\theta F$, where proj_θ denotes orthogonal projection onto the line through the origin at angle θ with $c=\tan\theta$. In particular,

$$\dim_H(L(F)\cap L_c)=\dim_H(\text{proj}_\theta F) \tag{12.1}$$

and

$$\mathcal{L}(L(F)\cap L_c)=0 \qquad \text{if and only if} \qquad \mathcal{L}(\text{proj}_\theta F)=0 \tag{12.2}$$

where $\mathcal{L}$ denotes length. In this way, duality relates the projections of F (for which we have the theory of Chapter 6) to the intersections of the line set $L(F)$ with vertical lines.

Projecting onto the y-axis also has an interpretation. The gradient of the line $L(a,b)$ is just $b=\text{proj}_{\pi/2}(a,b)$, so, for any F, the set of gradients of the lines in

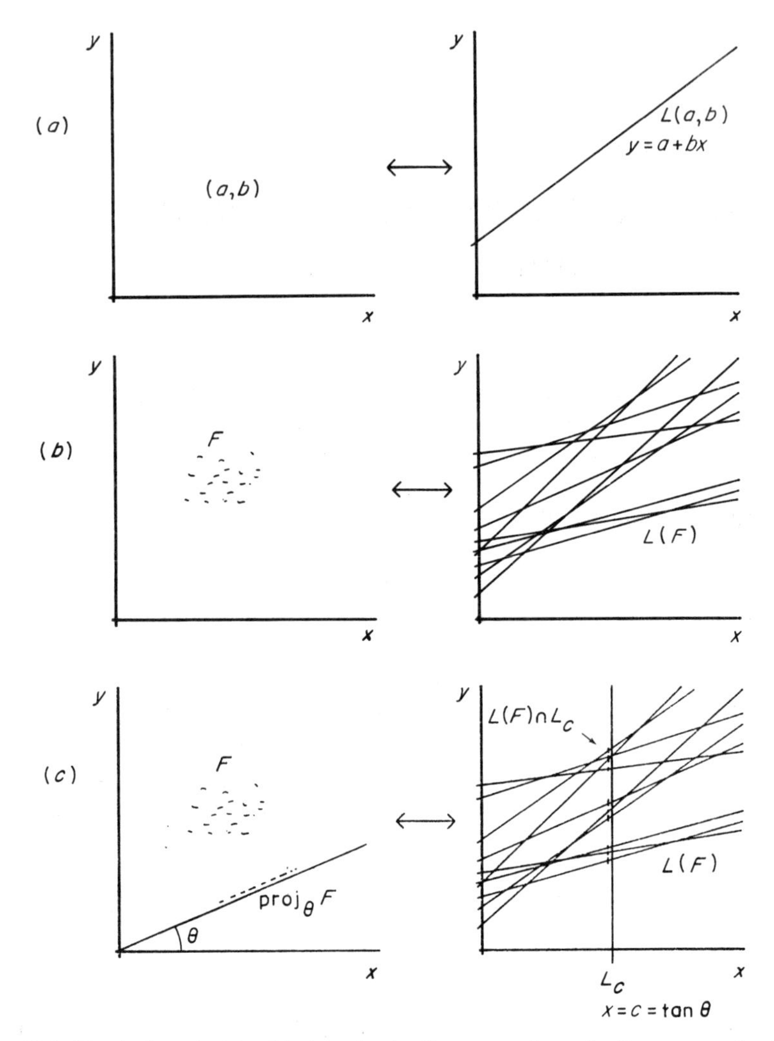

Figure 12.1 The duality principle: (*a*) the point (a, b) corresponds to the line $y = a + bx$; (*b*) the set F corresponds to the line set $L(F)$; (*c*) the projection $\mathrm{proj}_\theta F$ is geometrically similar to $L(F) \cap L_c$, where $c = \tan\theta$

the line set $L(F)$ is given by $\mathrm{proj}_{\pi/2} F$.

If F is a fractal its line set $L(F)$ often has a fractal structure, albeit a highly fibrous one. (In fact, $L(F)$ need not be a Borel set if F is Borel, though it will be if F is compact. We ignore the minor technical difficulties that this introduces.) We have the following dimensional relationship.

Proposition 12.1

Let $L(F)$ be the line set of a Borel set $F \subset \mathbb{R}^2$. Then

(*a*) $\dim_{\mathrm{H}} L(F) \geqslant \min\{2, 1 + \dim_{\mathrm{H}} F\}$, *and*
(*b*) *if F is a 1-set then* $\mathrm{area}(L(F)) = 0$ *if and only if F is irregular.*

Proof.

(*a*) By the Projection theorem 6.1, $\dim_H(\text{proj}_\theta F) = \min\{1, \dim_H F\}$ for almost all $\theta \in [0, \pi)$, so from (12.1) $\dim_H(L(F) \cap L_c) = \min\{1, \dim_H F\}$ for almost all $-\infty < c < \infty$. Part (*a*) now follows from Corollary 7.10.

(*b*) Let F be a 1-set. Corollary 6.5 tells us that if F is irregular then $\mathscr{L}(\text{proj}_\theta F) = 0$ for almost all θ, otherwise $\mathscr{L}(\text{proj}_\theta F) > 0$ for almost all θ. Using (12.2) we get the dual statement that if F is irregular then $\mathscr{L}(L(F) \cap L_c) = 0$ for almost all c, otherwise $\mathscr{L}(L(F) \cap L_c) > 0$ for almost all c. Since $\text{area}(L(F)) = \int_{-\infty}^{\infty} \mathscr{L}(L(F) \cap L_c)\, dc$, part (*b*) follows. □

In 1917 Kakeya posed the problem of finding the plane set of least area inside which a unit line segment could be reversed, i.e. manoeuvred continuously and without leaving the set to reach its original position but rotated through 180°. Essentially, this problem reduces to that of finding the smallest region that contains a unit line segment in every direction; certainly any set in which a segment can be reversed must have this property. By 1928 Besicovitch had found a surprising construction of a set of arbitrarily small area inside which a unit segment could be reversed. Only many years later did he realize that the method of duality gave a short and elegant solution to the problem.

Proposition 12.2

There is a plane set of zero area which contains a line in every direction. Any Borel set with this property must have Hausdorff dimension 2.

Proof. Let F be any irregular 1-set such that the projection of F onto the y-axis, $\text{proj}_{\pi/2} F$, contains the interval $[0, 1]$. (The set examined in Examples 2.6 and 6.7 certainly meets this requirement.) Since F is irregular, $L(F)$ has zero area, by Proposition 12.1(*b*). However, since $[0, 1] \subset \text{proj}_{\pi/2} F$, the set $L(F)$ contains lines that cut the x-axis at all angles between 0 and $\pi/4$. Taking $L(F)$, together with copies rotated through $\pi/4$, $\pi/2$ and $3\pi/4$, gives a set of area 0 containing a line in every direction.

For the second part, suppose that E contains a line in every direction. If

$$F = \{(a, b) : L(a, b) \subset E\}$$

then $\text{proj}_{\pi/2} F$ is the entire y-axis. Projection does not increase dimension (see (6.1)), so $\dim_H F \geqslant 1$. By Proposition 12.1(*a*) $\dim_H L(F) = 2$; since $L(F) \subset E$ it follows that $\dim_H E = 2$. □

Sets of this type provide important examples in functional analysis. Given a function $g: \mathbb{R}^2 \to \mathbb{R}$, write $G(\theta, t)$ for the integral of g along the line making angle θ with the x-axis and perpendicular distance t from the origin. Let F be a set of zero area containing a line in every direction, and let $g(x, y) = 1$ if (x, y) is a point of F and $g(x, y) = 0$ otherwise. It is clear that $G(\theta, t)$ is not continuous in t for any fixed value of θ. This example becomes significant when contrasted with the 3-dimensional situation. If $g: D \to \mathbb{R}$ is a bounded function on a bounded domain D in $\mathbb{R}^3$, and $G(\theta, t)$ is the integral of f over the plane perpendicular to the unit vector θ and perpendicular distance t from the origin, it may be shown that $G(\theta, t)$ must be continuous in t for almost all unit vectors θ.

The Kakeya construction may be thought of as a packing of lines in all directions into a set of area zero. Similar problems may be considered for packings of other collections of curves. For example, there are sets of zero area that contain the circumference of a circle of every radius. However, it has recently been shown that any set that contains some circle circumference centred at each point in the plane necessarily has positive area.

12.2 Vitushkin's conjecture

A longstanding conjecture of Vitushkin in complex potential theory was recently disproved using a fractal construction.

Let F be a compact subset of the complex plane. We say that F is a *removable* set if, given any bounded open domain V containing F and any bounded analytic (i.e. differentiable in the complex sense) function f on the complement $V \backslash F$, then f has an analytic extension to the whole of V. Thus the functions that are bounded and analytic on V are essentially the same as those that are bounded and analytic on $V \backslash F$; removing F makes no difference.

The problem of providing a geometrical characterisation of removable sets dates back many years. The removability, or otherwise, of F has been established in the following cases:

Removable	Not Removable
$\dim_{\mathrm{H}} F < 1$	$\dim_{\mathrm{H}} F > 1$
$0 < \mathscr{H}^1(F) < \infty$ and F irregular	$0 < \mathscr{H}^1(F) < \infty$ and F not irregular

This table should remind the reader of the projection theorems of Chapter 6. According to Theorem 6.1 and Corollary 6.5, if $\dim_{\mathrm{H}} F < 1$ or if F is an irregular 1-set then the projection $\mathrm{proj}_\theta F$ has length 0 for almost all θ. On the other hand, if $\dim_{\mathrm{H}} F > 1$ or if F is a 1-set that is not irregular, $\mathrm{proj}_\theta F$ has positive length for almost all θ. The apparent correspondence between removability and almost all projections having length 0, together with a considerable amount of further evidence in the delicate cases where $\dim_{\mathrm{H}} F = 1$ and $\mathscr{H}^1(F) = \infty$, led to Vitushkin's conjecture: F is removable if and only if $\mathrm{proj}_\theta F = 0$ for almost all $\theta \in [0, \pi)$.

A fractal construction shows that Vitushkin's conjecture cannot be true. Let V be an open domain in $\mathbb{C}$ and let $\phi: V \to \phi(V)$ be a conformal mapping (i.e. analytic bijection) on V that is not linear, so that straight lines are typically mapped onto (non-straight) curves; V as the unit disc and $\phi(z) = (z+2)^2$ would certainly be suitable. It is possible to construct a compact subset F of V such that $\mathrm{proj}_\theta F$ has zero length for almost all θ but $\mathrm{proj}_\theta \phi(F)$ has positive length

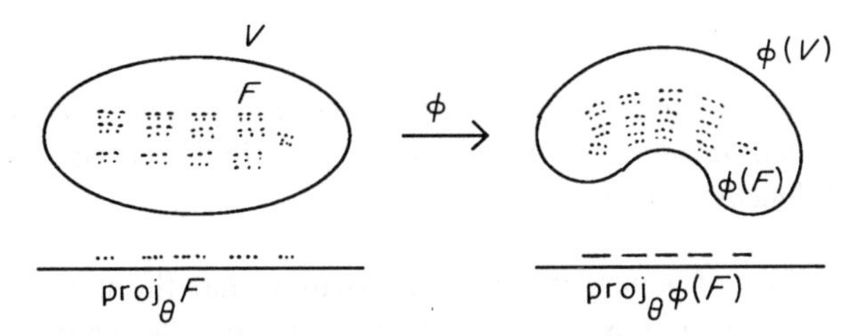

Figure 12.2 ϕ is an analytic mapping such that $\text{proj}_\theta F$ has zero length for almost all θ but $\text{proj}_\theta \varphi(F)$ has positive length for almost all θ

for almost all θ. This may be achieved using a version of the 'iterated Venetian blind' construction, outlined in the proof of Theorem 6.9—it may be shown that the 'slats' can be arranged so that they generally miss straight lines in V, but tend to intersect the inverse images under ϕ of straight lines in $\phi(V)$ (see figure 12.2). It follows that the property '$\text{proj}_\theta F$ has zero length for almost all θ' is not invariant under conformal transformations, since it can hold for F but not $\phi(F)$. However, removability is conformally invariant since the function $f(z)$ is analytic on $\phi(V)$ (respectively on $\phi(V\backslash F)$) if and only if $f(\phi(z))$ is analytic on V (respectively on $V\backslash F$). Therefore, the property of having almost all projections of zero length cannot be equivalent to removability.

One of the curious features of this particular argument is that it leaves us none the wiser as to whether sets with almost all projections of zero length must be removable or vice versa. All we can deduce is that both cannot be true.

Very recently, a non-removable set with almost all projections of zero length has been obtained using an iterated construction. The converse is still unresolved.

12.3 Convex surfaces

A continuous function $f:\mathbb{R}^2 \to \mathbb{R}$ is *convex* if

$$f(\lambda x + (1-\lambda)y) \geqslant \lambda f(x) + (1-\lambda) f(y)$$

for all $x, y \in \mathbb{R}^2$ and $0 \leqslant \lambda \leqslant 1$. Geometrically, if $S = \{(x, f(x)) : x \in \mathbb{R}^2\}$ is the surface in $\mathbb{R}^3$ representing the graph of f, then f is convex if the line segment joining any two points of S lies in or below S.

A convex function f need not be particularly smooth—there may be points where f is not differentiable. Dimension may be used to describe the size of the set of such 'singular' points. If f is not differentiable at x then the surface S supports more than one tangent plane at $(x, f(x))$. Notice that if P_1 and P_2 are distinct tangent planes at $(x, f(x))$ then there is a continuum of tangent planes through this point, namely those planes 'between P_1 and P_2' that contain the line $P_1 \cap P_2$.

Theorem 12.3

Let $f:\mathbb{R}^2 \to \mathbb{R}$ be a convex function. Then the set of points at which f is not differentiable is contained in a countable union of rectifiable curves, so in particular has Hausdorff dimension at most 1.

Proof. Without loss of generality, we may assume that the maximum value of f is strictly negative. Let S be the surface given by the graph of f and let $g:\mathbb{R}^2 \to S$ be the 'nearest point' mapping, so that if $x \in \mathbb{R}^2$ then $g(x)$ is that point of S for which the distance $|g(x)-x|$ is least. Convexity of f guarantees that this point is unique. If $x, y \in \mathbb{R}^2$ then the angles of the (possibly skew) quadrilateral $x, g(x), g(y), y$ at $g(x)$ and $g(y)$ must both be at least $\pi/2$; otherwise the segment $[g(x), g(y)]$ will contain a point on or below S that is nearer to x or y. It follows that g is distance decreasing, i.e.

$$|g(x)-g(y)| \leqslant |x-y| \qquad (x, y \in \mathbb{R}^2). \tag{12.3}$$

If f fails to be differentiable at x, then S supports more than one tangent plane at $(x, f(x))$. Thus $g^{-1}(x, f(x))$, the subset of the coordinate plane $\mathbb{R}^2$ mapped to this point by g, is the intersection of $\mathbb{R}^2$ with the normals to the tangent planes to S at $(x, f(x))$ and so contains a straight line segment. Let $\{L_1, L_2, \ldots\}$ be the (countable) collection of line segments in $\mathbb{R}^2$ with endpoints having rational coordinates. If f is not differentiable at x then $g^{-1}(x, f(x))$ contains a segment which must cut at least one of the L_i. Thus if $F = \{(x, f(x)): f \text{ is not differentiable at } x\}$ then $\bigcup_{i=1}^{\infty} g(L_i) \supset F$. Using (12.3) it follows that $g(L_i)$ is either a point or a rectifiable curve with $\mathscr{H}^1(g(L_i)) \leqslant \text{length}(L_i) < \infty$; see (2.9). Then $\bigcup_{i=1}^{\infty} g(L_i)$ is a countable union of rectifiable curves containing F, which in particular has dimension at most 1.

Since $|x-y| \leqslant |(x, f(x)) - (y, f(y))|$ for $x, y \in \mathbb{R}^2$, the set of points x at which

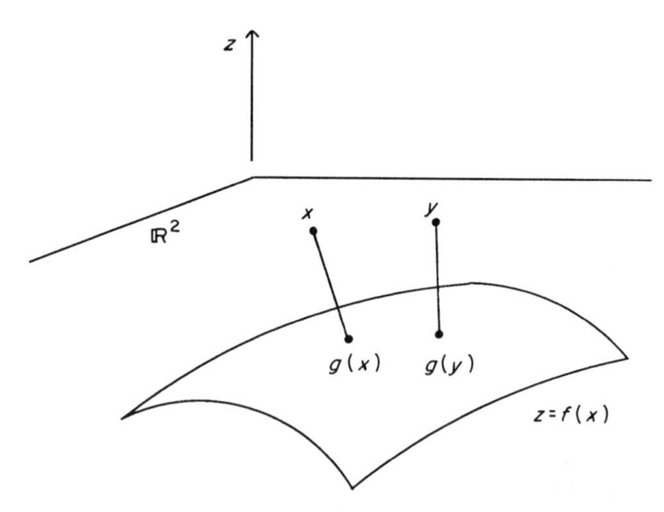

Figure 12.3 The 'nearest point mapping' g from $\mathbb{R}^2$ to a convex surface $z = f(x)$ is distance decreasing

f is non-differentiable is also of dimension at most 1 and is contained in a countable collection of rectifiable curves; again see (2.9). □

Hausdorff dimension has been used in various other ways to quantify the irregularity of surfaces. For example, a convex surface may contain line segments; however, the set of directions of such line segments may be shown to have dimension at most 1.

12.4 Groups and rings of fractional dimension

A subset F of $\mathbb{R}$ is a *subgroup* of the real numbers under the operation of addition if

(i) $0 \in F$,
(ii) $x + y \in F$ whenever $x \in F$ and $y \in F$, and
(iii) $-x \in F$ whenever $x \in F$.

The set F is a *subring* of $\mathbb{R}$ under addition and multiplication if, also,

(iv) $xy \in F$ whenever $x \in F$ and $y \in F$.

There are many simple examples of such structures: the integers, the rationals and the set of numbers $\{r + s\sqrt{2} : r, s \in \mathbb{Z}\}$ are all subrings (and therefore subgroups) of $\mathbb{R}$. These examples are countable sets and therefore have Hausdorff dimension 0. Do there exist subgroups and subrings of $\mathbb{R}$ of Hausdorff dimension s if $0 < s < 1$?

It is relatively easy to modify the earlier Example 4.7 to obtain a subgroup of any given dimension.

Example 12.4

Fix $0 < s < 1$. Let $n_1, n_2, \ldots$ be a rapidly increasing sequence of integers, say with $n_{k+1} \geqslant \max\{n_k^k, 3n_k^{1/s}\}$. For $r = 1, 2, \ldots$ let

$$F_r = \{x \in \mathbb{R} : |x - p/n_k| \leqslant r n_k^{-1/s} \text{ for some integer } p, \text{ for all } k\}$$

and let $F = \bigcup_{r=1}^{\infty} F_r$. Then $\dim_H F = s$, and F is a subgroup of $\mathbb{R}$ under addition.

Calculation. F_r is essentially the set of Example 4.7, so $\dim_H F_r = s$ for all r (it is easy to see that the value of r does not affect the dimension). Taking a countable union, $\dim_H F = s$.

Clearly $0 \in F_1 \subset F$. If $x, y \in F$ then $x, y \in F_r$ for some r, noting that $F_{r'} \subset F_r$ if $r \geqslant r'$. Thus, for each k, there are integers p, q such that

$$|x - p/n_k| \leqslant r n_k^{-1/s} \qquad |y - q/n_k| \leqslant r n_k^{-1/s}.$$

Adding,

$$|x + y - (p + q)/n_k| \leqslant 2r n_k^{-1/s}$$

so $x + y \in F_{2r} \subset F$. Clearly, if $x \in F_r$ then $-x \in F_r$, so F satisfies conditions (i)–(iii) above. □

Subrings are rather harder to analyse. One geometrical approach depends on estimating the dimension of the set of distances realized by a plane set. If E is a subset of $\mathbb{R}^2$, we define the *distance set* of E by

$$D(E) = \{|x - y| : x, y \in E\} \subset \mathbb{R}.$$

Theorem 12.5

Let $E \subset \mathbb{R}^2$ be a Borel set. Then

$$\dim_H D(E) \geqslant \min\{1, \dim_H E - \tfrac{1}{2}\}. \tag{12.4}$$

Note on proof. The potential theoretic proof of this theorem is a little complicated. Fourier transforms and the convolution theorem are used to examine the circles with centres in E that intersect E. It is unlikely that (12.4) is the best inequality possible. □

Assuming this theorem, it is not difficult to show that there are no subrings of dimension s if $\frac{1}{2} < s < 1$.

Theorem 12.6

Let F be a subring of $\mathbb{R}$ under addition and multiplication. Then, if F is a Borel set, it is not possible to have $\frac{1}{2} < \dim_H F < 1$.

Proof. Using (x, y) coordinates in $\mathbb{R}^2$, if (x_1, y_1), $(x_2, y_2) \in F \times F \subset \mathbb{R}^2$, then $|(x_1, y_1) - (x_2, y_2)|^2 = (x_1 - x_2)^2 + (y_1 - y_2)^2 \in F$, since F is a subring. Thus, if $D^2(F \times F)$ denotes the set of squares of distances between points of $F \times F$, we have $D^2(F \times F) \subset F$. Since the mapping $t \to t^2$ preserves Hausdorff dimension (see Exercise 2.5) we have

$$\begin{aligned} \dim_H F \geqslant \dim_H D^2(F \times F) &= \dim_H D(F \times F) \\ &\geqslant \min\{1, \dim_H (F \times F) - \tfrac{1}{2}\} \\ &\geqslant \min\{1, 2\dim_H F - \tfrac{1}{2}\} \end{aligned}$$

using Theorem 12.5 and Product formula 7.2. This inequality is satisfied if and only if $\dim_H F = 1$ or $\dim_H F \leqslant \frac{1}{2}$. □

It is an unsolved problem whether there exist (Borel) subrings of $\mathbb{R}$ of dimension between $\frac{1}{2}$ and 1, though it seems rather unlikely.

12.5 Notes and references

More detailed accounts of the Kakeya problem and its variants are given in Besicovitch (1963), Cunningham (1971, 1974) and Falconer (1985a). The dual approach was introduced by Besicovitch (1964).

For problems related to Vitushkin's conjecture see Harin, Hruščëv and Nikol'skii (1984). The construction outlined is due to Mattila (1986), and the recent counterexample is due to Jones and Murai (1988).

For a general introduction to convex geometry see Eggleston (1958). The result given here is due to Anderson and Klee (1952). For more recent results involving Hausdorff dimension and convexity see Dalla and Larman (1980).

Examples of groups of fractional dimension were given by Erdös and Volkmann (1966). They also raised the question of rings, which was addressed by Falconer (1985c).

Exercises

12.1 Construct a plane set of zero area that contains a line at every perpendicular distance from the origin between 0 and 1. (Hint: consider the image of the set obtained in Theorem 12.2 under the transformation $(x, y) \to (x(1 + y^2)^{1/2}, y)$.)

12.2 By transforming the set obtained in the previous exercise by the mapping given in polar coordinates by $(r, \theta) \to (1/r, \theta)$, show that there exists a plane set of zero area that contains a circle of radius r for all $r > 0$.

12.3 Show that there is a subset of the plane of area 0 that contains a different straight line through every point on the x-axis.

12.4 Let A be a (Borel) subset of $[0, \pi)$. Let F be a subset of the plane that contains a line running in direction θ for every $\theta \in A$. Show that $\dim_H F \geqslant 1 + \dim_H A$.

12.5 Dualize Theorem 6.9 to show that any Borel set of finite area a may be completely covered by a collection of straight lines of total area a.

12.6 Show that if a compact subset F of $\mathbb{C}$ supports a mass distribution μ such that $f(z) = \int_F (z - w)^{-1} \mathrm{d}\mu(w)$ is bounded then F is not removable in the sense of Section 12.2. Show that this is the case if $1 < \dim_H F \leqslant 2$. (Hint: see the proof of Theorem 4.13(*b*).)

12.7 Let $f: \mathbb{R} \to \mathbb{R}$ be a convex function. Show that the set of points at which f is not differentiable is finite or countable.

12.8 Show that any subgroup of $\mathbb{R}$ under addition has box dimension 0 or 1.

Chapter 13 Dynamical systems

Recently, there has been an explosion of interest in dynamical systems. This is largely due to the availability of powerful computers, which has allowed theoretical analysis to proceed alongside numerical investigation. It is also partly because of the advent of 'topological' methods for studying the qualitative behaviour of systems, such methods augmenting the more traditional quantitative approach. The subject is receiving impetus from an increasingly diverse range of applications—dynamical systems are now used to model phenomena in biology, geography and economics as well as in the traditional disciplines of engineering and physics. Volumes have been written on dynamical systems and chaos. We make no attempt to provide a general account, which would require excursions into ergodic theory, bifurcation theory and many other areas, but we illustrate various ways in which fractals can occur in dynamical systems.

Let D be a subset of $\mathbb{R}^n$ (often $\mathbb{R}^n$ itself), and let $f: D \to D$ be a continuous mapping. As usual, f^k denotes the kth iterate of f, so that $f^0(x) = x$, $f^1(x) = f(x)$, $f^2(x) = f(f(x))$, etc.; note that $f^k(x)$ is in D for all k if x is a point of D. Typically, $x, f(x), f^2(x), \ldots$ are the values of some quantity at times $0, 1, 2, \ldots$. Thus the value at time $k+1$ is given in terms of the value at k by the function f. Examples include biological populations, the value of an investment subject to certain interest and tax conditions, and the position of a fluid particle in a steady flow.

An iterative scheme $\{f^k\}$ is called a *discrete dynamical system*. We are interested in the behaviour of the sequence of iterates, or *orbits*, $\{f^k(x)\}_{k=1}^{\infty}$ for various initial points $x \in D$, particularly for large k. For example, if $f(x) = \cos x$, the sequence $f^k(x)$ converges to 0.739 as $k \to \infty$ for any initial x: try pressing the cosine button on a calculator repeatedly and see! Sometimes the distribution of iterates appears almost random. Alternatively, $f^k(x)$ may converge to a *fixed point* w, i.e. a point of D with $f(w) = w$. More generally, $f^k(x)$ may converge to an orbit of *period-p points* $\{w, f(w), \ldots, f^{p-1}(w)\}$, where p is the least positive integer with $f^p(w) = w$, in the sense that $|f^k(x) - f^i(w)| \to 0$ as $k \to \infty$. Sometimes, however, $f^k(x)$ may appear to move about at random, but always remaining close to a certain set, which may be a fractal. In this chapter we examine several ways in which such 'fractal attractors' or 'strange attractors' can occur.

Roughly speaking, an attractor is a set to which all nearby orbits converge. However, as frequently happens in dynamical systems theory, the precise definition varies between authors. We shall call a subset F of D an *attractor* for f if F is a closed set that a *invariant* under f (i.e. $f(F)=F$) such that the distance from $f^k(x)$ to F converges to zero as k tends to infinity for all x in an open set V containing F. The set V is called the *basin of attraction* of F. It is usual to require that F is minimal in the sense that it has no proper subset satisfying these conditions. Similarly, a closed invariant set F from which all nearby points not in F are iterated away is called a *repeller*; this is roughly equivalent to F being an 'attractor' for the (perhaps multivalued) inverse f^{-1}. An attractor or repeller may just be a single point or a period-p orbit. However, even relatively simple maps f can have fractal attractors.

Note that the set $F=\bigcap_{k=1}^{\infty} f^k(D)$ is invariant under f. Since $f^k(x)\in\bigcap_{i=1}^{k} f^i(D)$ for any $x\in D$, the iterates $f^k(x)$ approach F as $k\to\infty$, and F is often an attractor of f.

Very often, if f has a fractal attractor or repeller F, then f exhibits 'chaotic' behaviour on F. There are various definitions of chaos; f would certainly be regarded as chaotic on F if the following are all true.

(i) The orbit $\{f^k(x)\}$ is dense in F for some $x\in F$.
(ii) The periodic points of f in F (points for which $f^p(x)=x$ for some positive integer p) are dense in F.
(iii) f has *sensitive dependence on initial conditions*; that is, there is a number $\delta>0$ such that for any x in F there are points y in F arbitrarily close to x such that $|f^k(x)-f^k(y)|\geqslant\delta$ for some k. Thus points that are initially close to each other do not remain close under iterates of f.

Condition (i) implies that F cannot be decomposed into smaller closed invariant sets, (ii) suggests a skeleton of regularity in the structure of F, and (iii) reflects the unpredictability of iterates of points on F. In particular, (iii) implies that accurate long-term numerical approximation to orbits of f is impossible. Often conditions that give rise to fractal attractors also lead to chaotic behaviour.

Dynamical systems are naturally suited to computer investigation. Roughly speaking, attractors are the sets that are seen when orbits are plotted on a computer. For some initial point x one plots the iterates $f^k(x)$ for $k\geqslant 100$, say, on the assumption that they are indistinguishable from any attractor. If an attractor appears fractal, a 'box-counting' method can be used to estimate its dimension. However, computer pictures can be misleading, in that the distribution of $f^k(x)$ across an attractor can be very uneven, with certain parts of the attractor visited very rarely.

13.1 Repellers and iterated function schemes

Under certain circumstances, a repeller in a dynamical system concides with the invariant set of a related iterated function scheme. This is best seen by an example. The mapping $f:\mathbb{R}\to\mathbb{R}$ given by

$$f(x)=\tfrac{3}{2}(1-|2x-1|)$$

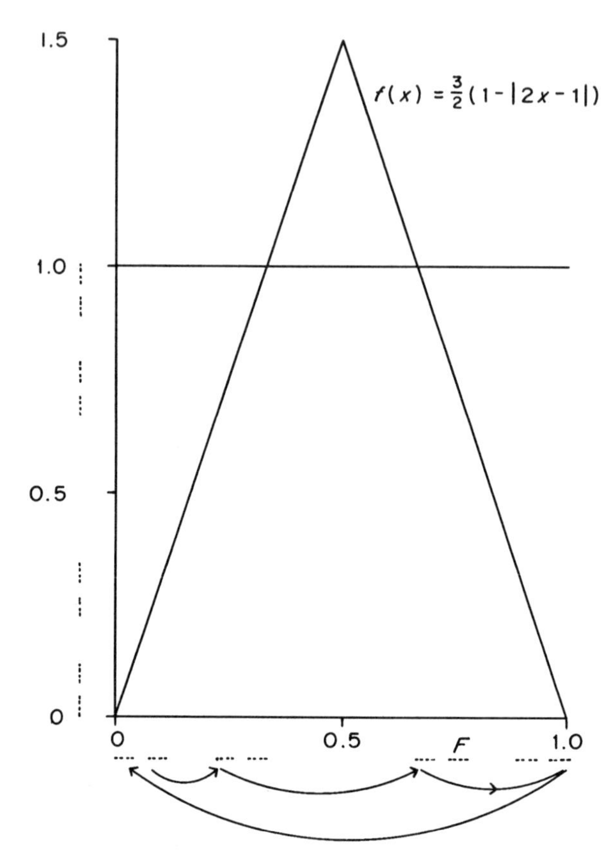

Figure 13.1 The tent map f. Notice that the middle third Cantor set F is mapped onto itself by f and is an invariant repeller. Notice, also, the chaotic nature of f on F: the iterates of a point are indicated by the arrows

is called the *tent map* because of the form of its graph; see figure 13.1. Clearly, f maps $\mathbb{R}$ in a two-to-one manner onto $(-\infty, \frac{3}{2})$. Defining $S_1, S_2: [0,1] \to [0,1]$ by

$$S_1(x) = \tfrac{1}{3}x \qquad S_2(x) = 1 - \tfrac{1}{3}x$$

we see that

$$f(S_1(x)) = f(S_2(x)) = x \qquad (0 \leqslant x \leqslant 1).$$

Thus S_1 and S_2 are the two branches of f^{-1}. Since S_1 and S_2 are contractions on $[0, 1]$ Theorem 9.1 implies that there is a unique compact invariant set F satisfying

$$F = S_1(F) \cup S_2(F) \tag{13.1}$$

which is given by $F = \bigcap_{k=0}^{\infty} S^k([0,1])$ (writing $S(E) = S_1(E) \cup S_2(E)$ for any set

E). It is easy to see that F is the middle third Cantor set of Hausdorff and box dimensions $\log 2/\log 3$.

It follows from (13.1) that $f(F) = F$. To see that f is a repeller, observe that if $x < 0$ then $f(x) < 3x$, so $f^k(x) \to -\infty$ as $k \to \infty$. If $x > 1$ then $f(x) < 0$ and again $f^k(x) \to -\infty$. If $x \in [0,1] \backslash F$ then for some k, we have $x \notin \cup \{S_{i_1} \circ \cdots \circ S_{i_k}[0,1] : i_j = 1,2\}$ so $f^k(x) \notin [0,1]$, and again $f^k(x) \to -\infty$ as $k \to \infty$. All points outside F are iterated to $-\infty$ so F is a repeller.

With the notation of Section 9.1, the chaotic nature of f on F is readily apparent. Denoting the points of F by $x_{i_1, i_2, \ldots}$ with $i_j = 1, 2$, as in (9.7), $|x_{i_1, i_2, \ldots} - x_{i'_1, i'_2, \ldots}| \leq 3^{-k}$ if $i_1 = i'_1, \ldots, i_k = i'_k$. Since $x_{i_1, i_2, \ldots} = S_{i_1}(x_{i_2, i_3, \ldots})$, it follows that $f(x_{i_1, i_2, \ldots}) = x_{i_2, i_3, \ldots}$. Suppose that $(i_1, i_2, \ldots)$ is an infinite sequence with every finite sequence of 0s and 1s appearing as a consecutive block of terms; for example,

$$(0, 1, 0{,}0, 0{,}1, 1{,}0, 1{,}1, 0{,}0{,}0, 0{,}0{,}1, \ldots)$$

where the spacing is just to indicate the form of the sequence. Then, for any point $x_{i'_1, i'_2, \ldots}$ in F and any integer q, we may find k such that $(i'_1, i'_2, \ldots, i'_q) = (i_{k+1}, \ldots, i_{k+q})$. Thus $f^k(x_{i_1, i_2, \ldots}) = x_{i_{k+1}, i_{k+2}, \ldots}$ comes arbitrarily close to each point of F for suitable large k, so that f has dense orbits in F. Moreover, since $x_{i_1, \ldots, i_k, i_1, \ldots, i_k, i_1, \ldots}$ is a periodic point of period k, the periodic points of f are also dense in F. The iterates have sensitive dependence on initial conditions, since $f^k(x_{i_1, \ldots, i_k, 1, \ldots}) \in [0, \frac{1}{3}]$ but $f^k(x_{i_1, \ldots, i_k, 2, \ldots}) \in [\frac{2}{3}, 1]$. We conclude that F is a chaotic repeller for f. (The study of f by its effect on points of F represented by sequences $(i_1, i_2, \ldots)$ is known as *symbolic dynamics*.)

In exactly the same way, the invariant sets of other iterated function scheme correspond to repellers of functions. If $S_1, \ldots, S_m$ is a set of bijective contractions on a domain D with invariant set F such that $S_1(F), \ldots, S_m(F)$ are disjoint, then F is a repeller for any mapping f such that $f(x) = S_i^{-1}(x)$ when x is near $S_i(F)$. Again, by examining the effect of f on the point $x_{i_1, i_2, \ldots}$, it may be shown that f acts chaotically on F.

13.2 The logistic map

The logistic map $f : \mathbb{R} \to \mathbb{R}$ is given by

$$f_\lambda(x) = \lambda x(1 - x) \tag{13.2}$$

where λ is a positive constant. This mapping was introduced to model the population development of certain species—if the population is x at the end of any year, it is assumed to be $f_\lambda(x)$ at the end of the following year. Nowadays the logistic map is studied intensively as an archetypal 1-dimensional dynamical system. We content ourselves here with an analysis when λ is large, and a brief discussion when λ is small.

For a given $\lambda > 2 + \sqrt{5}$ write $a = \frac{1}{2} - \sqrt{\frac{1}{4} - 1/\lambda}$ and $1 - a = \frac{1}{2} + \sqrt{\frac{1}{4} - 1/\lambda}$ for the roots of $f(x)=1$. Each of the intervals $[0,a]$ and $[1-a,1]$ is mapped bijectively onto $[0,1]$ by f_λ. The mappings $S_1:[0,1]\to[0,a]$ and $S_2:[0,1]\to[1-a,1]$ given by

$$S_1(x)=\tfrac{1}{2}-\sqrt{\tfrac{1}{4}-x/\lambda} \qquad\qquad S_2(x)=\tfrac{1}{2}+\sqrt{\tfrac{1}{4}-x/\lambda}$$

are the restrictions of the inverse f_λ^{-1} to $[0,a]$ and $[1-a,1]$ with $f_\lambda(S_1(x))=f_\lambda(S_2(x))=x$ for each x in $[0,1]$. For $i=1,2$ we have

$$|S_i'(x)|=\frac{1}{2\lambda}(\tfrac{1}{4}-x/\lambda)^{-1/2}$$

so

$$\frac{1}{\lambda}\leqslant|S_i'(x)|\leqslant\frac{1}{2\lambda}(\tfrac{1}{4}-1/\lambda)^{-1/2}=\tfrac{1}{2}(\lambda^2/4-\lambda)^{-1/2}$$

if $0\leqslant x\leqslant 1$. By the mean-value theorem

$$\frac{1}{\lambda}|x-y|\leqslant|S_i(x)-S_i(y)|\leqslant\tfrac{1}{2}(\lambda^2/4-\lambda)^{-1/2}|x-y| \qquad (0\leqslant x,y\leqslant 1). \quad (13.3)$$

Thus if $\lambda>2+\sqrt{5}$ the mappings S_1 and S_2 are contractions on $[0,1]$, so by Theorem 9.1 there is a unique non-empty compact subset F of $[0,1]$ satisfying

$$F=S_1(F)\cup S_2(F),$$

and it follows that $f_\lambda(F)=F$. Since this union is disjoint, F is totally disconnected. In exactly the same way as for the tent map, F is a repeller, and f is chaotic on F.

To estimate the dimension of F we proceed as in Example 9.8. Using Theorem 9.6 and 9.7, it follows from (13.3) that

$$\frac{\log 2}{\log\lambda}\leqslant\dim_{\mathrm{H}}F\leqslant\underline{\dim}_{\mathrm{B}}F\leqslant\overline{\dim}_{\mathrm{B}}F\leqslant\frac{\log 2}{\log(\lambda(1-4/\lambda)^{1/2})}.$$

Thus, if λ is large, the dimension of F is close to $\log 2/\log\lambda$.

For smaller values of λ, the dynamics of the logistic map (13.2) are subtle. If $0<\lambda\leqslant 4$, the function f_λ maps $[0,1]$ into itself, and we can restrict attention to the interval $[0,1]$. If x is a period-p point of f, i.e. $f^p(x)=x$ and p is the least positive integer with this property, we say that x is *stable* or *unstable* according to whether $|(f^p)'(x)|<1$ or >1. Stable periodic points attract nearby orbits, unstable periodic points repel them. If $0<\lambda\leqslant 1$, then f_λ has a fixed point at 0 which is attractive, in the sense that $f_\lambda^k(x)\to 0$ for all $x\in[0,1]$. For $1<\lambda<3$, the function f_λ has an unstable fixed point 0, and a stable fixed point $1-1/\lambda$, so $f_\lambda^k(x)\to 1-1/\lambda$ for all $x\in(0,1)$. As λ increases through the value $\lambda_1=3$, the fixed point at $1-1/\lambda$ becomes unstable, splitting into a stable orbit of period 2 to which all but countably many points of $(0,1)$ are attracted (see figure 13.2). When λ reaches $\lambda_2=1+\sqrt{6}$, the period-2 orbit becomes unstable and is replaced by a stable period-4 orbit. As λ is increased further, this period doubling continues with a stable orbit of period 2^q appearing at $\lambda=\lambda_q$; this

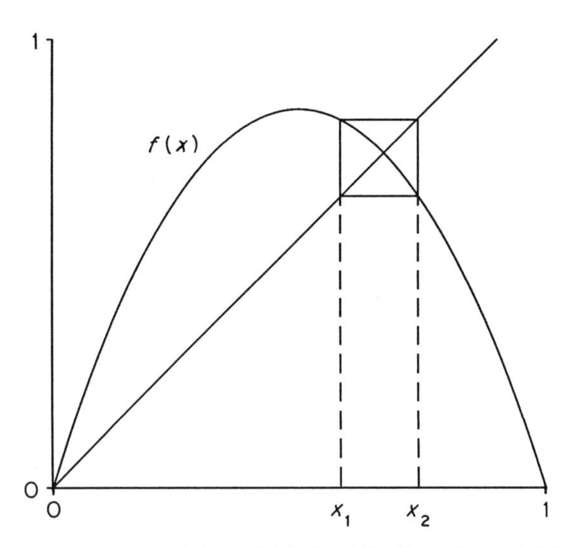

Figure 13.2 The logistic map $f(x) = \lambda x(1-x)$ for $\lambda = 3.38$. Note the period-2 orbit x_1, x_2 with $f(x_1) = x_2$ and $f(x_2) = x_1$

orbit attracts all but countably many initial points in (0, 1). One of the surprising features of this process is that the period doubling occurs more and more frequently as λ increases, and $q \to \infty$ as $\lambda \to \lambda_\infty$ where $\lambda_\infty \simeq 3.570$. As λ approaches λ_∞, the repeated splitting of stable orbits of period 2^q into nearby stable orbits of period 2^{q+1} provides a sequence of attracting orbits which approximate to a Cantor set; see figure 13.3. When $\lambda = \lambda_\infty$ the attractor F actually *is* a set of Cantor type. Then F is invariant under f_{λ_∞} with all except a countable number of points of $[0, 1]$ approaching F under iteration by f_{λ_∞} (the exceptional points are those that iterate onto the unstable periodic orbits). The effect of f_{λ_∞} on F can be analysed by extrapolating from the periodic orbits of f_{λ_q} when q is large. There are dense orbits but no sensitive dependence on initial conditions. It is possible to show that F is invariant in the sense of (9.1) under a pair of contractions, and, using the method of Example 9.8, the Hausdorff dimension may be estimated as 0.538.... A complete analysis of the structure of this fractal attractor is beyond the scope of this book.

For $\lambda_\infty < \lambda < 4$ several types of behaviour occur. There is a set K such that if $\lambda \in K$ then f_λ has a truly chaotic attractor of positive length. Moreover, K itself has positive Lebesgue measure. However, in the gaps or 'windows' of K, period doubling again occurs. For example, when $\lambda \simeq 3.83$ there is a stable period-3 orbit; as λ increases it splits first into a stable period-6 orbit, then into a stable period-12 orbit, and so on. When λ reaches about 3.855 the 'limit' of these stable orbits gives a Cantor-like attractor. Similarly there are other windows where period doubling commences with a 5-cycle, a 7-cycle and so on.

One of the most fascinating features of this subject is that the behaviour of the logistic map as λ increases is qualitatively the same as that of *any* family of transformations of an interval $f_\lambda(x) = \lambda f(x)$, provided that f is unimodal (i.e.

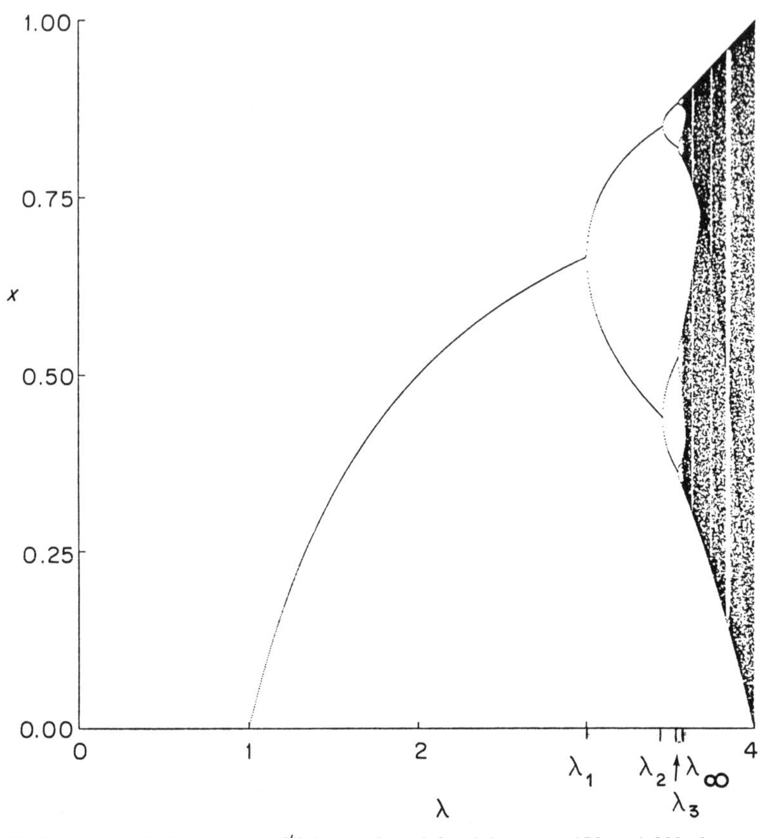

Figure 13.3 For each λ the iterates $f^k(x)$ are plotted for k between 150 and 300, for a suitable initial x. The intersection of the plot with vertical lines shows the periodic attractors for $\lambda < \lambda_\infty$. As λ approaches λ_∞, repeated splitting of the periodic orbits results in an attractor of Cantor-set form at $\lambda = \lambda_\infty$

has a single maximum). Although the values $\lambda_1, \lambda_2, \ldots$ at which period doubling occurs depend on f, the rate at which these values approach λ_∞ is universal, i.e. $\lambda_\infty - \lambda_k \simeq c\delta^{-k}$ where $\delta = 4.6692\ldots$ is the Feigenbaum constant and c depends on f. Moreover, the Hausdorff dimension of the fractal attractor of f_{λ_∞} is $0.538\ldots$, this same value occurring for any differentiable and unimodal f.

Mappings which have been used to model biological populations and which exhibit similar features include the following:

$$f_\lambda(x) = \lambda \sin \pi x$$
$$f_\lambda(x) = x \exp \lambda(1 - x)$$
$$f_\lambda(x) = x(1 + \lambda(1 - x))$$
$$f_\lambda(x) = \lambda x/(1 + ax)^5.$$

13.3 Stretching and folding transformations

One of the simplest planar dynamical systems with a fractal attractor is the 'baker's' transformation, so-called because it resembles the process of repeatedly stretching a piece of dough and folding it in two. Let $E = [0, 1] \times [0, 1]$ be the unit square. For fixed $0 < \lambda < \frac{1}{2}$ we define the *baker's transformation* $f: E \to E$ by

$$f(x, y) = \begin{cases} (2x, \lambda y) & (0 \leqslant x \leqslant \frac{1}{2}) \\ (2x - 1, \lambda y + \frac{1}{2}) & (\frac{1}{2} < x \leqslant 1). \end{cases} \tag{13.4}$$

This transformation may be thought of as stretching E into a $2 \times \lambda$ rectangle, cutting it into two $1 \times \lambda$ rectangles and placing these above each other with a gap of $\frac{1}{2} - \lambda$ in between; see figure 13.4. Then $E_k = f^k(E)$ is a decreasing sequence of sets, with E_k comprising 2^k horizontal strips of height λ^{-k} separated by gaps of at least $(\frac{1}{2} - \lambda)\lambda^{1-k}$. Since $f(E_k) = E_{k+1}$, the compact limit set $F = \bigcap_{k=0}^{\infty} E_k$ satisfies $f(F) = F$. If $(x, y) \in E$ then $f^k(x, y) \in E_k$, so $f^k(x, y)$ lies within distance λ^{-k} of F. Thus all points of E are attracted to F under iteration by f.

If the initial point (x, y) has $x = \cdot a_1 a_2 \ldots$ in base 2, then it is easily checked that

$$f^k(x, y) = (\cdot a_{k+1} a_{k+2} \cdots, y_k)$$

where y_k is some point in the strip of E_k numbered $a_k a_{k-1} \ldots a_1$ (base 2) counting from the bottom. Thus when k is large the position of $f^k(x, y)$ depends largely on the base-2 digits a_i of x with i close to k. By choosing an x with base 2 expansion containing all finite sequences, we can arrange for $f^k(x, y)$ to be dense in F for certain initial (x, y), just as in the case of the tent map.

Further analysis along these lines shows that f has sensitive dependence on

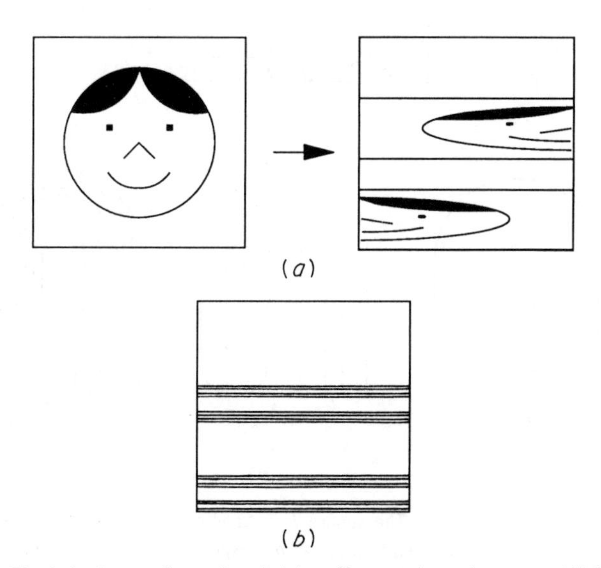

Figure 13.4 The baker's transformation: (*a*) its effect on the unit square; (*b*) its attractor

initial conditions, and that the periodic points of f are dense in F, so that F is a chaotic attractor for f. Certainly F is a fractal—it is the product $[0, 1] \times F_1$, where F_1 is a uniform Cantor set obtained by repeatedly replacing intervals I by a pair of subintervals of lengths $\lambda|I|$. Example 4.5 gives $\dim_H F_1 = \dim_B F_1 = \log 2/-\log\lambda$, so $\dim_H F = 1 + \log 2/-\log\lambda$, using Corollary 7.6.

The baker's transformation is rather artificial, being piecewise linear and discontinuous. However, it does serve to illustrate how the 'stretching and cutting' procedure results in a fractal attractor.

The closely related process of 'stretching and folding' *can* occur for continuous functions on plane regions. Let $E = [0, 1] \times [0, 1]$ and suppose that f maps E in a one-to-one manner onto a horseshoe-shaped region $f(E)$ contained in E. Then f may be thought of as stretching E into a long thin rectangle which is then bent in the middle. This figure is repeatedly stretched and bent by f so that $f^k(E)$ consists of an increasing number of side-by-side strips; see figure 13.5. We have $E \supset f(E) \supset f^2(E) \supset \cdots$, and the compact set $F = \bigcap_{k=1}^{\infty} f^k(E)$ attracts all points of E. Locally, F looks like the product of a Cantor set and an interval.

A variation on this construction gives a transformation with rather different characteristics; see figure 13.6. If D is a plane domain containing the unit square

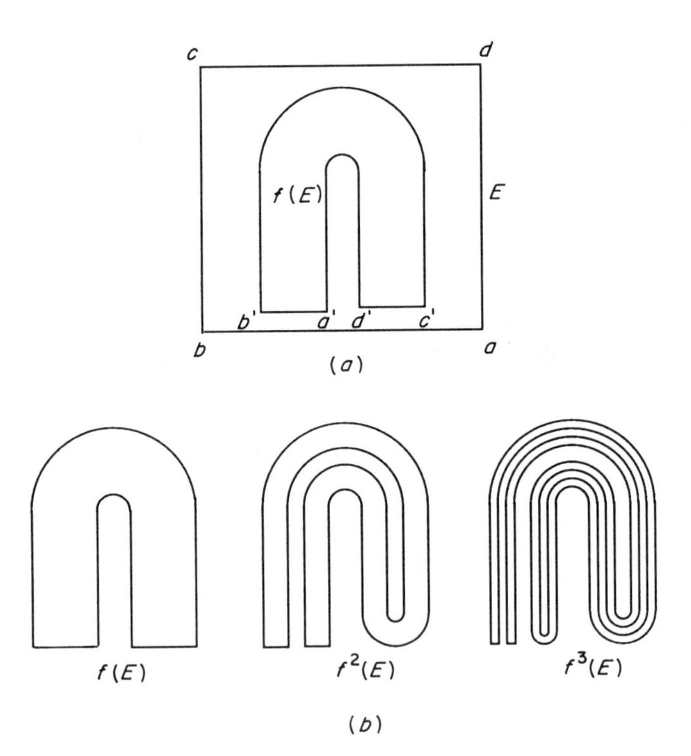

Figure 13.5 A horseshoe map. (*a*) The square E is transformed, by stretching and bending, to the horseshoe $f(E)$, with a, b, c, d mapped to a', b', c', d', respectively. (*b*) The iterates of E under f form a set that is locally a product of a line segment and a Cantor set

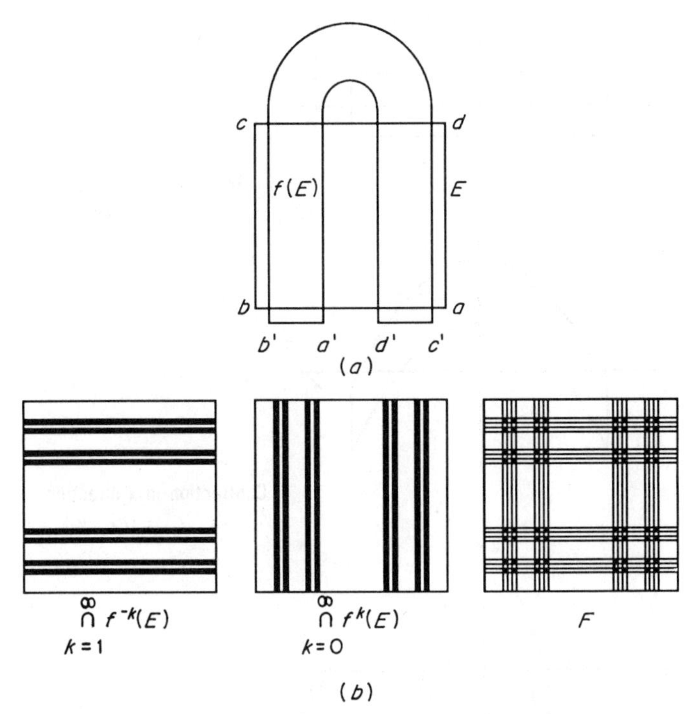

Figure 13.6 An alternative horseshoe map. (*a*) The square E is transformed so that the 'arch' and 'ends' of $f(E)$ lie outside E. (*b*) The sets $\bigcap_{k=1}^{\infty} f^{-k}(E)$ and $\bigcap_{k=0}^{\infty} f^{k}(E)$ are both products of a Cantor set and a unit interval. Their intersection F is an unstable invariant set for f

E and $f:D \to D$ is such that $f(E)$ is a horseshoe with 'ends' and 'arch' lying in a part of D outside E that is never iterated back into E, then almost all points of the square E (in the sense of plane measure) are eventually iterated outside E by f. If $f^k(x, y) \in E$ for all positive k, then $(x, y) \in \bigcap_{k=1}^{\infty} f^{-k}(E)$. With f suitably defined, $f^{-1}(E)$ consists of two horizontal bars across E, so $\bigcap_{k-1}^{\infty} f^{-k}(E)$ is the product of $[0, 1]$ and a Cantor set. The set $F = \bigcap_{k=-\infty}^{\infty} f^k(E) = \bigcap_{k=0}^{\infty} f^k(E) \cap \bigcap_{k=1}^{\infty} f^{-k}(E)$ is compact and invariant for f, and is the product of two Cantor sets. However, F is not an attractor, since points arbitrarily close to F are iterated outside E.

A specific example of a 'stretching and folding' transformation is the Hénon map $f:\mathbb{R}^2 \to \mathbb{R}^2$

$$f(x, y) = (y + 1 - ax^2, bx) \tag{13.5}$$

where a and b are constants. (The values $a = 1.4$ and $b = 0.3$ are usually chosen for study. For these values there is a quadrilateral D for which $f(D) \subset D$ to which we can restrict attention.) This mapping has Jacobian $-b$ for all (x, y), so it contracts area at a constant rate throughout $\mathbb{R}^2$; to within a linear change of coordinates, (13.5) is the most general quadratic mapping with this property.

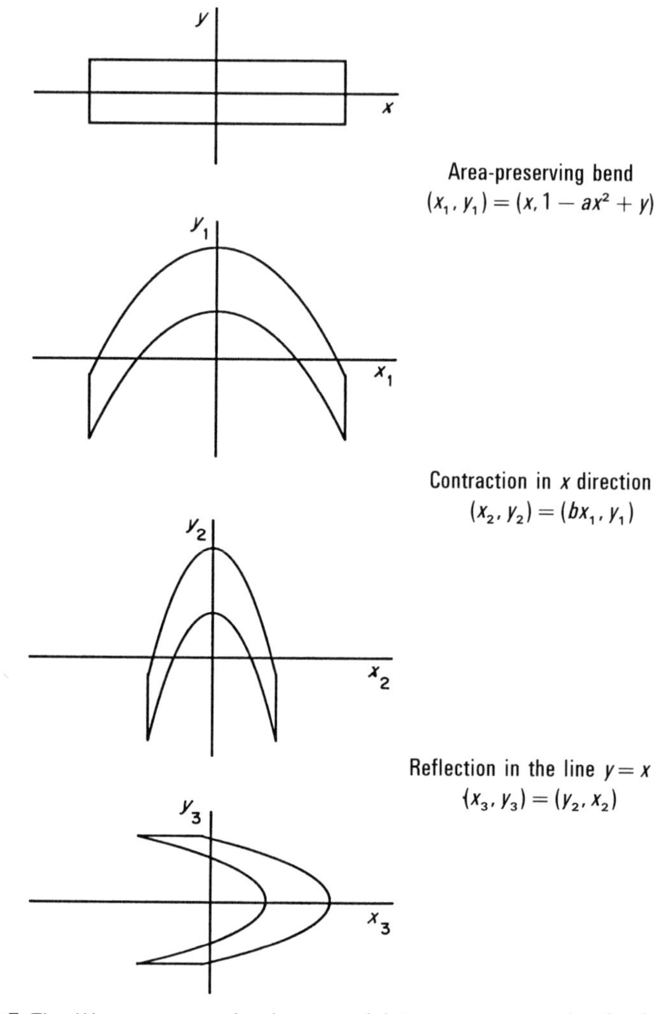

Figure 13.7 The Hénon map may be decomposed into an area-preserving bend, followed by a contraction, followed by a reflection in the line $y = x$. The diagrams show the effect of these successive transformations on a rectangle

The transformation (13.5) may be decomposed into an (area-preserving) bend, a contraction, and a reflection, the net effect being 'horseshoe-like'; see figure 13.7. This leads us to expect f to have a fractal attractor, and this is borne out by computer pictures. Detailed pictures show banding indicative of a set that is locally the product of a line segment and a Cantor-like set. Numerical estimates suggest that the attractor has box dimension of about 1.26 when $a = 1.4$ and $b = 0.3$.

Precise analysis of the Hénon map is quite difficult, and its dynamics are still

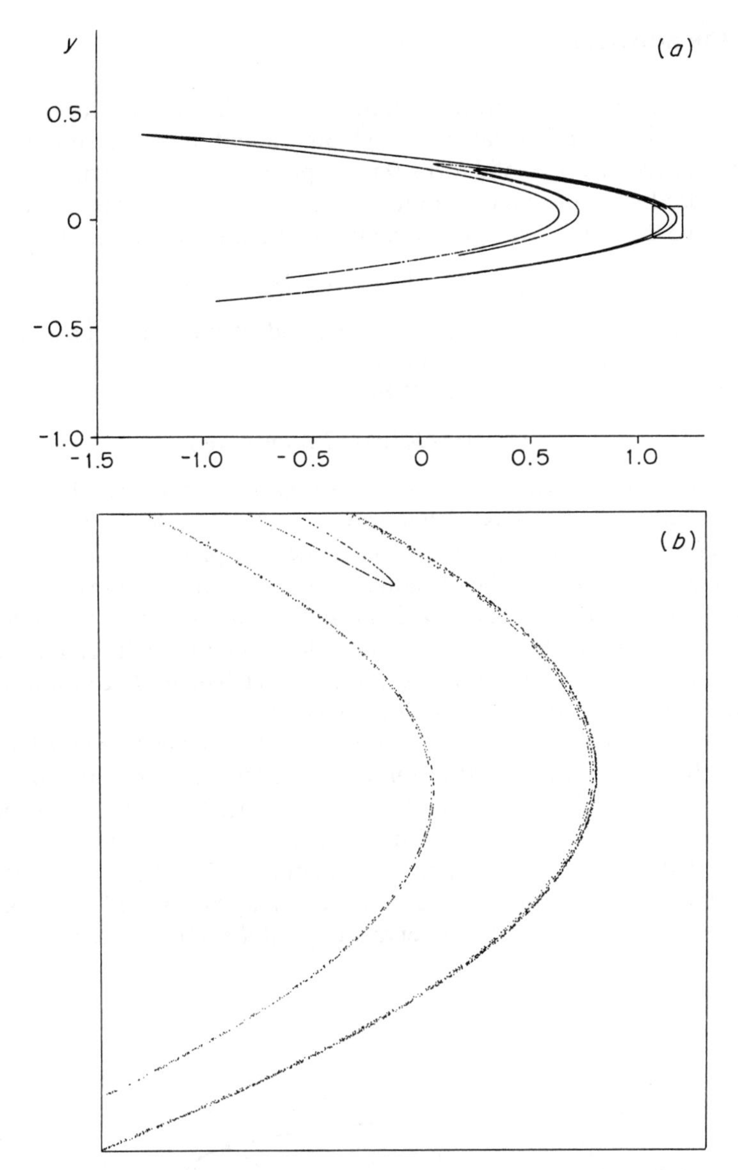

Figure 13.8 Iterates of a point under the Hénon map (13.5) showing the form of the attractor. In (*b*), a magnification of the square in (*a*), banding is becoming apparent

not fully understood. In particular the qualitative changes in behaviour (bifurcations) that occur as a and b vary are highly complex.

Many other types of 'stretching and folding' are possible. Transformations can fold several times or even be many-to-one; for example the ends of a horseshoe might cross. Such transformations often have fractal attractors, but their analysis tends to be difficult.

13.4 The solenoid

Our next example is of a transformation of a 3-dimensional region—a solid torus. If a unit disc B is rotated through 360° about an axis L in the plane of, but not intersecting, B, a solid torus D is swept out. The torus D may be thought of as the product of the circle C, of radius $r > 1$, obtained by rotating the centre of B around L, and B. This gives a convenient parametrization of D as

$$\{(\phi, w) \in C \times B : 0 \leqslant \phi < 2\pi, |w| \leqslant 1\}$$

where the angle ϕ specifies a point on C, and where w is a position vector relative to the centre of B; see figure 13.9.

Fix $0 < a < \frac{1}{4}$ and define $f: D \to D$ by

$$f(\phi, w) = (2\phi (\mathrm{mod}\, 2\pi), aw + \tfrac{1}{2}\hat{\phi}) \tag{13.6}$$

where $\hat{\phi}$ is the unit vector in B at angle ϕ to the outwards axis. Then f maps D onto a solid tube of radius a that traverses D twice. Note that (ϕ, w) and $(\phi + \pi, w)$ are mapped to points in the same half-plane bounded by L. The second iterate $f^2(D)$ is a tube of radius a^2 going round $f(D)$ twice, so around D four times; $f^3(D)$ traverses D eight times, and so on. The intersection $F = \bigcap_{k=1}^{\infty} f^k(D)$ is highly fibrous—locally it looks like a bundle of line segments that cut any cross section of D in a Cantor-like set. The set F, called a *solenoid*, is invariant under f, and attracts all points of D.

We may find the dimension of F by routine methods. Let P_ϕ be the half-plane bounded by L and cutting C at ϕ. Observe that $f^k(C)$ is a smooth closed curve traversing the torus 2^k times, with total length at most $2^k c$ where c is independent of k ($f^k(C)$ cannot oscillate too wildly—the angle between every curve $f^k(C)$ and every half-plane P_ϕ has a positive lower bound). The set $f^k(D)$ is a 'fattening' of the curve $f^k(C)$ to a tube of radius a^k, so it may be covered by a collection of balls of radius $2a^k$ spaced at intervals of a^k along $f^k(C)$. Clearly $2 \times 2^k c a^{-k}$

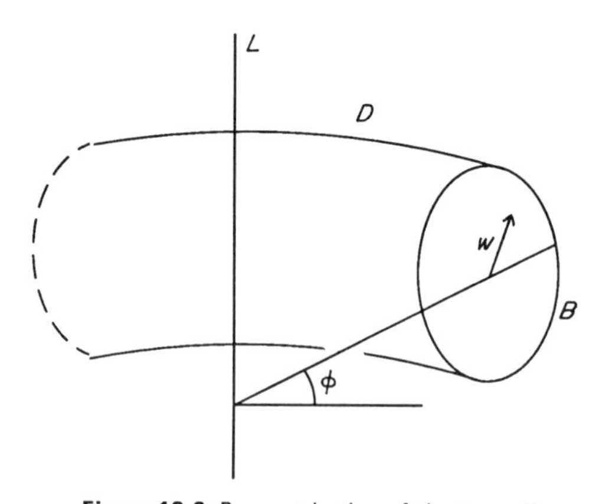

Figure 13.9 Parametrization of the torus D

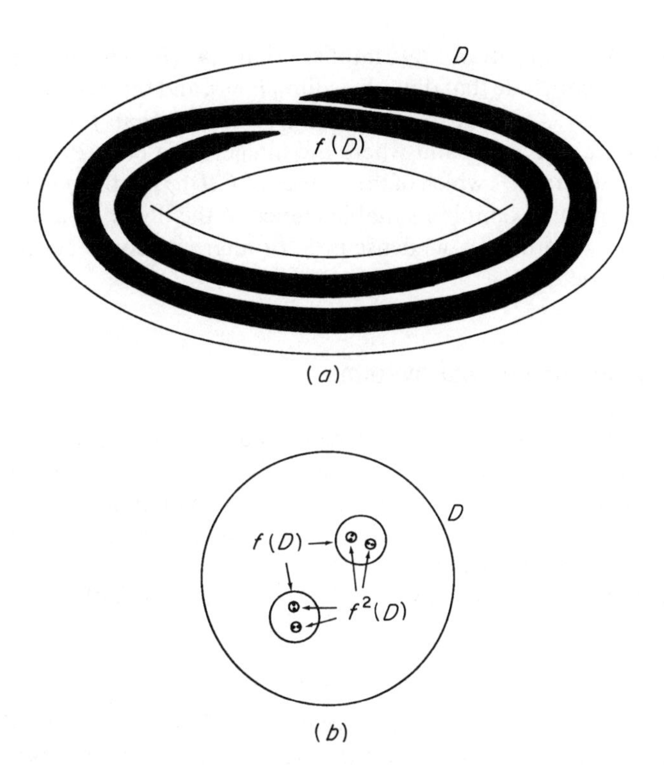

Figure 13.10 The solenoid. (*a*) The torus *D* and its image under *f*. (*b*) A plane section through *D* intersects *F* in a Cantor-like set

balls will suffice, so in the usual way we get $\dim_H F \leqslant \overline{\dim}_B F \leqslant s$ and $\mathscr{H}^s(F) < \infty$ for $s = 1 + \log 2/-\log a$.

To get a lower estimate for the dimension, we examine the sections $F \cap P_\phi$ for each ϕ. The set $f(D) \cap P_\phi$ consists of two discs of radius a situated diametrically opposite each other with centres $\frac{1}{2}$ apart inside $D \cap P_\phi$. Each of these discs contains two discs of $f^2(D) \cap P_\phi$ of radius a^2 and with centres $\frac{1}{2}a$ apart, and so on. We may place a mass distribution μ on $F \cap P_\phi$ in such a way that each of the 2^k discs of $f^k(D) \cap P_\phi$ has mass 2^{-k}. If $U \subset P_\phi$ satisfies

$$a^k(\tfrac{1}{2} - 2a) \leqslant |U| < a^{k-1}(\tfrac{1}{2} - 2a)$$

then U intersects at most one the discs of $f^k(D) \cap P_\phi$, so

$$\mu(U) \leqslant 2^{-k} = a^{k(\log 2/-\log a)} \leqslant c_1 |U|^{\log 2/-\log a}$$

where c_1 is independent of $|U|$. It follows from the Mass distribution principle 4.2 that

$$\mathscr{H}^{\log 2/-\log a}(F \cap P_\phi) \geqslant c_1^{-1}.$$

Since F is built up from sections $F \cap P_\phi$ $(0 \leqslant \phi < 2\pi)$, a higher-dimensional modification of Proposition 7.9 implies that $\mathscr{H}^s(F) > 0$, where $s = 1 + \log 2 / -\log a$. We conclude that $\dim_{\mathrm{H}} F = \dim_{\mathrm{B}} F = s$, and that $0 < \mathscr{H}^s(F) < \infty$.

If $\phi/2\pi = 0{\cdot}a_1 a_2 \ldots$ to base 2, it follows from (13.6) that $f^k(\phi, w) = (\phi_k, v_k)$ where $\phi_k/2\pi = 0{\cdot}a_{k+1}a_{k+2}\ldots$ and where the integer with base-2 representation $a_k a_{k-1} \cdots a_{k-d+1}$ determines which of the 2^d discs of $f^d(D) \cap P_{\phi_k}$ the point v_k belongs to. Just as in previous examples, suitable choice of the digits $a_1, a_2, \ldots$ leads to initial points (ϕ, w) with $f^k(\phi, w)$ dense in F, or alternatively to periodic orbits, so that f is chaotic on F.

13.5 Continuous dynamical systems

A discrete dynamical system may be thought of as a formula relating the value of a quantity at successive discrete time intervals. If the time interval is allowed to tend to 0, then the formula becomes a differential equation in the usual way. Thus it is natural to regard an autonomous (time-independent) differential equation as a continuous dynamical system.

Let D be a domain in $\mathbb{R}^n$ and let $f: D \to \mathbb{R}^n$ be a smooth function. The differential equation

$$\dot{x}(t) = \mathrm{d}x/\mathrm{d}t = f(x) \tag{13.7}$$

has a family of *solution curves* or *trajectories* which fill D. If an initial point $x(0)$ is given, the solution $x(t)$ remains on the unique trajectory that passes through $x(0)$ for all time t; the behaviour of $x(t)$ as $t \to \pm\infty$ may be found by following the trajectory. Given reasonable conditions on f, no two trajctories cross; otherwise the equations (13.7) would not determine the motion of x. Moreover, the trajectories vary smoothly across D except at points where $\dot{x}(t) = f(x) = 0$ and the trajectories are single points.

As in the discrete case, continuous dynamical systems give rise to attractors and repellers. A closed subset F of D might be termed an attractor with basin of attraction V containing F if, for all initial points $x(0)$ in the open set V, the trajectory $x(t)$ through $x(0)$ approaches F as t tends to infinity. Of course, we require F to be invariant, so that if $x(0)$ is a point of F then $x(t)$ is in F for $-\infty < t < \infty$ implying that F is a union of trajectories. We also require F to be minimal, in the sense that there is some point $x(0)$ such that $x(t)$ is dense in F.

When D is a plane domain, the range of attractors for continuous systems is rather limited. The only attractors possible are isolated points (x for which $f(x) = 0$ in (13.7)) or closed loops. More complicated attractors cannot occur. To demonstrate this, suppose that $x(t)$ is a dense trajectory in an attractor and that for t near t_2 it runs close to, but distinct from, its path when t is near t_1. Since the trajectories vary smoothly, the directions of $x(t)$ at t_1 and t_2 are almost parallel (see figure 13.11). Thus for $t > t_2$ the trajectory $x(t)$ is 'blocked' from ever getting too close to $x(t_1)$ so that $x(t_1)$ cannot in fact be a point on an attractor. (The precise formulation of this fact is known as the Poincaré–Bendixson theorem.)

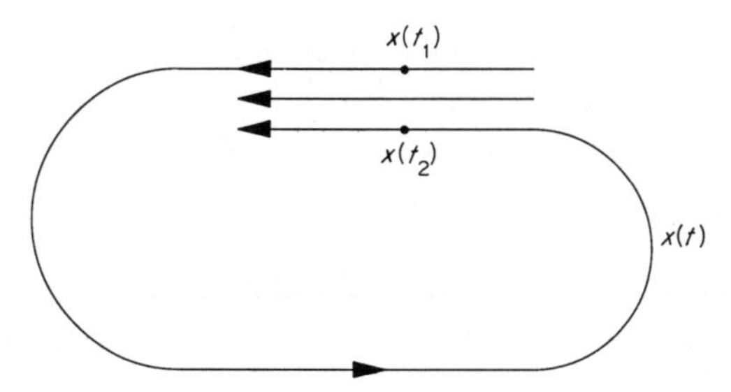

Figure 13.11 A trajectory of a continuous dynamical system in the plane. Assuming that the trajectories vary smoothly, the trajectory shown is 'cut off' from returning too close to $x(t_1)$ any time after t_2

Consequently, to find continuous dynamical systems with fractal attractors we need to look at systems in 3 or more dimensions. Linear differential equations (with $f(x)$ a linear function of x in (13.7)) can be solved completely by classical methods, the solutions involving periodic or exponential terms. However, even simple non-linear terms can lead to trajectories of a highly intricate from. Non-linear differential equations, particularly in higher dimensions, are notoriously difficult to analyse, and present knowledge stems from a combination of qualitative mathematical analysis and numerical study. One standard approach is to reduce a 3-dimensional continuous system to a 2-dimensional discrete system by looking at plane 'cross sections'. If P is a plane region transverse to the trajectories, we may define the 'first return' map $g: P \to P$ by taking $g(x)$ as the point at which the trajectory through x next intersects P; see figure 13.12. Then g is a discrete dynamical system on P. If g has an attractor E in P it follows that the union of trajectories through the points of E is an attractor F of f. Locally F looks like a product of E and a line segment, and typically $\dim_H F = 1 + \dim_H E$, by a variation on Corollary 7.4.

Perhaps the best known example of a continuous dynamical system with a

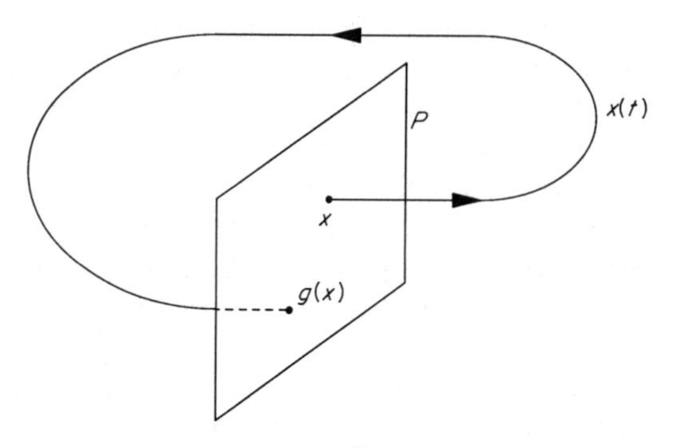

Figure 13.12 A continuous dynamical system in $\mathbb{R}^3$ induces a discrete dynamical system on the plane P by the 'first return' map g

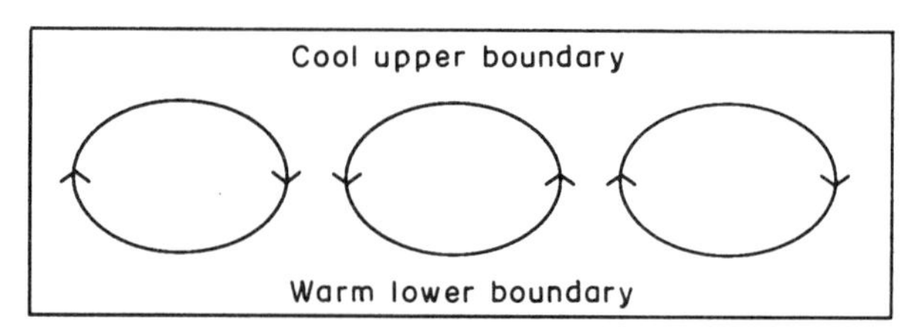

Figure 13.13 The Lorenz equations describe the behaviour of one of the rotating cylindrical rolls of heat-conducting viscous fluid

fractal attractor is the Lorenz system of equations. Lorenz studied thermal convention of a horizontal layer of fluid heated from below: the warm fluid may rise owing to its buoyancy and circulate in cylindrical rolls. Under certain conditions these cells are a series of parallel rotating cylindrical rolls; see figure 13.13. Lorenz used the continuity equation and Navier–Stokes equations from fluid dynamics, together with the heat conduction equation to describe the behaviour of one of these rolls. A series of approximations and simplifications lead to the *Lorenz equations*

$$\begin{aligned}\dot{x} &= \sigma(y - x)\\ \dot{y} &= rx - y - xz\\ \dot{z} &= xy - bz.\end{aligned} \tag{13.8}$$

The term x represents the rate of rotation of the cylinder, z represents the deviation from a linear vertical temperature gradient, and y corresponds to the difference in temperature at opposite sides of the cylinder. The constant σ is the Prandtl number of the air (the Prandtl number involves the viscosity and thermal conductivity), b depends on the width-to-height ratio of the layer, and r is a control parameter representing the fixed temperature difference between the bottom and top of the system. The non-linearity in the second and third equations results from the non-linearity of the equations of flow.

Working in (x, y, z)-space, the first thing to notice is that the system (13.8) contracts volumes at a constant rate. The differences in velocity between pairs of opposite faces of a small coordinate box of side δ are approximately $\delta(\partial\dot{x}/\partial x)$, $\delta(\partial\dot{y}/\partial y)$, $\delta(\partial\dot{z}/\partial z)$, so the rate of change of volume of the box is $\delta^3((\partial\dot{x}/\partial x) + (\partial\dot{y}/\partial y) + (\partial\dot{z}/\partial z)) = -(\sigma + b + 1)\delta^3 < 0$. Nevertheless, with $\sigma = 10$, $b = \frac{8}{3}$, $r = 28$ (the values usually chosen for study) the trajectories are concentrated onto an attractor of a highly complex form. This *Lorenz attractor* consists of two 'discs' each made up of spiralling trajectories (figure 13.14). Certain trajectories leave each of the discs almost perpendicularly and flow into the other disc. If a trajectory $x(t)$ is computed, the following behaviour is typical. As t increases, $x(t)$ circles around one of the discs a number of times and then 'flips' over to the other disc. After a few loops round this second disc, it flips back to the original disc. This pattern continues, with an apparently random number of circuits before leaving each disc. The motion seems to be chaotic;

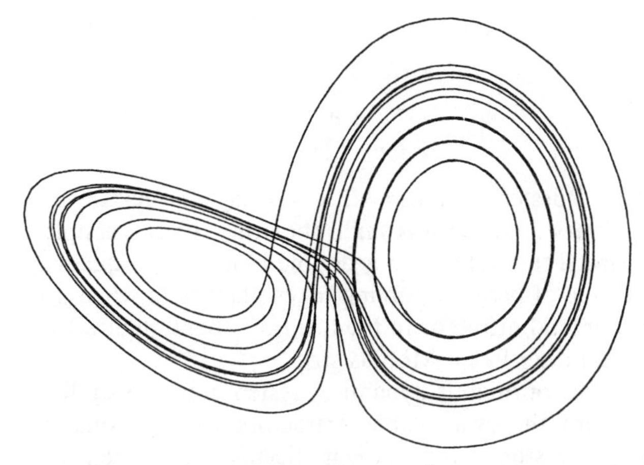

Figure 13.14 A view of the Lorenz attractor for $\sigma = 10$, $b = \frac{8}{3}$, $r = 28$. Note the spiralling round the two discs and the 'jumps' from one disc to the other

in particular points that are initially close together soon have completely different patterns of residence in the two discs of the attractor. One interpretation of this sensitive dependence on initial conditions is that long-term weather prediction is impossible.

The Lorenz attractor appears to be a fractal with numerical estimates suggesting a dimension of 2.06 when $\sigma = 10$, $b = \frac{8}{3}$, $r = 28$.

Other systems of differential equations also have fractal attractors. The

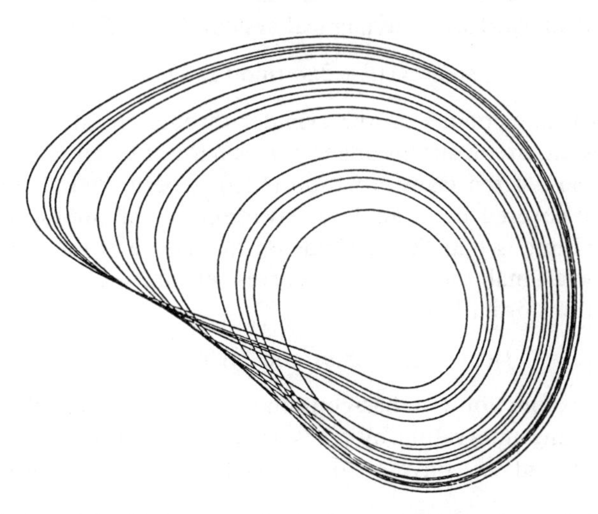

Figure 13.15 A view of the Rössler band attractor for $a = 0.375$, $b = 2$, $c = 4$. Note the banding, suggestive of a set that is locally the product of a Cantor set and a line segment

equations

$$\dot{x} = -y - z$$
$$\dot{y} = x + ay$$
$$\dot{z} = b + z(x - c)$$

were studied by Rössler. Fixing $b = 2$, $c = 4$, say, the nature of the attractor changes as a is varied. When a is small the attractor is a simple closed curve, but on increasing a this splits into a double loop, then a quadruple loop, and so on. Thus a type of period doubling takes place, and when a reaches about 0.375 there is a fractal attractor in the form of a band (figure 13.15). The band has a twist in it, rather like a Möbius strip.

At present, each continuous dynamical system must be studied individually; there is little general theory available. Attractors of continuous systems are well suited to computer study, and mathematicians are frequently challenged to explain 'strange' attractors that are observed on computer screens.

*13.6 Small divisor theory

There are a number of important dynamical systems dependent on a parameter ω, which are, in some sense, stable provided that ω is 'not too close to a rational number', in other words if ω is badly approximable in the sense of Section 10.3. By Jarník's theorem 10.3 the well-approximable numbers are fractal sets, so the stable parameters lie in sets with fractal complement.

The following simple example indicates how badly approximable parameters can result in stable systems.

Let C be the infinite cylinder of unit radius $\{(\theta, y): 0 \leqslant \theta < 2\pi, -\infty < y < \infty\}$. Fix $\omega \in \mathbb{R}$ and define a discrete dynamical system $f: C \to C$ by

$$f(\theta, y) = (\theta + 2\pi\omega \pmod{2\pi}, y). \tag{13.9}$$

Clearly, f just rotates points on the cylinder through an angle $2\pi\omega$, and the circles $y = \text{constant}$ are invariant under f. It is natural to ask if these invariant curves are stable—if the transformation (13.9) is perturbed slightly, will the cylinder still be covered by a family of invariant closed curves (figure 13.16)? The surprising thing is that this depends on the nature of the number ω: if ω is 'sufficiently irrational' then invariant curves remain.

We modify transformation (13.9) to

$$f(\theta, y) = (\theta + 2\pi\omega \pmod{2\pi}, y + g(\theta)) \tag{13.10}$$

where g is a C^∞ function (i.e. has continuous derivatives of all orders). It is easy to show, using integration by parts, that a function is C^∞ if and only if its Fourier coefficients a_k converge to 0 faster than any power of k, i.e. if

$$g(\theta) = \sum_{-\infty}^{\infty} a_k e^{ik\theta}$$

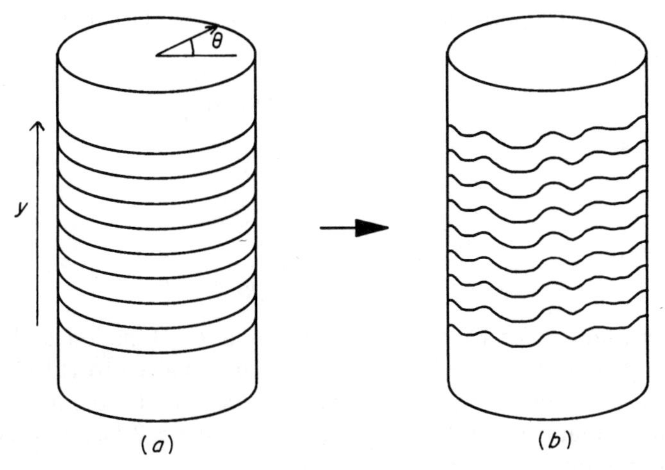

Figure 13.16 (*a*) Invariant circles for the mapping $f(\theta, y) = (\theta + 2\pi\omega \pmod{2\pi}, y)$. (*b*) If ω is not 'too rational', a small perturbation of the mapping to $f(\theta, y) = (\theta + 2\pi\omega \pmod{2\pi}, y + g(\theta))$ distorts the circles to a family of smooth invariant curves

is the Fourier series of g, then for every positive integer m there is a constant c such that for $k \neq 0$

$$|a_k| \leqslant c|k|^{-m}.$$

Suppose that $y(\theta)$ has Fourier series

$$y(\theta) = \sum_{-\infty}^{\infty} b_k \mathrm{e}^{\mathrm{i}k\theta}.$$

If $y(\theta)$ is an invariant curve, the point $(\theta + 2\pi\omega \pmod{2\pi}, y(\theta) + g(\theta))$ must lie on the curve whenever $(\theta, y(\theta))$ does; hence

$$y(\theta + 2\pi\omega \pmod{2\pi}) = y(\theta) + g(\theta)$$

or

$$\sum_{-\infty}^{\infty} b_k \mathrm{e}^{\mathrm{i}k(\theta + 2\pi\omega)} = \sum_{-\infty}^{\infty} b_k \mathrm{e}^{\mathrm{i}k\theta} + \sum_{-\infty}^{\infty} a_k \mathrm{e}^{\mathrm{i}k\theta}.$$

Equating terms in $\mathrm{e}^{\mathrm{i}k\theta}$ we get that

$$b_k = \frac{a_k}{\mathrm{e}^{2\pi\mathrm{i}k\omega} - 1} \qquad (k \neq 0)$$

with b_0 arbitrary if ω is irrational. Thus the invariant curves are given by

$$y(\theta) = b_0 + \sum_{k \neq 0} \frac{a_k}{\mathrm{e}^{2\pi\mathrm{i}k\omega} - 1} \mathrm{e}^{\mathrm{i}k\theta} \tag{13.11}$$

provided that this Fourier series converges to a continuous function. This will happen if the denominators $\mathrm{e}^{2\pi\mathrm{i}k\omega} - 1$ are not 'too small too often'. Suppose that ω is not α-well-approximable for some $\alpha > 2$; see (10.4). Then there is a constant c_1 such that

$$|e^{2\pi i k\omega} - 1| \geqslant \|k\omega\| \geqslant c_1 |k|^{1-\alpha}$$

for all $k \neq 0$, so

$$\left|\frac{a_k}{e^{2\pi i k\omega} - 1}\right| \leqslant c_1^{-1} \frac{|a_k|}{|k|^{1-\alpha}}$$

$$\leqslant c c_1^{-1} |k|^{-m-1+\alpha}$$

for some constant c for each m. Thus if g is a C^∞ function and ω is not α-well-approximable for some $\alpha > 2$, the function $y(\theta)$ given by (13.11) is C^∞, so that f has a family of C^∞ invariant curves. We saw in Theorem 10.3 that the set of numbers that are α-well-approximable for all $\alpha > 2$ has dimension 0, so for 'most' ω the invariant curves are stable.

The above example is a special case of a much more general class of transformations of the cylinder known as twist maps. Define $f: C \to C$ by

$$f(\theta, y) = (\theta + 2\pi\omega(y) \pmod{2\pi}, y). \tag{13.12}$$

Again the circles $y = \text{constant}$ are invariant, but this time the angle of rotation $\omega(y)$ is allowed to vary smoothly with y. We perturb f to

$$f(\theta, y) = (\theta + 2\pi\omega(y) + \varepsilon h(\theta, y) \pmod{2\pi}, y + \varepsilon g(\theta, y)) \tag{13.13}$$

where h and g are smooth functions and ε is small, and ask if the invariant curves round C are preserved. Moser's twist theorem, a very deep result, roughly says that the invariant circles $y = \text{constant}$ of (13.12), for which $\omega(y) = \omega$, will deform into differentiable closed invariant curves of (13.13) if ε is small enough, provided that $\|k\omega\| \geqslant c_1 |k|^{-3/2}$ for all $k \neq 0$ for some constant c_1. Thus the exceptional set of frequencies ω has dimension $\frac{4}{5}$, by Theorem 10.3. Typically C is filled by invariant curves corresponding to badly approximable ω, where the motion is regular, and regions in between where the motion is chaotic. The chaotic regions grow as ε increases.

Perhaps the most important application of small divisor theory is to the stability of Hamiltonian systems. Consider a 4-dimensional space parametrized by $(\theta_1, \theta_2, j_1, j_2)$. A Hamiltonian function $H(\theta_1, \theta_2, j_1, j_2)$ determines a conservative (volume-preserving) dynamical system according to the differential equations

$$\dot{\theta}_1 = \partial H/\partial j_1 \qquad \dot{\theta}_2 = \partial H/\partial j_2 \qquad \dot{j}_1 = -\partial H/\partial \theta_1 \qquad \dot{j}_2 = -\partial H/\partial \theta_2.$$

Thus if $H(\theta_1, \theta_2, j_1, j_2) = H_0(j_1, j_2)$ is independent of θ_1, θ_2, we get the solution

$$\theta_1 = \omega_1 t + c_1 \qquad \theta_2 = \omega_2 t + c_2 \qquad j_1 = c_3 \qquad j_2 = c_4$$

where ω_1 and ω_2 are angular frequencies (which may depend on j_1, j_2) and $c_1, \ldots, c_4$ are constants. A trajectory of the system remains on the same 2-dimensional torus, $(j_1, j_2) = \text{constant}$, for all time; such tori are called invariant.

It is important to know whether such invariant tori are stable under small perturbations of the system. If the Hamiltonian is replaced by

$$H_0(j_1,j_2) + \varepsilon H_1(\theta_1,\theta_2,j_1,j_2)$$

where ε is small, do the trajectories of this new system stay on new invariant tori expressible as $(j'_1,j'_2) = \text{constant}$, after a suitable coordinate transformation $(\theta_1,\theta_2,j_1,j_2) \mapsto (\theta'_1,\theta'_2,j'_1,j'_2)$? In other words, do the invariant tori of the original system distort slightly to become invariant tori for the new system, or do they break up altogether? The celebrated Kolmogorov–Arnold–Moser (KAM) theorem gives an answer to this question—essentially a torus is stable under sufficiently small perturbations provided that the frequency ratio ω_1/ω_2 is badly approximable by rationals; more precisely it is stable if for some $c > 0$ we have $|(\omega_1/\omega_2) - (p/q)| \geqslant c/q^{2.5}$ for all positive integers p, q. The set of numbers that fails to satisfy this condition is a fractal of dimension $\frac{4}{5}$ by Theorem 10.3, so, in particular, almost all frequency ratios (in the sense of Lebesgue measure) have tori that are stable under sufficiently small perturbations. (In fact, the condition can be weakened to $|(\omega_1/\omega_2) - (p/q)| \geqslant c/q^\alpha$ for any $\alpha > 2$.)

There is some astronomical evidence for small divisor theory. For example, the angular frequencies ω of asteroids tend to avoid values for which the ratio ω/ω_J is close to p/q where q is a small integer, where ω_J is the angular frequency of Jupiter, the main perturbing influence. On the assumptions that orbits in the Solar System are stable (which, fortunately, seems to be the case) and that we can consider a pair of orbiting bodies in isolation (a considerable oversimplification), this avoidance of rational frequency ratios is predicted by KAM theory.

*13.7 Liapounov exponents and entropies

So far we have looked at attractors of dynamical systems largely from a geometric point of view. However, a dynamical system f provides a much richer structure than a purely geometric one. In this section we outline some properties of f that often go hand in hand with fractal attractors.

The concept of invariant measures is fundamental in dynamical systems theory. A measure μ on D is *invariant* for a mapping $f: D \to D$ if for every subset A of D we have

$$\mu(f^{-1}(A)) = \mu(A). \tag{13.14}$$

We assume that μ has been normalized so that $\mu(D) = 1$. Any attractor F supports at least one invariant measure: for x in the basin of attraction of F and A a Borel set, write

$$\mu(A) = \lim_{m\to\infty} \frac{1}{m} \#\{k : 1 \leqslant k \leqslant m \text{ and } f^k(x) \in A\} \tag{13.15}$$

for the proportion of iterates in A. It may be shown using ergodic theory that this limit exists and is the same for μ-almost all points in the basin of attraction under very general circumstances. Clearly, $\mu(A \cup B) = \mu(A) + \mu(B)$ if A and B

are disjoint, and $f^k(x)\in A$ if and only if $f^{k-1}(x)\in f^{-1}(A)$, giving (13.14). The measure (13.15) is concentrated on the set of points to which $f^k(x)$ comes arbitrarily close infinitely often; thus μ is supported by an attractor of f. The measure $\mu(A)$ reflects the proportion of the iterates that lie in A, and may be thought of as the distribution that is seen when a large number of iterates $f^k(x)$ are plotted on a computer screen. As far as the size of an attractor is concerned, it is often the dimension of the set occupied by the invariant measure μ that is relevant, rather than the entire attractor. With this in mind, we define the *Hausdorff dimension of a measure* μ for which $\mu(D)=1$ as

$$\dim_{\mathrm{H}}\mu=\inf\{\dim_{\mathrm{H}}E:\mu(E)=1\}. \tag{13.16}$$

If μ is supported by F then clearly $\dim_{\mathrm{H}}\mu\leqslant\dim_{\mathrm{H}}F$, but we may have strict inequality; see Exercise 13.8. However, if there are numbers $s>0$ and $c>0$ such that for any set U

$$\mu(U)\leqslant c|U|^s \tag{13.17}$$

then the Mass distribution principle 4.2 implies that for any set E with $0<\mu(E)$ we have $\mathscr{H}^s(E)\geqslant\mu(E)/c>0$, so that $\dim_{\mathrm{H}}E\geqslant s$. Hence if (13.17) holds

$$\dim_{\mathrm{H}}\mu\geqslant s. \tag{13.18}$$

Once f is equipped with an invariant measure μ several other dynamical constants may be defined. For convenience, we assume that D is a domain in $\mathbb{R}^2$ and $f:D\to D$ is differentiable. The derivative $(f^k)'(x)$ is a linear mapping; we write $a_k(x)$ and $b_k(x)$ for the lengths of the major and minor semi-axes of the ellipse $(f^k)'(x)(B)$ where B is the unit ball. Thus the image under f^k of a small ball of radius r and centre x approximates to an ellipse with semi-axes of lengths $ra_k(x)$ and $rb_k(x)$. We define the *Liapounov exponents* as the average logarithmic rate of growth with k of these semi-axes:

$$\lambda_1(x)=\lim_{k\to\infty}\frac{1}{k}\log a_k(x)\qquad\qquad \lambda_2(x)=\lim_{k\to\infty}\frac{1}{k}\log b_k(x). \tag{13.19}$$

Techniques from ergodic theory show that if μ is invariant for f, these exponents exist and have the same values λ_1,λ_2 for μ-almost all x. Hence in a system with an invariant measure, we refer to λ_1 and λ_2 as *the* Liapounov exponents of the system. The Liapounov exponents represent the 'average' rates of expansion of f. If B is a disc of small radius r, then $f^k(B)$ will 'typically' be close to an ellipse with semi-axes of lengths $re^{\lambda_1 k}$ and $re^{\lambda_2 k}$; see figure 13.17.

A related dynamical idea is the entropy of a mapping $f:D\to D$. Write

$$V(x,\varepsilon,k)=\{y\in D:|f^i(x)-f^i(y)|<\varepsilon\text{ for }0\leqslant i\leqslant k\} \tag{13.20}$$

for the set of points with their first k iterates within ε of those of x. If μ is an invariant measure for f, we define the *μ-entropy* of f as

$$h_\mu(f)=\lim_{\varepsilon\to 0}\varlimsup_{k\to\infty}\left(-\frac{1}{k}\log\mu(V(x,\varepsilon,k))\right). \tag{13.21}$$

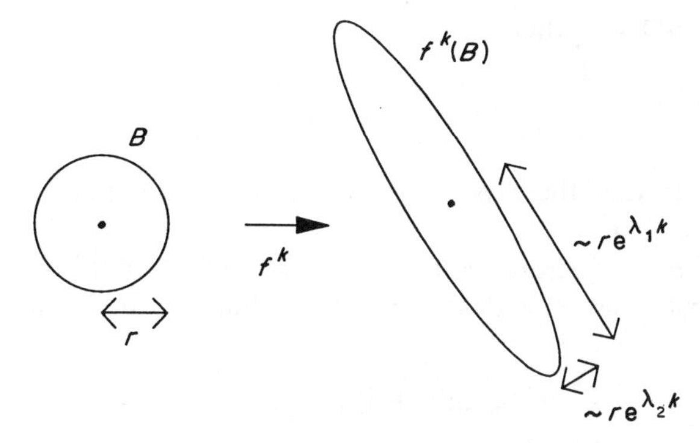

Figure 13.17 The definition of the Liapounov exponents λ_1 and λ_2

Under reasonable conditions, this limit exists and has a constant value for μ-almost all x. The entropy $h_\mu(f)$ reflects the rate at which nearby points spread out under iteration by f, or alternatively, the amount of extra information about an orbit $\{f^k(x)\}_{k=1}^{\infty}$ that is gained from knowing the position of an additional point on the orbit.

The baker's transformation (13.4) provides a simple illustration of these ideas. There is a natural invariant mass distribution μ on the attractor F such that each of the 2^k strips of E_k has mass 2^{-k}, with this mass spread uniformly across the width of the E. Just as in Example 4.3 we get that $\mu(U) \leqslant c|U|^s$ where $s = 1 + \log 2/(-\log \lambda)$ so by (13.17) and (13.18) $s \leqslant \dim_H \mu \leqslant \dim_H F = s$.

The Liapounov exponents are also easily found. The derivative of (13.4) is $f'(x, y) = \begin{bmatrix} 2 & 0 \\ 0 & \lambda \end{bmatrix}$ (provided $x \neq \frac{1}{2}$) so $(f^k)'(x, y) = \begin{bmatrix} 2^k & 0 \\ 0 & \lambda^k \end{bmatrix}$ (except where $x = p/2^k$ for non-negative integers p and k). Hence $a_k(x, y) = 2^k$ and $b_k(x, y) = \lambda^k$. By (13.19) $\lambda_1(x, y) = \log 2$, $\lambda_2(x, y) = \log \lambda$ for μ-almost all (x, y), and the Liapounov exponents of the system are $\lambda_1 = \log 2$ and $\lambda_2 = \log \lambda$.

Since f^k stretches by a factor 2^k horizontally and contracts by a factor λ^k vertically, we get, using (13.20) and ignoring the 'cutting' effect of f, that $V((x, y), \varepsilon, k)$ is approximately a rectangle with sides $2^{-k}\varepsilon$ and ε, which has μ-measure approximately $\varepsilon^s 2^{-k}$, if $(x, y) \in F$. Thus

$$h_\mu(f) = \lim_{\varepsilon \to 0} \varliminf_{k \to \infty} \left(-\frac{1}{k} \log(\varepsilon^s 2^{-k}) \right) = \log 2.$$

The Hausdorff and box dimensions, Liapounov exponents and entropies of an invariant measure of a given dynamical system can be estimated computationally or experimentally and are often useful when comparing different systems. However, the very nature of these quantities suggests that they may not be completely independent of each other. One relationship that has been derived rigorously applies to a smooth bijective transformation f on a 2-dimensional surface. If μ is an invariant measure for f with Liapounov exponents $\lambda_1 > 0 > \lambda_2$ then

$$\dim_{\mathrm{H}} \mu = h_\mu(f)\left(\frac{1}{\lambda_1} - \frac{1}{\lambda_2}\right). \tag{13.22}$$

It is easily seen that the exponents calculated for the baker's transformation satisfy this formula.

The following conjectured relationship is known to hold in many cases; if f is a plane transformation with attractor F and Liapounov exponents $\lambda_1 > 0 > \lambda_2$, then

$$\dim_{\mathrm{B}} F \leqslant 1 - (\lambda_1/\lambda_2). \tag{13.23}$$

An argument to support this runs as follows. Let $N_\delta(F)$ be the least number of discs of radius δ that can cover F. If $\{U_i\}$ are $N_\delta(F)$ such discs, then $f^k(F)$ is covered by the $N_\delta(F)$ sets $f^k(U_i)$ which are approximately elliptical with semi-axis lengths $\delta\exp(\lambda_1 k)$ and $\delta\exp(\lambda_2 k)$. These ellipses may be covered by about $\exp((\lambda_1 - \lambda_2)k)$ discs of radii $\delta\exp(\lambda_2 k)$. Hence

$$N_{\delta\exp(\lambda_2 k)}(F) \leqslant \exp((\lambda_1 - \lambda_2)k) N_\delta(F)$$

so

$$\frac{\log N_{\delta\exp(\lambda_2 k)}(F)}{-\log(\delta\exp(\lambda_2 k))} \leqslant \frac{\log(\exp((\lambda_1 - \lambda_2)k)N_\delta(F))}{-\log(\delta\exp(\lambda_2 k))}$$

$$= \frac{(\lambda_1 - \lambda_2)k + \log N_\delta(F)}{-\lambda_2 k - \log\delta}.$$

Letting $k \to \infty$ gives $\overline{\dim}_{\mathrm{B}} F \leqslant 1 - (\lambda_1/\lambda_2)$. The drawback with this argument is that it assumes that the Liapounov exponents are constant across the domain D, which need not be the case.

The relationship between these and other dynamical parameters is not yet fully understood. However, it is clear that such concepts are closely interrelated with the chaotic properties of f and the fractal nature of the attractor.

Recently, the theory of multifractal measures has been introduced to analyse measures such as the invariant measures of dynamical systems. This is discussed in Chapter 17.

13.8 Notes and references

The literature on dynamical systems is enormous and growing rapidly. The books by Guckenheimer and Holmes (1983), Bergé, Pomeau and Vidal (1984), Schuster (1984), Devaney (1986) and Thompson and Stewart (1986) provide accounts of differing aspects at a fairly basic level; see also Ruelle (1980). The collections of papers edited by Cvitanović (1984), Fischer and Smith (1985), Holden (1986), Barnsley and Demko (1986) and Bedford and Swift (1988) highlight a variety of relevant aspects. The dimension of attractors is considered by Farmer, Ott and Yorke (1983). Accounts of the logistic map are contained in May (1976) and Devaney (1986). The horseshoe attractor was introduced in the fundamental paper by Smale (1967) and the Hénon attractor in Hénon and Pomeau (1976).

The book by Sparrow (1982) contains a full account of the Lorenz equations, and the paper by Holden and Muhamad (1986) has pictures of attractors of a variety of continuous dynamical systems.

The main theory and applications of small divisor theory are brought together in the collected papers on Hamiltonian dynamical systems edited by MacKay and Meiss (1987). For results relating Liapounov exponents to dimensions see the papers by Young (1982) and by Frederickson, Kaplan, Yorke and Yorke (1983) and those in Mayer Kress (1986).

Exercises

13.1 Find a fractal invariant set F for the 'tent-like' map $f:\mathbb{R}\to\mathbb{R}$ given by $f(x)=2(1-|2x-1|)$. Show that F is a repeller for f and that f is chaotic on F.

13.2 Investigate the iterates $f_\lambda^k(x)$ of x in $[0,1]$ under the logistic mapping (13.2) for various values of λ and initial points x. Show that if the sequence of iterates converges then it converges either to 0 or to $1-1/\lambda$. Show that if $\lambda=\frac{1}{2}$ then, for all x in $(0,1)$, the iterates converge to 0, but that if $\lambda=2$ they converge to $\frac{1}{2}$. Show that if $\lambda=4$, then there are infinitely many values of x in $(0,1)$ such that $f_\lambda^k(x)$ converges to 0, infinitely many x in $(0,1)$ for which $f_\lambda^k(x)$ converges to $\frac{3}{4}$, and infinitely many x in $(0,1)$ for which $f_\lambda^k(x)$ does not converge. Use a programmable calculator or computer to investigate the behaviour of the orbits for other values of λ. Investigate other transformations listed at the end of Section 13.2 in a similar way.

13.3 In the cases $\lambda=2$ and $\lambda=4$ it is possible to obtain a simple formula for the iterates of the logistic map f_λ on $[0,1]$. For a given $x=x_0$ we write $x_k=f_\lambda^k(x)$.

(i) Show that if $\lambda=2$ and a is chosen so that $x=\frac{1}{2}(\exp a-1)$, then the iterates are given by $x_k=\frac{1}{2}(\exp(2^k a)-1)$.

(ii) Show that if $\lambda=4$ and $0\leqslant a<1$ is chosen so that $x=\sin^2(\pi a)$, then $x_k=\sin^2(2^k\pi a)$. By writing $a=0\cdot a_1a_2\ldots$ in binary form, show that f_4 has an unstable orbit of period p for all positive integers p and also has a dense orbit.

13.4 Consider the modified baker's transformation $f:E\to E$, where E is the unit square, given by

$$f(x,y)=\begin{cases}(2x,\lambda y) & (0\leqslant x\leqslant\frac{1}{2})\\ (2x-1,\mu y+\frac{1}{2}) & (\frac{1}{2}<x\leqslant 1)\end{cases}$$

where $0<\lambda,\mu<\frac{1}{2}$. Show that there is a set F that attracts all points of E, and find the Hausdorff dimension of F.

13.5 Consider the Hénon mapping (13.5) with $a=1.4$ and $b=0.3$. Show that the quadrilateral D with vertices $(1.32,0.133)$, $(-1.33,0.42)$, $(-1.06,-0.5)$ and

$(1.245, -0.14)$ is mapped into itself by f. Use a computer to plot the iterates of a typical point in D.

13.6 With notation as in Section 13.4 consider the transformation f of the solid torus D given by

$$f(\phi, w) = (3\phi \,(\mathrm{mod}\, 2\pi), aw + \tfrac{1}{2}\hat{\phi})$$

where $0 < a < \frac{1}{10}$. Show that f has an attractor F of Hausdorff and box dimensions equal to $1 + \log 3/-\log a$, and verify that f is chaotic on F.

13.7 Let $g: \mathbb{R} \to \mathbb{R}$ be differentiable, and let $h: \mathbb{R}^2 \to \mathbb{R}^2$ be given by

$$h(t, x) = (\lambda t, \lambda^{2-s}(x - g(t)))$$

where $\lambda > 1$ and $0 < s < 2$. Show that graph f is a repeller for h, where f is the function

$$f(t) = \sum_{k=0}^{\infty} \lambda^{(s-2)k} g(\lambda^k t).$$

Thus functions of Weierstrass type (see (11.7)) can occur as invariant sets in dynamical systems.

13.8 Give an example of a mass distribution μ on $[0, 1]$ for which $\dim_{\mathrm{H}} \mu < \dim_{\mathrm{H}} F$, where F is the support of μ. (Hint: see Section 10.1.)

13.9 Consider the mapping $f: E \to E$, where E is the unit square, given by

$$f(x, y) = (x + y \,(\mathrm{mod}\, 1), x + 2y \,(\mathrm{mod}\, 1)).$$

(This mapping has become known as Arnold's cat map.) Show that plane Lebesgue measure is invariant for f (i.e. f is area preserving), and find the Liapounov exponents of f.

13.10 Write a computer program that plots the orbits of a point x under iteration by a mapping of a region in the plane. Use it to study the attractors of the baker's transformation, the Hénon mapping and experiment with other functions.

13.11 Write a computer program to draw trajectories of the Lorenz equations (13.8). See how the trajectories change as σ, r and b are varied. Do a similar study for the Rössler equations.

Chapter 14 Iteration of complex functions—Julia sets

Julia sets provide some of the most striking illustrations of how an apparently simple process can lead to highly intricate sets. Functions on the complex plane $\mathbb{C}$ as simple as $f(z) = z^2 + c$, with c a constant, can give rise to fractals of an exotic appearance—look ahead to figure 14.7.

Julia sets arise in connection with the iteration of a function of a complex variable f, so are related to the dynamical systems discussed in the previous chapter—in general a Julia set is a dynamical repeller. However, by specializing to functions that are analytic on the complex plane (i.e. differentiable in the sense that $f'(z) = \lim_{w \to 0}(f(z+w) - f(z))/w$ exists as a complex number, where $z, w \in \mathbb{C}$) we can use the powerful techniques of complex variable theory to obtain much more detailed information about the structure of such repelling sets.

14.1 General theory of Julia sets

For convenience of exposition, we take $f:\mathbb{C} \to \mathbb{C}$ to be a polynomial of degree $n \geqslant 2$ with complex coefficients, $f(z) = a_0 + a_1 z + \cdots + a_n z^n$. Note that with minor modifications the theory remains true if f is a rational function $f(z) = p(z)/q(z)$ (where p, q are polynomials) on the extended complex plane $\mathbb{C} \cup \{\infty\}$, and much of it holds if f is any meromorphic function (that is a function that is analytic on $\mathbb{C} \cup \{\infty\}$ except at a finite number of poles).

As usual we write f^k for the k-fold composition $f \circ \cdots \circ f$ of the function f, so that $f^k(w)$ is the kth iterate $f(f(\cdots(f(w))))$ of w. As before, if $f(w) = w$ we call w a *fixed point* of f, and if $f^p(w) = w$ for some integer $p \geqslant 1$ we call w a *periodic point* of f; the least p such that $f^p(w) = w$ is called the *period* of w. We call $w, f(w), \ldots, f^p(w)$ a period p *orbit*. Let w be a periodic point of period p, with $(f^p)'(w) = \lambda$, where the prime denotes complex differentiation. The point w is called

superattractive	if $\lambda = 0$
attractive	if $0 \leqslant \lvert\lambda\rvert < 1$
indifferent	if $\lvert\lambda\rvert = 1$
repelling	if $\lvert\lambda\rvert > 1$.

The *Julia set* $J(f)$ of f may be defined as the closure of the set of repelling periodic points of f. (We write J for $J(f)$ when the function is clear.) The complement of the Julia set is called the *Fatou set* or *stable set* $F(f)$. This chapter investigates the geometry and fractal nature of the Julia sets of polynomials. We show that $J(f)$ is both forwards and backwards invariant under f, i.e. $J = f(J) = f^{-1}(J)$, and that J is non-empty and compact. Moreover, f behaves 'chaotically' on J, and J is usually a fractal.

For the simplest example, let $f(z) = z^2$, so that $f^k(z) = z^{2^k}$. The points satisfying $f^p(z) = z$ are $\{\exp(2\pi i q/(2^p - 1)) : 0 \leqslant q < 2^p - 2\}$, which are certainly repelling, since $|(f^p)'(z)| = 2$ at such points. Thus the Julia set $J(f)$ is the unit circle $|z| = 1$. Clearly $J = f(J) = f^{-1}(J)$, with $f^k(z) \to 0$ as $k \to \infty$ if $|z| < 1$ and $f^k(z) \to \infty$ if $|z| > 1$, but with $f^k(z)$ remaining on J for all k if $|z| = 1$. The Julia set J is the boundary between the sets of points which iterate to 0 and ∞. Of course, in this special case J is not a fractal.

Suppose that we modify this example slightly, taking $f(z) = z^2 + c$ where c is a small complex number. It is easy to see that we still have $f^k(z) \to w$ if z is small, where w is the fixed point of f close to 0, and that $f^k(z) \to \infty$ if z is large. Again, the Julia set is the boundary between these two types of behaviour, but it turns out that now J *is* a fractal curve; see figure 14.1.

To establish the basic properties of Julia sets we cannot avoid the idea of normal families of analytic functions and Montel's theorem.

*[Readers who wish to omit this quite technical work involving complex variable theory should skip to Summary 14.12.]

Let U be an open set in $\mathbb{C}$, and let $g_k: U \to \mathbb{C}$ be a family of complex analytic functions (i.e. functions differentiable on U in the complex sense). The family $\{g_k\}$ is said to be *normal* on U if every sequence of functions selected from $\{g_k\}$ has a subsequence which converges uniformly on every compact subset of U, either to a bounded analytic function or to ∞. Notice that by standard complex

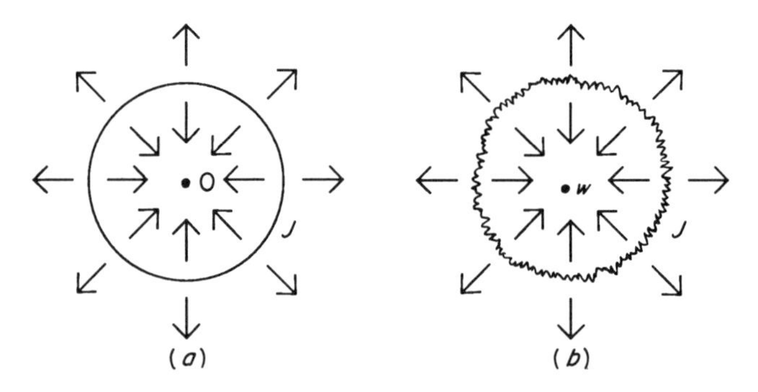

Figure 14.1 (*a*) The Julia set of $f(z) = z^2$ is the circle $|z| = 1$, with the iterates $f^k(z) \to 0$ if z is inside J, and $|f^k(z)| \to \infty$ if z is outside J. (*b*) If f is perturbed to the function $f(z) = z^2 + c$ for small c this picture distorts slightly, with a curve J separating those points z for which $f^k(z)$ converges to the fixed point w of f near 0 from those points z with $|f^k(z)| \to \infty$. The curve J is now a fractal

variable theory this means that the subsequence converges either to a finite analytic function or to ∞ on each connected component of U. In the former case the derivatives of the subsequence must converge to the derivative of the limit function. The family $\{g_k\}$ is *normal at the point* w of U if there is some open subset V of U containing w such that $\{g_k\}$ is a normal family on V. Observe that this is equivalent to there being a neighbourhood V of w on which every sequence from $\{g_k\}$ has a subsequence uniformly convergent to a bounded analytic function or to ∞.

The fundamental result on which the theory of Julia sets hangs is due to Montel. This deep theorem asserts that non-normal families of functions take all except possibly one complex value near every point.

Montel's theorem 14.1

Let $\{g_k\}$ be a family of complex analytic functions on an open domain U. If $\{g_k\}$ is not a normal family, then for all $w \in \mathbb{C}$ with at most one exception we have $g_k(z) = w$ for some $z \in U$ and some k.

Proof. Consult the literature on complex function theory. □

We examine the normality of the iterates of a complex polynomial f. Define

$$J_0(f) = \{z \in \mathbb{C}: \text{the family } \{f^k\}_{k \geq 0} \text{ is not normal at } z\}. \tag{14.1}$$

Using Montel's theorem, we shall show that $J_0(f)$ is the same as the closure of the repelling periodic points, $J(f)$. In fact (14.1) is often taken as the definition of the Julia set. Although our definition of $J(f)$ is intuitively more appealing, $J_0(f)$ is rather easier to work with, since complex variable techniques are more readily applicable. We derive some basic properties of $J_0(f)$, with the eventual aim of showing that $J(f) = J_0(f)$.

Observe that the complement

$$\begin{aligned} F_0(f) &\equiv \mathbb{C} \backslash J_0(f) \\ &= \{z \in \mathbb{C} \text{ such that there is an open set } V \text{ with} \\ &\qquad z \in V \text{ and } \{f^k\} \text{ normal on } V\} \end{aligned} \tag{14.2}$$

is trivially an open set.

Proposition 14.2

If f is a polynomial, then $J_0(f)$ is compact.

Proof. By the above remark $J_0(f)$ has open complement, so is closed. Since f is a polynomial of degree at least 2, we may find r such that $|f(z)| \geq 2|z|$ if $|z| \geq r$, implying that $|f^k(z)| > 2^k r$ if $|z| > r$. Thus $f^k(z) \to \infty$ uniformly on the open set $V = \{z: |z| > r\}$. By definition, $\{f^k\}$ is normal on V, so that $V \subset \mathbb{C} \backslash J_0(f)$. Thus $J_0(f)$ is bounded, and so is compact. □

(Note that if $f:\mathbb{C}\cup\{\infty\}\to\mathbb{C}\cup\{\infty\}$ is a rational function then J_0 must be closed, but need not be bounded. Indeed, it is possible for J_0 to be the whole complex plane; for example, if $f(z)=((z-2)/z)^2$.)

Proposition 14.3

$J_0(f)$ is non-empty.

Proof. Suppose $J_0(f)=\varnothing$. Then, for every $r>0$, the family $\{f^k\}$ is normal on the open disc $B_r^\circ(0)$ with centre the origin and radius r (since the closed disc $B_r(0)$ is compact, it may be covered by a finite number of open sets on which f^k is normal). Since f is a polynomial, taking r large enough ensures that $B_r^\circ(0)$ contains a point z for which $|f^k(z)|\to\infty$ and also contains a fixed point w of f with $f^k(w)=w$ for all k. Thus it is impossible for any subsequence of $\{f^k\}$ to converge uniformly either to a bounded function or to infinity on any compact subset of $B_r^\circ(0)$ which contains both z and w, contradicting the normality of $\{f^k\}$. □

Proposition 14.4

$J_0(f)$ is forward and backward invariant, i.e. $J_0=f(J_0)=f^{-1}(J_0)$.

Proof. We show, equivalently, that the complement $F_0(f)$ is invariant. Let V be an open set with $\{f^k\}$ normal on V. Since f is continuous, $f^{-1}(V)$ is open. Let $\{f^{k_i}\}$ be a subsequence of $\{f^k\}$. Then $\{f^{k_i+1}\}$ has a subsequence $\{f^{k_i'+1}\}$ that is uniformly convergent on compact subsets of V. Thus if D is a compact subset of $f^{-1}(V)$, then $\{f^{k_i'+1}\}$ is uniformly convergent on the compact set $f(D)$, so $\{f^{k_i'}\}$ is uniformly convergent on D. Thus $\{f^k\}$ is normal on $f^{-1}(V)$, so $F_0\subset f^{-1}(F_0)$. The other inclusions required may be obtained in a similar way, using that a polynomial $f:\mathbb{C}\to\mathbb{C}$ is an open mapping, i.e. that $f(V)$ is open whenever V is open. □

Proposition 14.5

$J_0(f^p)=J_0(f)$ for every positive integer p.

Proof. Again we work with the complement F_0. Clearly, if every subsequence of $\{f^k\}$ has a subsequence uniformly convergent on a given set, the same is true of $\{f^{pk}\}_{k\geqslant 1}$. Thus $F_0(f)\subset F_0(f^p)$.

If D is compact and $\{g_k\}$ is a family of functions uniformly convergent on D either to a bounded function or to ∞, then the same is true of $\{h\circ g_k\}$ for any polynomial h. Thus if $\{f^{pk}\}_{k\geqslant 1}$ is normal on an open set V, so is $\{f^{pk+r}\}_{k\geqslant 1}$ for $r=0,1,\ldots,p-1$. But any subsequence of $\{f^k\}_{k\geqslant 1}$ contains an infinite subsequence of $\{f^{pk+r}\}_{k\geqslant 1}$ for some integer r with $0\leqslant r\leqslant p-1$, which has a subsequence that is uniformly convergent on compact subsets of V. Hence $\{f^k\}$ is normal, so $F_0(f)\supset F_0(f^p)$. □

Our next result tells us that f is 'mixing'; that is, neighbourhoods of points in J_0 are spread right across the complex plane by iterates of f.

Lemma 14.6

Let f be a polynomial, let $w \in J_0(f)$ and let U be any neighbourhood of w. Then $W \equiv \bigcup_{k=1}^{\infty} f^k(U)$ is the whole of $\mathbb{C}$, except possibly for a single point. Any such exceptional point is not in $J_0(f)$, and is independent of w and U.

Proof. By definition of J_0, the family $\{f^k\}$ is not normal at w, so the first part is immediate from Montel's theorem 14.1.

Suppose $v \notin W$. If $f(z) = v$, then, since $f(W) \subset W$, it follows that $z \notin W$. As $\mathbb{C} \setminus W$ consists of at most one point, then $z = v$. Hence f is a polynomial of degree n such that the only solution of $f(z) - v = 0$ is v, which implies that $f(z) - v = c(z - v)^n$ for some constant c.

If z is sufficiently close to v, then $f^k(z) - v \to 0$ as $k \to \infty$, convergence being uniform on, say, $\{z : |z - v| < (2c)^{-1/(n-1)}\}$. Thus $\{f^k\}$ is normal at v, so the exceptional point $v \notin J_0(f)$. Clearly v only depends on the polynomial f. (In fact, if W omits a point v of $\mathbb{C}$, then $J_0(f)$ is the circle centre v and radius $c^{-1/(n-1)}$.) □

The following corollary is the basis of many computer pictures of Julia sets; see page 215.

Corollary 14.7

(*a*) *The following holds for all $z \in \mathbb{C}$ with, at most, one exception: if U is an open set intersecting $J_0(f)$ then $f^{-k}(z)$ intersects U for infinitely many values of k.*

(*b*) *If $z \in J_0(f)$ then $J_0(f)$ is the closure of $\bigcup_{k=1}^{\infty} f^{-k}(z)$.*

Proof.

(*a*) Provided z is not the exceptional point of Lemma 14.6, $z \in f^k(U)$ so $f^{-k}(z)$ intersects U for some k. Using this repeatedly we get an infinite sequence of k with $f^{-k}(z)$ intersecting U.

(*b*) If $z \in J_0(f)$ then $f^{-k}(z) \subset J_0(f)$, by Property 14.4, so that $\bigcup_{k=1}^{\infty} f^{-k}(z)$ and, therefore, its closure is contained in the closed set $J_0(f)$. On the other hand, if U is an open set containing $z \in J_0(f)$, then $f^{-k}(z)$ intersects U for some k, by part (*a*); z cannot be the exceptional point by Lemma 14.6. □

Another immediate consequence of Lemma 14.6 is that $J_0(f)$ cannot be 'too thick'.

Corollary 14.8

If f is a polynomial, $J_0(f)$ has empty interior.

Proof. Suppose $J_0(f)$ contains an open set U. Then $J_0(f) \supset f^k(U)$ for all k, by Proposition 14.4, so $J_0(f) \supset \bigcup_{k=1}^{\infty} f^k(U)$. By Lemma 14.6 $J_0(f)$ is all of $\mathbb{C}$ except possibly for one point, contradicting Property 14.2, that $J_0(f)$ is bounded. □

Proposition 14.9

$J_0(f)$ is a perfect set (i.e. closed and with no isolated points) and is therefore uncountable.

Proof. Let $v \in J_0(f)$ and let U be a neighbourhood of v. We must show that U contains other points of $J_0(f)$. We consider three cases separately.

(i) v is not a fixed or periodic point of f. By Corollary 14.7(*b*) and Property 14.4, U contains a point of $f^{-k}(v) \subset J_0(f)$ for some $k \geqslant 1$, and this point must be different from v.

(ii) $f(v) = v$. If $f(z) = v$ has no solution other than v, then, just as in the proof of Lemma 14.6, $v \notin J_0(f)$. Thus, there exists $w \neq v$ with $f(w) = v$. By Corollary 14.7(*b*), U contains a point of $f^{-k}(w)$ for some $k \geqslant 1$. Any such point is in $J_0(f)$ by backward invariance and is distinct from v, since $f^k(v) = v$.

(iii) $f^p(v) = v$ for some $p > 1$. By Proposition 14.5, $J_0(f) = J_0(f^p)$, so by applying (ii) to f^p we see that U contains points of $J_0(f^p) = J_0(f)$ other than v.

Thus $J_0(f)$ has no isolated points; since it is closed, it is perfect. □

We can now prove the main result of this section, that $J_0(f)$, the set of points of non-normality of $\{f^k\}$ is exactly the same as $J(f)$, the closure of the repelling periodic points of f.

Theorem 14.10

If f is a polynomial, $J(f) = J_0(f)$.

Proof. Let w be a repelling periodic point of f of period p, so w is a repelling fixed point of $g = f^p$. Suppose that $\{g^k\}$ is normal at w; then w has an open neighbourhood V on which a subsequence $\{g^{k_i}\}$ coverges to a finite analytic function g_0 (it cannot converge to ∞ since $g^k(w) = w$ for all k). By a standard result from complex analysis, the derivatives also converge, $(g^{k_i})'(z) \to g_0'(z)$ if $z \in V$. However, by the chain rule, $|(g^{k_i})'(w)| = |(g'(w))^{k_i}| \to \infty$ since w is a repelling fixed point and $|g'(w)| > 1$. This contradicts the finiteness of $g_0'(w)$, so $\{g^k\}$ cannot be normal at w. Thus $w \in J_0(g) = J_0(f^p) = J_0(f)$, by Proposition 14.5. Since $J_0(f)$ is closed, it follows that $J(f) \subset J_0(f)$.

Let $K = \{w \in J_0(f)$ such that there exists $z \neq w$ with $f(z) = w$ and $f'(z) \neq 0\}$. Suppose that $w \in K$. Then there is an open neighbourhood V of w on which we may find a local analytic inverse $f^{-1}: V \to \mathbb{C} \setminus V$ so that $f(f^{-1}(z)) = z$ for $z \in V$

(just choose values of $f^{-1}(z)$ in a continuous manner). Define a family of analytic functions $\{h_k\}$ on V by

$$h_k(z) = \frac{(f^k(z) - z)}{(f^{-1}(z) - z)}.$$

Let U be any open neighbourhood of w with $U \subset V$. Since $w \in J_0(f)$ the family $\{f^k\}$ and thus, from the definition, the family $\{h_k\}$ is not normal on U. By Montel's theorem 14.1, $h_k(z)$ must take either the value 0 or 1 for some k and $z \in U$. In the first case $f^k(z) = z$ for some $z \in U$; in the second case $f^k(z) = f^{-1}(z)$ so $f^{k+1}(z) = z$ for some $z \in U$. Thus U contains a periodic point of f, so $w \in J(f)$.

We have shown that $K \subset J(f)$; taking closures $\bar{K} \subset \overline{J(f)} = J(f)$. However, K contains all of $J_0(f)$ except for a finite number of points. Since $J_0(f)$ contains no isolated points, by Property 14.9, $J_0(f) = \bar{K} \subset J(f)$, as required. □

If w is an attractive fixed point of f, we write

$$A(w) = \{z \in \mathbb{C} : f^k(z) \to w \quad \text{as} \quad k \to \infty\}$$

for the *basin of attraction* of w. We define the basin of attraction of infinity, $A(\infty)$, in the same way. Since w is attractive, there is an open set V containing w in $A(w)$ (if $w = \infty$, we may take $\{z : |z| > r\}$, for sufficiently large r). This implies that $A(w)$ is open, since if $f^k(z) \in V$ for some k, then $z \in f^{-k}(V)$, which is open. The following characterization of J as the boundary of any basin of attraction is extremely useful in determining Julia sets. Recall the notation ∂A for the boundary of the set A.

Lemma 14.11

Let w be an attractive fixed point of f. Then $\partial A(w) = J(f)$. The same is true if $w = \infty$.

Proof. If $z \in J(f)$ then $f^k(z) \in J(f)$ for all k so cannot converge to an attractive fixed point, and $z \notin A(w)$. However, if U is any neighbourhood of z, the set $f^k(U)$ contains points of $A(w)$ for some k by Lemma 14.6, so there are points arbitrarily close to z that iterate to w. Thus $z \in \overline{A(w)}$ and so $z \in \partial A(w)$.

Suppose $z \in \partial A(w)$ but $z \notin J(f) = J_0(f)$. Then z has a connected open neighbourhood V on which $\{f^k\}$ has a subsequence convergent either to an analytic function or to ∞. The subsequence converges to w on $V \cap A(w)$, which is open and non-empty, and therefore on V, since an analytic function is constant on a connected set if it is constant on any open subset. All points of V are mapped into $A(w)$ by iterates of f, so $V \subset A(w)$, contradicting that $z \in \partial A(w)$. □

For an example of this lemma, recall the case $f(z) = z^2$. The Julia set is the unit circle, which is the boundary of both $A(0)$ and $A(\infty)$.

We collect together the main points of this section.

Summary 14.12

The Julia set $J(f)$ is the closure of the repelling periodic points of the polynomial f. It is an uncountable compact set containing no isolated points and is invariant under f and f^{-1}. If $z \in J(f)$, then $J(f)$ is the closure of $\bigcup_{k=1}^{\infty} f^{-k}(z)$. The Julia set is the boundary of the basin of attraction of each attractive fixed point of f, including ∞, and $J(f) = J(f^p)$ for each positive integer p.

Proof. This collects together the results of this section, using that $J(f) = J_0(f)$. □

It is possible to discover a great deal more about the dynamics of f on the Julia set. It may be shown that 'f acts chaotically on J' (see Chapter 13). Periodic points of f are dense in J, by definition. On the other hand, J contains points z with iterates $f^k(z)$ that are dense in J. Moreover, f has 'sensitive dependence on initial conditions' on J; thus $|f^k(z) - f^k(w)|$ will be large for certain k regardless of how close $z, w \in J$ are, making accurate computation of iterates impossible.

14.2 Quadratic functions—the Mandelbrot set

We now specialize to the case of quadratic functions on $\mathbb{C}$. We study Julia sets of polynomials of the form

$$f_c(z) = z^2 + c. \tag{14.3}$$

This is not as restrictive as it first appears: if $h(z) = \alpha z + \beta\ (\alpha \neq 0)$ then

$$h^{-1}(f_c(h(z))) = (\alpha^2 z^2 + 2\alpha\beta z + \beta^2 + c - \beta)/\alpha.$$

By choosing appropriate values of α, β and c we can make this expression into any quadratic function f that we please. Then $h^{-1} \circ f_c \circ h = f$, so $h^{-1} \circ f_c^k \circ h = f^k$ for all k. This means that the sequence of iterates $\{f^k(z)\}$ of a point z under f is just the image under h^{-1} of the sequence of iterates $\{f_c^k(h(z))\}$ of the point $h(z)$ under f_c. The mapping h transforms the dynamical picture of f to that of f_c. In particular, z is a period-p point of f if and only if $h(z)$ is a period-p point of f_c; thus the Julia set of f is the image under h^{-1} of the Julia set of f_c.

The transformation h is called a *conjugacy* between f and f_c. Any quadratic function is conjugate to f_c for some c, so by studying the Julia sets of f_c for $c \in \mathbb{C}$ we effectively study the Julia sets of all quadratic polynomials. Since h is a similarity transformation, the Julia set of any quadratic polynomial is geometrically similar to that of f_c for some $c \in \mathbb{C}$.

It should be borne in mind throughout this section that $f_c^{-1}(z)$ takes two distinct values $\pm(z - c)^{1/2}$, called the two *branches* of $f_c^{-1}(z)$, except when $z = c$. Thus if U is a small open set with $c \notin U$, then the pre-image $f_c^{-1}(U)$ has two parts, both of which are mapped bijectively and smoothly by f_c onto U.

We define the *Mandelbrot set M* to be the set of parameters c for which the

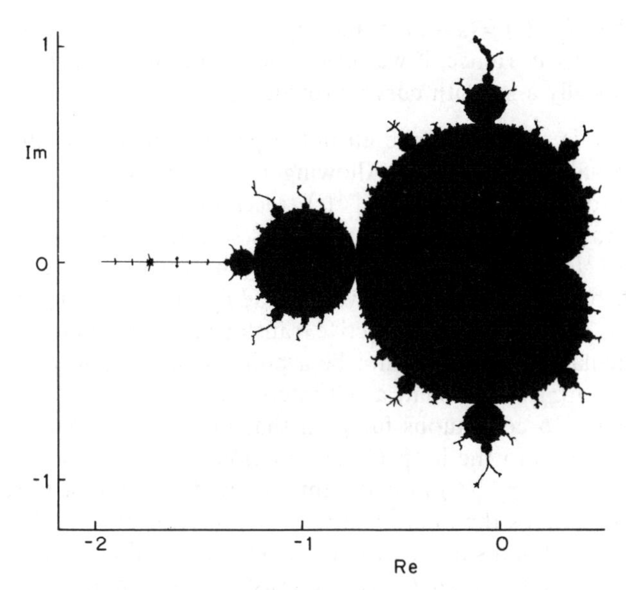

Figure 14.2 The Mandelbrot set M in the complex plane

Julia set of f_c is connected

$$M = \{c \in \mathbb{C} : J(f_c) \text{ is connected}\}. \tag{14.4}$$

At first, M appears to relate to one rather specific property of $J(f_c)$. In fact, as we shall see, M contains an enormous amount of information about the structure of Julia sets.

The definition (14.4) is awkward for computational purposes. We derive an equivalent definition that is much more useful for determining whether a parameter c lies in M and for investigating the extraordinarily intricate form of M; see figure 14.2.

To do this, we first need to know a little about the effect of the transformation f_c on smooth curves. For brevity, we term a smooth (i.e. differentiable), closed, simple (i.e. non-self-intersecting) curve in the complex plane a *loop*. We refer to the parts of $\mathbb{C}$ inside and outside such a curve as the *interior* and *exterior* of the loop. A *figure of eight* is a smooth closed curve with a single point of self-intersection.

Lemma 14.13

Let C be a loop in the complex plane.

- (*a*) *If c is inside C then $f_c^{-1}(C)$ is a loop, with the inverse image of the interior of C as the interior of $f_c^{-1}(C)$.*
- (*b*) *If c lies on C then $f_c^{-1}(C)$ is a figure of eight, such that the inverse image of the interior of C is the interior of the two loops.*

Proof. Note that $f_c^{-1}(z)=(z-c)^{1/2}$ and $(f_c^{-1})'(z)=\frac{1}{2}(z-c)^{-1/2}$, which is finite and non-zero if $z\neq c$. Hence, if we select one of the two branches of f_c^{-1}, the set $f_c^{-1}(C)$ is locally a smooth curve, provided $c\notin C$.

(*a*) Suppose c is inside C. Take an initial point w on C and choose one of the two values for $f_c^{-1}(w)$. Allowing $f_c^{-1}(z)$ to vary continuously as z moves around C, the point $f_c^{-1}(z)$ traces out a smooth curve. When z returns to w, however, $f_c^{-1}(w)$ takes its second value. As z traverses C again, $f_c^{-1}(z)$ continues on its smooth path, which closes as z returns to w the second time. Since $c\notin C$, we have $0\notin f_c^{-1}(C)$, so $f_c'(z)\neq 0$ on $f_c^{-1}(C)$. Thus f_c is locally a smooth bijective transformation near points on $f_c^{-1}(C)$. In particular $z\in f_c^{-1}(C)$ cannot be a point of self-intersection of $f_c^{-1}(C)$, otherwise $f_c(z)$ would be at a self-intersection of C.

Since f_c is a continuous function that maps the loop $f_c^{-1}(C)$ and no other points onto the loop C, the polynomial f_c must map the interior and exterior of $f_c^{-1}(C)$ into the interior and exterior of C respectively. Hence f_c^{-1} maps the interior of C to the interior of $f_c^{-1}(C)$.

(*b*) This is proved in a similar way to (*a*), noting that if C_0 is a smooth piece of curve through c, then $f_c^{-1}(C_0)$ consists of two smooth pieces of curve through 0 which cross at right angles, providing the self-intersection of the figure of eight. □

We now obtain the alternative characterization of the Mandelbrot set in terms of the iterates of f_c.

Theorem 14.14

$$M=\{c\in\mathbb{C}:\{f_c^k(0)\}_{k\geqslant 1} \text{ is bounded}\} \tag{14.5}$$

$$=\{c\in\mathbb{C}:f_c^k(0)\not\to\infty \text{ as } k\to\infty\}. \tag{14.6}$$

Proof. Since there is a number r such that $|f_c(z)|>2|z|$ if $|z|>r$, it is clear that $f_c^k(0)\not\to\infty$ if and only if $\{f_c^k(0)\}$ is bounded, so (14.5) and (14.6) are equal.

(*a*) We first show that if $\{f_c^k(0)\}$ is bounded then $J(f_c)$ is connected. Let C be a large circle in $\mathbb{C}$ such that all the points $\{f_c^k(0)\}$ lie inside C, such that $f_c^{-1}(C)$ is interior to C and such that points outside C iterate to ∞ under f_c^k. Since $c=f_c(0)$ is inside C, Lemma 14.13(*a*) gives that $f_c^{-1}(C)$ is a loop contained in the interior of C. Also, $f_c(c)=f_c^2(0)$ is inside C and f_c^{-1} maps the exterior of C onto the exterior of $f_c^{-1}(C)$, so c is inside $f_c^{-1}(C)$. Thus $f_c^{-2}(C)$ is a loop contained in the interior of $f_c^{-1}(C)$. Proceeding in this way, we see that $\{f_c^{-k}(C)\}$ consists of a sequence of loops, each containing the next in its interior (figure 14.3(*a*)). Let K denote the closed set of points that are on or inside the loops $f_c^{-k}(C)$ for all k. If $z\in\mathbb{C}\backslash K$ some iterate $f_c^k(z)$ lies outside C and so $f_c^k(z)\to\infty$. Thus

$$A(\infty)=\{z:f_c^k(z)\to\infty \quad \text{as} \quad k\to\infty\}=\mathbb{C}\backslash K.$$

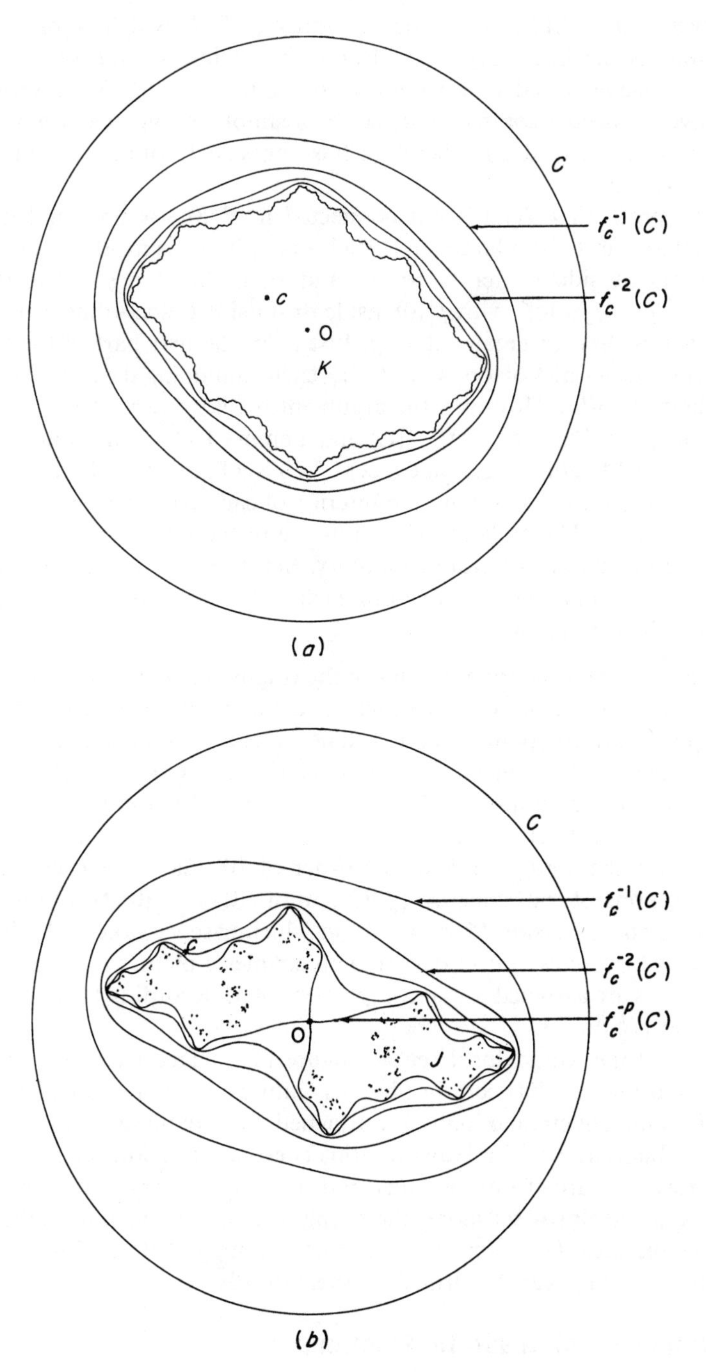

Figure 14.3 Inverse iterates of a circle c under f_c, illustrating the two parts of the proof of Theorem 14.14: (*a*) $c = -0.3 + 0.3i$; (*b*) $c = -0.9 + 0.5i$

By Lemma 14.11 $J(f_c)$ is the boundary of $\mathbb{C}\backslash K$ which is, of course, the same as the boundary of K. But K is the intersection of a decreasing sequence of closed simply connected sets (i.e. sets that are connected and have connected complement), so, by a simple topological argument, K is simply connected and therefore has connected boundary. Thus $J(f_c)$ is connected.

(*b*) The proof that $J(f_c)$ is not connected if $\{f_c^k(0)\}$ is unbounded is fairly similar. Let C be a large circle such that $f_c^{-1}(C)$ is inside C, such that all points outside C iterate to ∞, and such that for some p, the point $f_c^{p-1}(c) = f_c^p(0) \in C$ with $f_c^k(0)$ inside or outside C according as to whether k is less than or greater than p. Just as in the first part of the proof, we construct a series of loops $\{f_c^{-k}(C)\}$, each containing the next in its interior (figure 14.3(*b*)). However, the argument breaks down when we get to the loop $f_c^{1-p}(C)$, since $c \in f_c^{1-p}(C)$ and Lemma 14.13(*a*) does not apply. By Lemma 14.13(*b*) we get that $E \equiv f^{-p}(C)$ is a figure of eight inside the loop $f_c^{1-p}(C)$, with f_c mapping the interior of each half of E onto the interior of $f_c^{1-p}(C)$. The Julia set $J(f_c)$ must lie in the interior of the loops of E, since other points iterate to infinity. Since $J(f_c)$ is invariant under f_c^{-1}, parts of it must be contained in each of the loops of E. Thus this figure of eight E disconnects $J(f_c)$. □

The reason for considering iterates of the origin in (14.5) and (14.6) is that the origin is the *critical point* of f_c for each c, i.e. the point for which $f_c'(z) = 0$. The critical points are the points where f_c fails to be a local bijection—a property that was crucial in distinguishing the two cases in the proof of Theorem 14.14.

The equivalent definition of M provided by (14.5) is the basis of computer pictures of the Mandelbrot set.

Choose numbers r, k_0, both of the order of 100, say. For each c compute successive terms of the sequence $\{f_c^k(0)\}$ until either $|f_c^k(0)| > r$, in which case c is deemed to be outside M, or $k = k_0$ in which case we take $c \in M$. Repeating this process for values of c across a region enables a picture of M to be drawn. Often colours are assigned to the complement of M according to the first integer k such that $|f_c^k(0)| > r$.

Pictures of the Mandelbrot set (see figure 14.2) suggest that it has a highly complicated form. It has certain obvious features: a main cardioid to which a series of prominent circular 'buds' are attached. Each of these buds is surrounded by further buds, and so on. However, this is not all. In addition, fine, branched 'hairs' grow outwards from the buds, and these hairs carry miniature copies of the entire Mandelbrot set along their length. It is easy to miss these hairs in computer pictures. However, accurate pictures suggest that M is a connected set, a fact that has been confirmed mathematically.

14.3 Julia sets of quadratic functions

In this section we will see a little of how the structure of the Julia set $J(f_c)$ changes as the parameter c varies across the complex plane. In particular, the significance of the various parts of the Mandelbrot set will start to become apparent.

The attractive periodic points of f_c are crucial to the form of $J(f_c)$. It may be shown (see Exercise 14.8) that if $w \neq \infty$ is an attractive periodic point of a polynomial f then there is a critical point z (a point with $f'(z)=0$) such that $f^k(z)$ is attracted to the periodic orbit containing w. Since the only critical point of f_c is 0, f_c can have at most one attractive periodic orbit. Moreover, if $c \notin M$ then, by Theorem 14.14, $f_c^k(0) \to \infty$, so f_c can have no attractive periodic orbit. It is conjectured, but not yet proved, that the set of c for which f_c has an attractive periodic orbit fills the interior of M.

It is natural to categorize f_c by the period p of the (finite) attractive orbit, if any; the values of c corresponding to different p may be identified as different regions of the Mandelbrot set M.

To begin with, suppose c lies outside M, so f_c has no attractive periodic points. By definition, $J(f_c)$ is not connected. In fact, $J(f_c)$ must be totally disconnected and expressible as the disjoint union $J = S_1(J) \cup S_2(J)$, where S_1 and S_2 are the two branches of f_c^{-1} on J. This means that J is invariant in the sense of (9.1). Basically, this follows from the second half of the proof of Theorem 14.14—we get that f_c maps the interior of each loop of a figure of eight E onto a region D containing E. The mappings S_1 and S_2 may be taken as the restrictions of f_c^{-1} to the interior of each loop. Since $S_1(J)$ and $S_2(J)$ are interior to the two halves of E, they are disjoint, so J must be totally disconnected; see page 116.

We look at this situation in more detail when c is large enough to allow some simplifications to be made.

Theorem 14.15

Suppose $|c| > \frac{1}{4}(5 + 2\sqrt{6})$. *Then* $J(f_c)$ *is totally disconnected, and is the invariant set (in the sense of* (9.1)) *of the contractions given by the two branches of* $f_c^{-1}(z)$

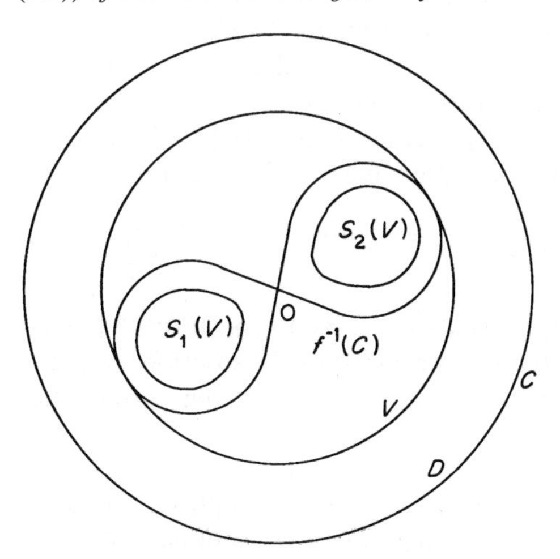

Figure 14.4 Proof of Theorem 14.15

for z near J. When $|c|$ is large

$$\dim_{\mathrm{B}} J(f_c) = \dim_{\mathrm{H}} J(f_c) \sim 2\log 2/\log|c|.$$

Proof. Let C be the circle $|z| = |c|$ and D its interior $|z| < |c|$. Then

$$f_c^{-1}(C) = \{(c\,\mathrm{e}^{i\theta} - c)^{1/2} : 0 \leqslant \theta \leqslant 4\pi\}$$

which is a figure of eight with self-intersection point at 0, with the loops on either side of a straight line through the origin (figure 14.4). Since $|c| > 2$ we have $f_c^{-1}(C) \subset D$. The interior of each of the loops of $f_c^{-1}(C)$ is mapped by f_c in a bijective manner onto D. If we define $S_1, S_2 : D \to D$ as the branches of $f_c^{-1}(z)$ inside each loop, then $S_1(D)$ and $S_2(D)$ are the interiors of the two loops.

Let V be the disc $\{z : |z| < |2c|^{1/2}\}$. We have chosen the radius of V so that V just contains $f_c^{-1}(C)$, so $S_1(D), S_2(D) \subset V \subset D$. Hence $S_1(V), S_2(V) \subset V$ with $S_1(\bar{V})$ and $S_2(\bar{V})$ disjoint. We have

$$S_1'(z) = S_2'(z) = (f_c^{-1})'(z) = \tfrac{1}{2}(z - c)^{-1/2}.$$

Hence if $z \in \bar{V}$,

$$\tfrac{1}{2}(|c| + |2c|^{1/2})^{-1/2} \leqslant |S_i'(z)| \leqslant \tfrac{1}{2}(|c| - |2c|^{1/2})^{-1/2}. \tag{14.7}$$

The upper bound is less than 1 if $|c| > \frac{1}{4}(5 + 2\sqrt{6})$, in which case S_1 and S_2 are contractions on the disc $\bar{V}$. By Theorem 9.1 there is a unique non-empty compact invariant set $F \subset \bar{V}$ satisfying

$$S_1(F) \cup S_2(F) = F. \tag{14.8}$$

Since $S_1(\bar{V})$ and $S_2(\bar{V})$ are disjoint, so are $S_1(F)$ and $S_2(F)$, implying that F is totally disconnected.

Of course, F is none other than the Julia set $J = J(f_c)$. One way to see this is to note that, since $\bar{V}$ contains at least one point z of J (for example, a repelling fixed point of f_c), we have $J = \mathrm{closure}(\bigcup_{k=1}^{\infty} f_c^{-k}(z)) \subset \bar{V}$, since $f_c^{-k}(\bar{V}) \subset \bar{V}$. Using further results from Summary 14.12, J is a non-empty compact subset of $\bar{V}$ satisfying $J = f_c^{-1}(J)$ or, equivalently, $J = S_1(J) \cup S_2(J)$. Thus $J = F$, the unique compact set satisfying (14.8).

Finally, we estimate the dimension of $J(f_c) = F$. It may be shown, using (14.7) and an appropriate complex 'mean-value theorem' that,

$$\tfrac{1}{2}(|c| + |2c|^{1/2})^{-1/2} \leqslant \frac{|S_i(z_1) - S_i(z_2)|}{|z_1 - z_2|} \leqslant \tfrac{1}{2}(|c| - |2c|^{1/2})^{-1/2}$$

if z_1, z_2 are distinct points of $\bar{V}$. By Propositions 9.6 and 9.7 lower and upper bounds for $\dim_{\mathrm{H}} J(f_c)$ are provided by the solutions of $2(\frac{1}{2}(|c| \pm |2c|^{1/2})^{-1/2})^s = 1$, i.e. by $s = 2\log 2/\log 4(|c| \pm |2c|^{1/2})$, which gives the stated asymptotic estimate. □

We next turn to the case where c is small. We know that if $c = 0$ then $J(f_c)$ is the unit circle. If c is small and z is small enough, then $f_c^k(z) \to w$ as $k \to \infty$, where w is the attractive fixed point $\frac{1}{2}(1-\sqrt{1-4c})$ close to 0. On the other hand, $f_c^k(z)\to\infty$ if z is large. It is not unreasonable to expect the circle to 'distort' into a simple closed curve (i.e. having no points of self-intersection) separating these two types of behaviour as c moves away from 0.

In fact, this is the case provided that f_c retains an attractive fixed point, i.e. if $|f_c'(z)|<1$ at one of the roots of $f_c(z)=z$. Simple algebra shows that this happens if c lies inside the cardioid $z=\frac{1}{2}e^{i\theta}(1-\frac{1}{2}e^{i\theta})\,(0\leqslant\theta\leqslant 2\pi)$—this is the main cardioid of the Mandelbrot set.

For convenience, we treat the case of $|c|<\frac{1}{4}$, but the proof is easily modified if f_c has any attractive fixed point.

Theorem 14.16

If $|c|<\frac{1}{4}$ then $J(f_c)$ is a simple closed curve.

Proof. Let C_0 be the curve $|z|=\frac{1}{2}$, which encloses both c and the attractive fixed point of f_c. Then by direct calculation the inverse image $f_c^{-1}(C_0)$ is a loop C_1 surrounding C_0. We may fill the annular region A_1 between C_0 and C_1 by a continuum of curves, which we call 'trajectories', which leave C_0 and reach C_1 perpendicularly; see figure 14.5(a). For each θ let $\psi_1(\theta)$ be the point on C_1 at the end of the trajectory leaving C_0 at $\psi_0(\theta)=\frac{1}{2}e^{i\theta}$. The inverse image $f_c^{-1}(A_1)$ is an annular region A_2 with outer boundary the loop $C_2=f_c^{-1}(C_1)$ and inner boundary C_1, with f_c mapping A_2 onto A_1 in a two-to-one manner. The inverse image of the trajectories joining C_0 to C_1 provides a family of trajectories joining C_1 to C_2. Let $\psi_2(\theta)$ be the point on C_2 at the end of the trajectory leaving C_1 at $\psi_1(\theta)$. We continue in this way to get a sequence of loops C_k, each surrounding its predecessor, and families of trajectories joining the points $\psi_k(\theta)$ on C_k to $\psi_{k+1}(\theta)$ on C_{k+1} for each k.

As $k\to\infty$, the curves C_k approach the boundary of the basin of attraction of w; by Lemma 14.11 this boundary is just the Julia set $J(f_c)$. Since $|f_c'(z)|>\gamma$ for some $\gamma>1$ outside C_1, it follows that f_c^{-1} is contracting near J. Thus the length of the trajectory joining $\psi_k(\theta)$ to $\psi_{k+1}(\theta)$ converges to 0 at a geometric rate as $k\to\infty$. Consequently $\psi_k(\theta)$ converges uniformly to a continuous function $\psi(\theta)$ as $k\to\infty$, and J is the closed curve given by $\psi(\theta)\,(0\leqslant\theta\leqslant 2\pi)$.

It remains to show that ψ represents a simple curve. Suppose that $\psi(\theta_1)=\psi(\theta_2)$. Let D be the region bounded by C_0 and the two trajectories joining $\psi_0(\theta_1)$ and $\psi_0(\theta_2)$ to this common point. The boundary of D remains bounded under iterates of f_c, so by the maximum modulus theorem (that the modulus of an analytic function takes its maximum on the boundary point of a region) D remains bounded under iteration of f. By Lemma 14.6 the interior of D cannot contain any points of J. Thus the situation of figure 14.5(b) cannot occur, so that $\psi(\theta)=\psi(\theta_1)=\psi(\theta_2)$ for all θ between θ_1 and θ_2. It follows that $\psi(\theta)$ has no point of self-intersection. $\square$

By an extension of this argument, if c is in the main cardioid of M, then $J(f_c)$

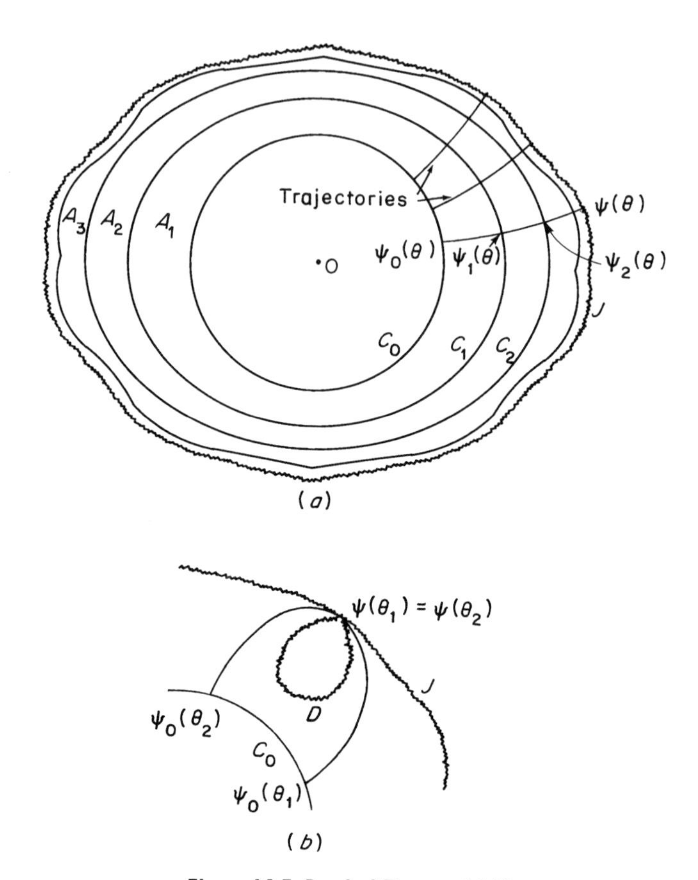

Figure 14.5 Proof of Theorem 14.16

is a simple closed curve; such curves are sometimes referred to as quasi-circles. Of course, $J(f_c)$ will be a fractal curve if $c > 0$. It may be shown that, for small c, its dimension is given by

$$s = \dim_B J(f_c) = \dim_H J(f_c) = 1 + |c|^2/4\log 2 + \text{terms in } |c|^3 \text{ and higher powers.} \tag{14.9}$$

Moreover, $0 < \mathscr{H}^s(J) < \infty$, with $\dim_B J(f_c) = \dim_H J(f_c)$ given by a real analytic function of c.

The next case to consider is when f_c has an attractive periodic orbit of period 2. By a straightforward calculation this occurs if $|c+1| < \frac{1}{4}$; that is, if z lies in the prominent circular disc of M abutting the cardioid.

Since f_c^2 is a polynomial of degree 4, f_c has two fixed points and two period-2 points. Let w_1 and w_2 be the points of the attractive period-2 orbit. It may be shown, as in the proof of Theorem 14.16, that the basin of attraction for w_i (i.e. $\{z: f_c^{2k}(z) \to w_i \text{ as } k \to \infty\}$) includes a region bounded by a simple closed curve

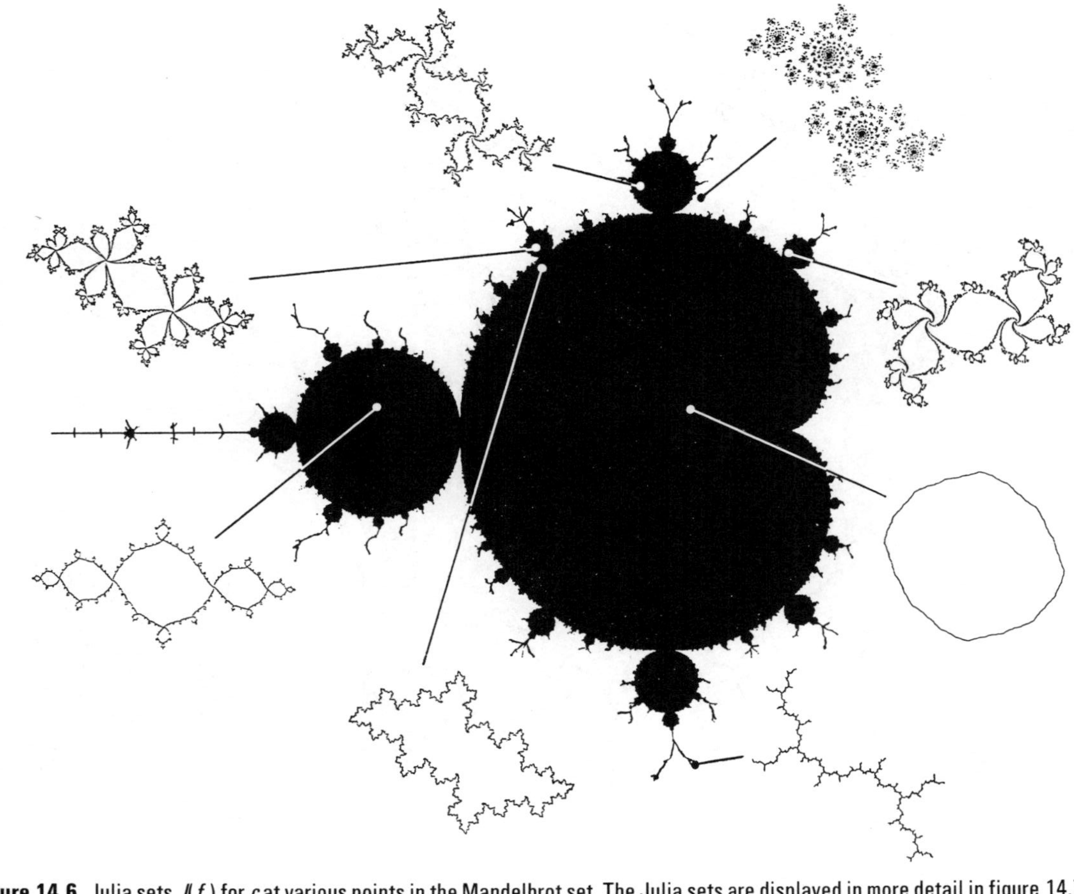

Figure 14.6 Julia sets $J(f_c)$ for c at various points in the Mandelbrot set. The Julia sets are displayed in more detail in figure 14.7

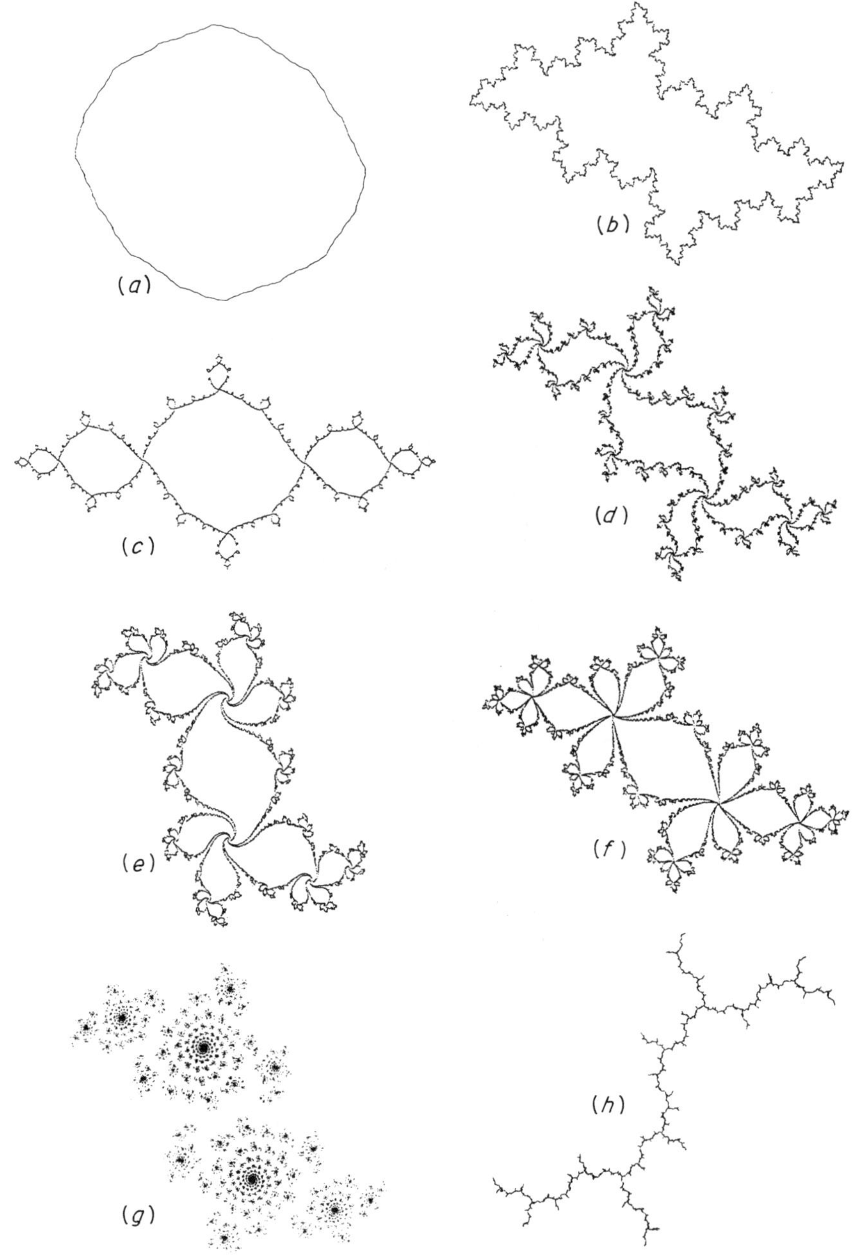

Figure 14.7 A selection of Julia sets of the quadratic function $f_c(z) = z^2 + c$. (*a*) $c = -0.1 + 0.1\mathrm{i}$; f_c has an attractive fixed point, and J is a quasi-circle. (*b*) $c = -0.5 + 0.5\mathrm{i}$; f_c has an attractive fixed point, and J is a quasi-circle. (*c*) $c = -1 + 0.05\mathrm{i}$; f_c has an attractive period-2 orbit. (*d*) $c = -0.2 + 0.75\mathrm{i}$, f_c has an attractive period-3 orbit. (*e*) $c = 0.25 + 0.52\mathrm{i}$; f_c has an attractive period-4 orbit, (*f*) $c = -0.5 + 0.55\mathrm{i}$; f_c has an attractive period-5 orbit. (*g*) $c = 0.66\mathrm{i}$; f_c has no attractive orbits and J is totally disconnected. (*h*) $c = -\mathrm{i}$, $f_c^2(0)$ is periodic and J is a dendrite

C_i surrounding w_i, for $i=1,2$. By Lemma 14.11 and Proposition 14.5, $C_i \subset J(f_c^2)=J(f_c)$. The curves C_i are mapped onto themselves in a two-to-one fashion by f_c^2, which implies that there is a fixed point of f_c^2 on each C_i. The period-2 points are strictly inside the C_i, so there is a fixed point of f_c on each C_i; since the C_i are mapped onto each other by f_c, the only possibility is for C_1 and C_2 to touch at one of the fixed points of f_c. The inverse function f_c^{-1} is two-valued on C_1. One of the inverse images is C_2 (which encloses w_2). However, the other branch of $f_c^{-1}(C_1)$ is a further simple closed curve enclosing the second value of $f_c^{-1}(w_1)$. We may continue to take inverse images in this way to find that $J(f_c)$ is made up of infinitely many simple closed which enclose the pre-images of w_1 and w_2 of all orders and touch each other in pairs at 'pinch points'—see figure 14.7(*c*). Thus we get fractal Julia sets that are topologically much more complicated than in the previous cases.

It is possible to use these sorts of ideas to analyse the case when f_c has an attractive periodic orbit of period $p>2$. The immediate neighbourhoods of the period-p points that are drawn in to the attractive orbits are bounded by simple closed curves which touch each other at a common point. The Julia set consists of these fractal curves together with all their pre-images under f^k.

A variety of examples are shown in figures 14.6 and 14.7. The 'buds' on the Mandelbrot set corresponding to attractive orbits of period p are indicated in figure 14.8.

The Julia sets $J(f_c)$ that are most intricate, and are mathematically hardest to analyse are at the 'exceptional' values of c on the boundary of M. If c is on the boundary of the cardioid or a bud of M, then f_c has an indifferent periodic point. If c is at a 'neck' where a bud touches a parent region, then $J(f_c)$ includes a series of 'tendrils' joining its boundary to the indifferent periodic points. For c elsewhere on the boundary of the cardioid the Julia set may contain 'Siegel discs'. The Julia set $J(f_c)$ consists of infinitely many curves bounding open regions, with f_c mapping each region into a 'larger' one, until the region containing the fixed point is reached. Inside this *Siegel disc*, f_c rotates points on invariant circles around the fixed point.

There are still further possibilities. If c is on one of the 'hairs' of M then $J(f_c)$ may be a *dendrite*, i.e. of tree-like form. This occurs if an iterate of the critical point 0 is periodic, i.e. if $f_c^k(0)=f_c^{k+q}(0)$ for positive integers k and q.

We have mentioned that there are miniature copies of M located in the hairs of M. If c belongs to one of these, then $J(f_c)$ will be of dendrite form, but with small copies of the Julia set from the corresponding value of c in the main part of M inserted at the 'vertices' of the dendrite.

A good way to explore the range of Julia sets and, indeed, the Julia sets of other functions, is using a computer. There are two usual methods of drawing Julia sets, based on properties that we have discussed.

For the first method, we choose a repelling periodic point z. For suitable k, we may compute the set of inverse images $J_k = f^{-k}(z)$. By Corollary 14.7(*b*) these 2^k points are in J, and should fill J as k becomes large. A difficulty with picturing J in this way is that the points of J_k need not be uniformly distributed

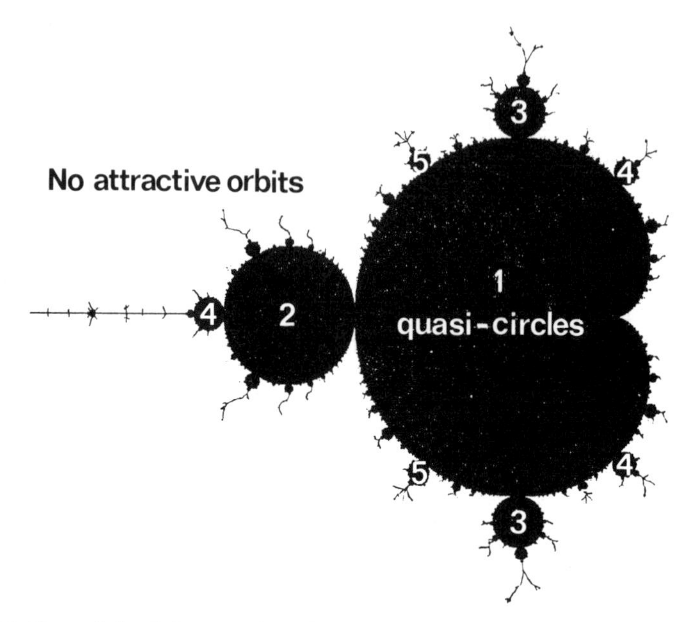

Figure 14.8 The periods of the attractive orbits of f_c for c in various parts of the Mandelbrot set M. If c is in the main cardioid, f_c has an attractive fixed point, and the Julia set $J(f_c)$ is a quasi-circle. For c in the buds of M, f_c has an attractive orbit with the period p shown, with p regions inside the Julia set $J(f_c)$ meeting at each pinch point. Outside M, the function f_c has no attractive orbits and $J(f_c)$ is totally disconnected

across J—they may tend to cluster in some parts of J and be sparse in other parts. Consequently, even with k quite large, parts of J can be missed altogether. (This tends to happen for f_c with c close to the boundary of M.) There are various ways of getting around this difficulty. For instance, with $J_0 = \{z\}$, instead of taking $J_k = f^{-1}(J_{k-1})$ for each k, we can choose a subset J_k of $f^{-1}(J_{k-1})$ by ignoring all but one of the points in every 'small' cluster. This ensures that we are working with a reasonably well distributed set of points of J at each step of the iteration, and also reduces the calculation involved.

A second method is to test individual points to see if they are close to the Julia set. Suppose, for example, that f has two or more attractive fixed points (now including ∞ if f is a polynomial). If z is a point of $J(f)$ then there are points arbitrarily close to z in the attractive basin of each attractive point by Lemma 14.11. To find J we divide a region of $\mathbb{C}$ into a fine mesh. We examine the ultimate destination under iteration by f of the four corners of each mesh square. If two of the corners are attracted to different points, we deem the mesh square to contain a point of J. Often, the other squares, the 'Fatou set', are coloured according to which point the vertices of the square is attracted to, perhaps with different shading according to how close the kth iterates are to the attractive point for some fixed k.

Both of these methods can be awkward to use in certain cases. A knowledge of the mathematical theory is likely to be beneficial in overcoming the difficulties that can occur.

14.4 Characterisation of quasi-circles by dimension

We saw in the previous section that, if c is in the main cardioid of the Mandelbrot set, then the Julia set of $f_c(z) = z^2 + c$ is a simple closed curve. By similar arguments, the Julia set of $f(z) = z^n + c$ is a simple closed curve for any integer $n \geqslant 2$ provided that c is small enough, and, indeed, the same is true for $f(z) = z^2 + g(z)$ for a wide variety of analytic functions g that are 'sufficiently small' near the origin. Thus all these functions have Julia sets that are topologically the same—they are all homeomorphic to a circle. The surprising thing is that they are essentially the same *as fractals*, in other words are Lipschitz equivalent, if and only if they have the same Hausdorff dimension. Of course, if two sets have different dimensions they cannot be Lipschitz equivalent (Corollary 2.4). However, in this particular situation the converse is also true.

We term a set F a *quasi-self-similar circle* or *quasi-circle* if the following conditions are satisfied.

(i) F is homeomorphic to a circle (i.e. F is a simple closed curve).
(ii) $0 < \mathscr{H}^s(F) < \infty$ where $s = \dim_H F$.
(iii) There are constants $a, b, r > 0$ such that for any subset U of F with $|U| \leqslant r$ there is a mapping $\varphi: U \to F$ such that

$$a|x - y| \leqslant |U||\varphi(x) - \varphi(y)| \leqslant b|x - y| \qquad (x, y \in F). \qquad (14.10)$$

The 'quasi-self-similar' condition (iii) says that arbitrarily small parts of F are 'roughly similar' to a large part of F.

The following theorem depends on using s-dimensional Hausdorff measure to measure 'distance' round a quasi-circle.

Theorem 14.17

Quasi-circles E and F are bi-Lipschitz equivalent if and only if $\dim_H E = \dim_H F$.

Sketch of proof. If there is a bi-Lipschitz mapping between E and F then $\dim_H E = \dim_H F$ by Corollary 2.4(*b*).

Suppose that $\dim_H E = \dim_H F$. Let $E(x, y)$ be the 'arc' of E between points $x, y \in E$, taken in the clockwise sense, with a similar notation for arcs of F. Conditions (ii) and (iii) imply that $\mathscr{H}^s(E(x, y))$ is continuous in $x, y \in E$ and is positive if $x \neq y$. We claim that there are constants $c_1, c_2 > 0$ such that

$$c_1 \leqslant \frac{\mathscr{H}^s(E(x, y))}{|x - y|^s} \leqslant c_2 \qquad (14.11)$$

whenever $E(x, y)$ is the 'shorter' arc, i.e. $\mathscr{H}^s(E(x, y)) \leqslant \mathscr{H}^s(E(y, x))$. Assume that

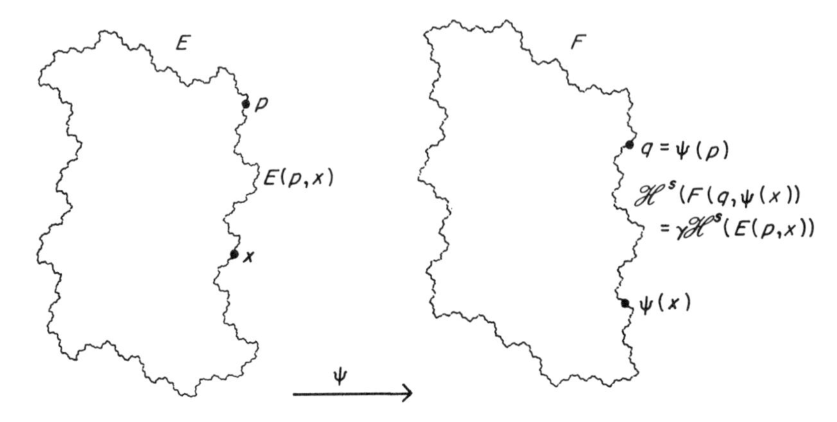

Figure 14.9 Setting up a bi-Lipschitz mapping ψ between two quasi-circles of Hausdorff dimension s

$\varepsilon > 0$ is sufficiently small. If $|x-y| \geqslant \varepsilon$ then (14.11) is true by a continuity argument for suitable constants. If $|x-y| < \varepsilon$ there is a mapping $\varphi: E(x,y) \to E$ satisfying (14.10) such that $|\varphi(x) - \varphi(y)| \geqslant \varepsilon$. Inequalities (14.10) and (2.8) and (2.9) imply that the ratio (14.11) changes by a bounded amount if x and y are replaced by $\varphi(x)$ and $\varphi(y)$, so (14.11) holds for suitable c_1 and c_2 for all $x, y \in E$.

Now choose base points $p \in E, q \in F$. Set $\gamma = \mathscr{H}^s(E)/\mathscr{H}^s(F)$ and define $\psi: E \to F$ by taking $\psi(x)$ to be the point of F such that

$$\mathscr{H}^s(E(p,x)) = \gamma \mathscr{H}^s(F(q, \psi(x)))$$

(see figure 14.9). Then ψ is a continuous bijection, and also

$$\mathscr{H}^s(E(x,y)) = \gamma \mathscr{H}^s(F(\psi(x), \psi(y))) \quad (x, y \in E).$$

Using (14.11) together with similar inequalities for arcs of F, this gives

$$c_3 \leqslant \frac{|\psi(x) - \psi(y)|}{|x-y|} \leqslant c_4$$

if $x \neq y$, so that ψ is bi-Lipschitz, as required. □

Corollary 14.18

Suppose that the Julia sets J_1 and J_2 of the polynomials f_1 and f_2 are simple closed curves. Suppose that f_i is strictly repelling on J_i (i.e. $|f_i'(z)| > 1$ for $i = 1, 2$). Then J_1 and J_2 are bi-Lipschitz equivalent if and only if $\dim_H J_1 = \dim_H J_2$.

Proof. It may be shown that if a polynomial f is strictly repelling on its Julia set J then $0 < \mathscr{H}^s(J) < \infty$, where $s = \dim_H J$. Moreover, given a subset U of J, we may choose k so that $f^k(U)$ has diameter comparable with that of J itself, and (14.10) holds taking $\varphi = f^k$ (this reflects the quasi-self-similarity of J). Thus J_1 and J_2 are quasi-circles to which Theorem 14.17 may be applied. □

14.5 Newton's method for solving polynomial equations

Anyone who has done any numerical analysis will have found roots of equations using Newton's method. Let $p(x)$ be a function with continuous derivative. If $f(x) = x - p(x)/p'(x)$ then the iterates $f^k(x)$ converge to a solution of $p(x) = 0$ provided that $p'(x) \neq 0$ at the solution and that the initial value of x is chosen appropriately. Cayley suggested investigating the method in the complex plane, and in particular which initial points of $\mathbb{C}$ iterate to which zero of p.

Let $p: \mathbb{C} \to \mathbb{C}$ be a polynomial with complex coefficients, and form the rational function $f: \mathbb{C} \cup \{\infty\} \to \mathbb{C} \cup \{\infty\}$

$$f(z) = z - p(z)/p'(z). \tag{14.12}$$

Then the fixed points of f, given by $p(z)/p'(z) = 0$, are the zeros of p together with ∞. Differentiating, we find that

$$f'(z) = p(z)p''(z)/p'(z)^2 \tag{14.13}$$

so a zero z of p is a superattractive fixed point of f, provided that $p'(z) \neq 0$. If $|z|$ is large, $f(z) \sim z(1 - 1/n)$, where n is the degree of p, so ∞ is a repelling point of f. As usual, we write

$$A(w) = \{z : f^k(z) \to w\} \tag{14.14}$$

for the basin of attraction of the zero w, i.e. the set of initial points which converge to w under Newton iteration. Since the zeros are attractive, the basin $A(w)$ includes an open region containing w. We shall see, however, that $A(w)$ can be remarkably complicated further away from w.

The theory of Julia sets developed for polynomials in Section 14.1 is almost the same for a rational function, provided that the point $\{\infty\}$ is included in the natural way. The main differences are that if f is a rational function $J(f)$ need not be bounded (though it must be closed) and it is possible for $J(f)$ to have interior points, in which case $J(f) = \mathbb{C} \cup \{\infty\}$. However, Lemma 14.11 remains true, so that $J(f)$ is the boundary of $A(w)$ for each attractive fixed point w. Thus $J(f)$ is likely to be important when analysing the domains of attraction of the roots in Newton's method.

A straightforward case is the quadratic polynomial

$$p(z) = z^2 - c$$

with zeros $\pm\sqrt{c}$ (as before, more general quadratic polynomials can be reduced to this form by a conjugacy). Newton's formula (14.12) becomes

$$f(z) = (z^2 + c)/2z.$$

Thus

$$f(z) \pm \sqrt{c} = (z \pm \sqrt{c})^2/2z$$

so

$$\frac{f(z) + \sqrt{c}}{f(z) - \sqrt{c}} = \left(\frac{z + \sqrt{c}}{z - \sqrt{c}}\right)^2. \tag{14.15}$$

It follows immediately that if $|z+\sqrt{c}|/|z-\sqrt{c}|<1$ then $|f^k(z)+\sqrt{c}|/|f^k(z)-\sqrt{c}| \to 0$ and $f^k(z) \to -\sqrt{c}$ as $k \to \infty$, and similarly if $|z+\sqrt{c}|/|z-\sqrt{c}|>1$ then $f^k(z) \to \sqrt{c}$. The Julia set $J(f)$ is the line $|z+\sqrt{c}|=|z-\sqrt{c}|$ (the perpendicular bisector of $-\sqrt{c}$ and $\sqrt{c}$) and $A(-\sqrt{c})$ and $A(\sqrt{c})$ are the half-planes on either side. (Letting $h(z)=(z+\sqrt{c})/(z-\sqrt{c})$ in (14.15) gives $f(z)=h^{-1}(h(z))^2$, so that f is conjugate to, and therefore has similar dynamics to, the mapping $g(z)=z^2$.) In this case the situation is very regular—any initial point is iterated by f to the nearest zero of p.

The quadratic example might lead us to hope that the domains of attraction under Newton iteration of the zeros of any polynomial are reasonably regular. However, for higher-order polynomials the situation is fundamentally different. Lemma 14.11 provides a hint that something very strange happens. If p has zeros $z_1, \ldots, z_n$ with $p'(z_i) \neq 0$, Lemma 14.11 tells us that the Julia set of f is the boundary of the domain of the attraction of *every* zero:

$$J(f) = \partial A(z_1) = \cdots = \partial A(z_n).$$

A point on the boundary of any one of the domains of attraction must be on the boundary of all of them; since $J(f)$ is uncountable, there are a great many such multiple boundary points. An attempt to visualize three or more disjoint sets with this property will convince the reader that they must be very complicated indeed.

Let us look at a specific example. The cubic polynomial

$$p(z) = z^3 - 1$$

has zeros $1, e^{i2\pi/3}, e^{i4\pi/3}$, and Newton function

$$f(z) = \frac{2z^3+1}{3z^2}.$$

The transformation $\rho(z) = z\,e^{i2\pi/3}$ is a rotation of $120°$ about the origin. It is easily checked that $f(\rho(z)) = \rho(f(z))$, in other words ρ is a conjugacy of f to itself. It follows that a rotation of $120°$ about the origin maps $A(w)$ onto $A(w\,e^{i2\pi/3})$ for each of the three zeros w, so that the Julia set has threefold symmetry about the origin. (Of course, these symmetries would be expected from the symmetric disposition of the three zeros of p.) If z is real then $f^k(z)$ remains real for all k, and, by elementary arguments, $f^k(z)$ converges to 1 except for countably many real z. Thus $A(1)$ contains the real axis except for a countable number of points, and, by symmetry, $A(e^{i2\pi/3})$ and $A(e^{i4\pi/3})$ contain the lines through the origin making $120°$ and $240°$ to the real axis, again except for countably many points. We also know that each $A(w)$ contains an open region round w, that any point on the boundary of one of the $A(w)$ is on the boundary of all three, and that there are uncountably many such 'triple points'. Most people require the insight of a computer drawing to resolve this almost paradoxical situation, see figure 14.10.

The domain $A(1)$ is shown in black in figure 14.10(b); note that the basins of attraction of the other two zeros, obtained by rotation of $120°$ and $240°$, key

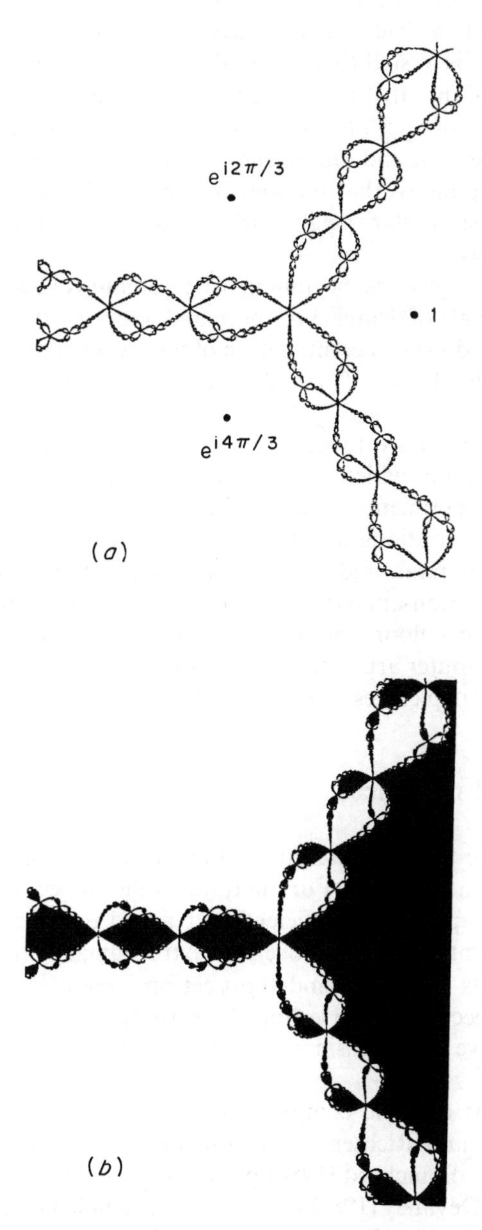

Figure 14.10 Newton's method for $p(z) = z^3 - 1$. The Julia set for the Newton function $f(z) = (z^2 + 1)/3z^2$ is shown in (*a*), and the domain of attraction of the zero $z = 1$ is shown in black in (*b*)

into the picture in a natural way. The Julia set shown in figure 14.10(a) is the boundary of the three basins and is made up of three 'chains' leading away from the origin. These fractal chains have a fine structure—arbitrarily close to each point of $J(f)$ is a 'slightly distorted' copy of the picture at the origin with six components of the $A(w)$ meeting at a point. This reflects Corollary 14.7(b): $J(f)$ is the closure of $\bigcup_{k=1}^{\infty} f^{-k}(0)$, so that if z is in $J(f)$ then there is a point w arbitrarily close to z, and an integer k, such that $f^k(w)=0$. But f^k is locally a conformal mapping, so that the local inverse $(f^k)^{-1}$ maps a neighbourhood of 0 to an 'almost similar' neighbourhood of w. The Julia set $J(f)$ exhibits quasi-self-similarity.

This, of course, is just the beginning. The domains of attraction of the zeros of other polynomials of degree 3 or more and, indeed, other analytic functions, may be investigated using a combination of theory and computer graphics. This leads to a wealth of sets of a highly intricate form that are still far from understood.

In this chapter we have touched on what is a complicated and fascinating area of mathematics in which fractals play a fundamental rôle. It is an area in which computer experiments often lead the way with mathematical theory trying to keep up. The variations are endless—we can investigate the Julia set of higher-order polynomials and of other analytic functions such as $\exp z$, as well as invariant sets of non-analytic transformations of the plane. With the advent of relatively cheap colour computer graphics, these ideas have become the basis of much computer art. A single function of simple form can lead to highly intricate yet regular pictures—often very beautiful, sometimes almost weird.

14.6 Notes and references

Much of the basic theory of iteration of complex functions was developed by Julia (1918) and Fatou (1919). For many years the subject lay almost dormant, until computer graphics was sufficiently advanced to reveal the intricate form of Julia sets. Recently, there has been an extraordinary interest in the subject. Drawing Julia sets and the Mandelbrot set on computers has almost become a craze, perhaps because of the feeling of creativity that it gives the programmer, but also there have been considerable advances in the mathematical theory of the subject.

For an account of basic complex variable theory, see Ahlfors (1979). The book by Peitgen and Richter (1986) provides a richly illustrated account of complex iteration. Blanchard (1984) provides a full survey of the mathematics and the book by Devaney (1986) contains some detailed mathematical analysis. Saupe (1987) discusses the computation of Julia sets. Mandelbrot (1980) introduced the set bearing his name; the fundamental Theorem 14.14 is given by Brolin (1965).

The formula (14.9) for the dimension of $J(f_c)$ is due to Ruelle (1982). For details of the characterization of quasi-circles by dimension, see Falconer and

Marsh (1989). Fractals associated with Newton's method are discussed in Peitgen and Richter (1986), Peitgen, Saupe and von Haeseler (1984) and Curry, Garnett and Sullivan (1983).

Exercises

14.1 Show, directly, that, if f is a complex polynomial, then the family $\{f^k\}_k$ is not normal at any repelling periodic point of f.

14.2 Describe the Julia set of $f(z) = z^2 + 4z + 2$.

14.3 Show that the Julia set of $f(z) = z^2 - 2$ is contained in the real interval $[-2, 2]$. Use Theorem 14.14 to deduce that the Julia set is connected, and hence that it *is* the interval $[-2, 2]$.

14.4 Show that f_c has an attractive fixed point precisely when c lies inside the main cardioid of the Mandelbrot set given by $z = \frac{1}{2}e^{i\theta}(1 - \frac{1}{2}e^{i\theta})$ where $(0 \leqslant \theta \leqslant 2\pi)$.

14.5 Show that if c is a non-real number with $|c| < \frac{1}{4}$ and $w = \frac{1}{2}(1 + (1 - 4c)^{1/2})$ is the repelling fixed point of $f_c(z) = z^2 + c$ then $f_c'(w)$ is not real. Deduce that the simple closed curve that forms the Julia set $J(f_c)$ cannot have a tangent at w. Hence deduce that the curve contains no differentiable arcs.

14.6 Show that if $|c| < 1$ then the Julia set of $f(z) = z^3 + cz$ is a simple closed curve.

14.7 Obtain an estimate for the Hausdorff dimension of the Julia set of $f(z) = z^3 + c$ when $|c|$ is large.

14.8 Show that if w is a (finite) attractive fixed point of f_c then the attractive basin $A(w)$ must contain the point c. (Hint: show that otherwise there is a small open set containing w on which the inverse iterates f^{-k} of f can be uniquely defined and form a normal family, which is impossible, since w is a repelling fixed point of f^{-1}.) Deduce that f_c can have at most one attractive fixed point. Generalize this to show that if w is an attractive fixed point of any polynomial f then $A(w)$ contains a point $f(z)$ for some z with $f'(z) = 0$.

14.9 Let f be a quadratic polynomial. Show, by applying Exercise 14.8 to f^p for positive integers p, that f can have at most one attractive periodic orbit.

14.10 Write a computer program to draw Julia sets of functions (see the end of Section 14.3). Try it out first on quadratic functions, then experiment with other polynomials and rational functions, and then other functions such as $\exp z$.

14.11 Use a computer to investigate the domains of attraction for the zeros of some other polynomials under Newton's method iteration; for example, for $p(z) = z^4 - 1$ or $p(z) = z^3 - z$.

Chapter 15 Random fractals

Many of the fractal constructions that have been encountered in this book have random analogues. For example, in the von Koch curve construction, each time we replace the middle third of an interval by the other two sides of an equilateral triangle, we might toss a coin to determine whether to position the new part 'above' or 'below' the removed segment. After a few steps, we get a rather irregular looking curve which nevertheless retains certain of the characteristics of the von Koch curve; see figure 15.1.

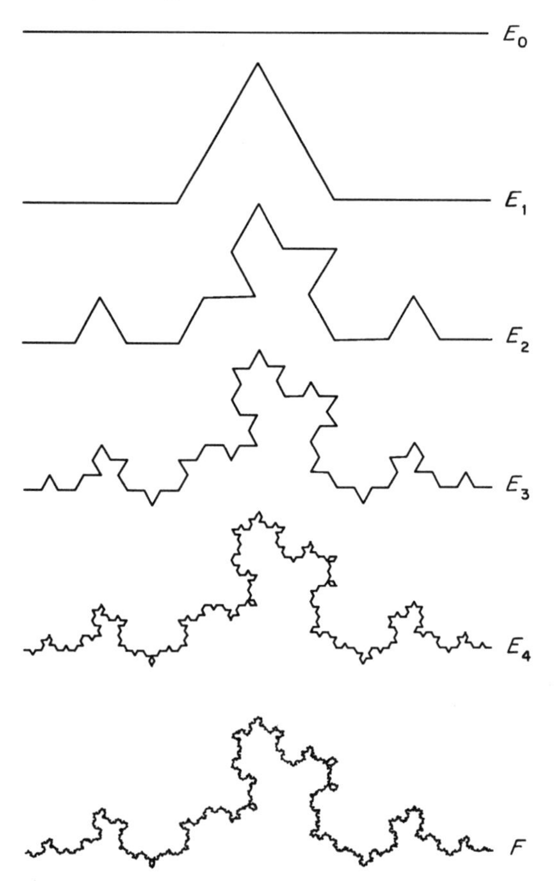

Figure 15.1 Construction of a 'random von Koch curve'. At each step a coin is tossed to determine on which side of the removed segment to place the new pair of segments

The middle third Cantor set construction may be randomized in several ways, as in figure 15.2. Each time we divide a segment into three parts we could, instead of always removing the middle segment, throw a die to decide which parts to remove. Alternatively, we might choose the interval lengths at each stage of the construction at random, so that at the kth stage we are left with 2^k intervals of differing lengths, resulting in a rather irregular looking fractal.

Whilst such 'random fractals' do not have the self-similarity of their non-random counterparts, their non-uniform appearance is often rather closer to natural phenomena such as coastlines, topographical surfaces or cloud boundaries. Indeed, random fractal constructions are the basis of many impressive computer-drawn landscapes or skyscapes.

Most fractals discussed in this book involve a sequence of approximations E_k, each obtained from its predecessor by modification in increasingly fine detail, with a fractal F as a limiting set. A random fractal worthy of the name should display randomness at all scales, so it is appropriate to introduce a random element at each stage of the construction. By relating the size of the random variations to the scale, we can arrange for the fractal to be *statistically self-similar* in the sense that enlargements of small parts have the same statistical distribution as the whole set. This compares with (non-random) self-similar sets (see Chapter 9) where enlargements of small parts are identical to the whole.

In order to describe fractal constructions involving infinitely many random

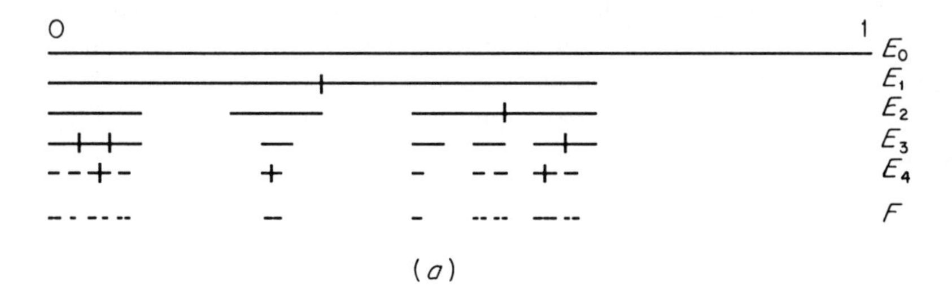

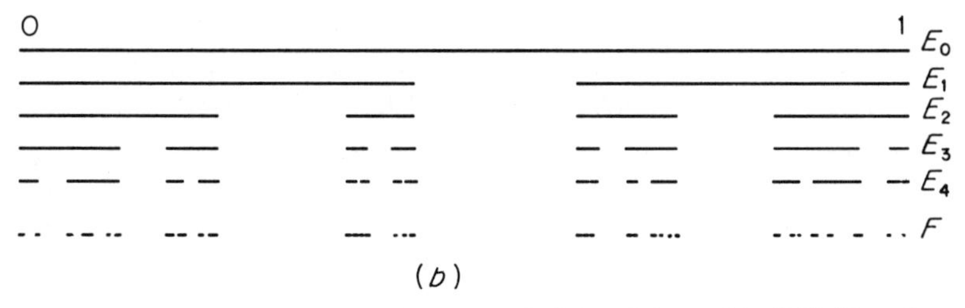

Figure 15.2 Two random versions of the Cantor set. In (*a*) each interval is divided into three equal parts from which some are selected at random. In (*b*) each interval is replaced by two subintervals of random lengths

steps with precision, we must use the language of probability theory, a brief survey of which is given in Section 1.4.

15.1 A random Cantor set

We give a detailed analysis of a specific statistically self-similar construction. It resembles that of the middle third Cantor set, except that the lengths of the intervals at each stage are random.

Intuitively, we consider a construction $F = \bigcap_{k=1}^{\infty} E_k$, where $[0, 1] = E_0 \supset E_1 \supset \cdots$ is a decreasing sequence of closed sets, with E_k a union of 2^k disjoint closed 'basic' intervals. We assume that each basic interval I in E_k contains two intervals I_{L} and I_{R} of E_{k+1}, abutting the left-and right-hand ends of I respectively. The lengths of the intervals are random, and we impose statistical self-similarity by the requirement that the length ratios $|I_{\mathrm{L}}|/|I|$ have the same probability distribution independently for every basic interval I of the construction, and similarly for the ratios $|I_{\mathrm{R}}|/|I|$. The 'random Cantor set' F is statistically self-similar, in that the distribution of the set $F \cap I$ is the same as that of F, but scaled by a factor $|I|$, for each I.

We describe this random construction in probabilistic terms. Let a, b be constants with $0 < a \leqslant b < \frac{1}{2}$. We let Ω denote the class of all decreasing sequences of sets $[0, 1] = E_0 \supset E_1 \supset E_2 \supset \cdots$ satisfying the following conditions. The set E_k comprises 2^k disjoint closed intervals $I_{i_1,\ldots,i_k}$, where $i_j = 1$ or 2 $(1 \leqslant j \leqslant k)$; see figure 15.3. The interval $I_{i_1,\ldots,i_k}$ of E_k contains the two intervals $I_{i_1,\ldots,i_k,1}$ and $I_{i_1,\ldots,i_k,2}$ of E_{k+1}, with the left-hand ends of $I_{i_1,\ldots,i_k}$ and $I_{i_1,\ldots,i_k,1}$ and the right-hand ends of $I_{i_1,\ldots,i_k}$ and $I_{i_1,\ldots,i_k,2}$ coinciding. We write $C_{i_1,\ldots,i_k} = |I_{i_1,\ldots,i_k}|/|I_{i_1,\ldots,i_{k-1}}|$, and suppose that $a \leqslant C_{i_1,\ldots,i_k} \leqslant b$ for all $i_1, \ldots, i_k$. We let $F = \bigcap_{k=1}^{\infty} E_k$.

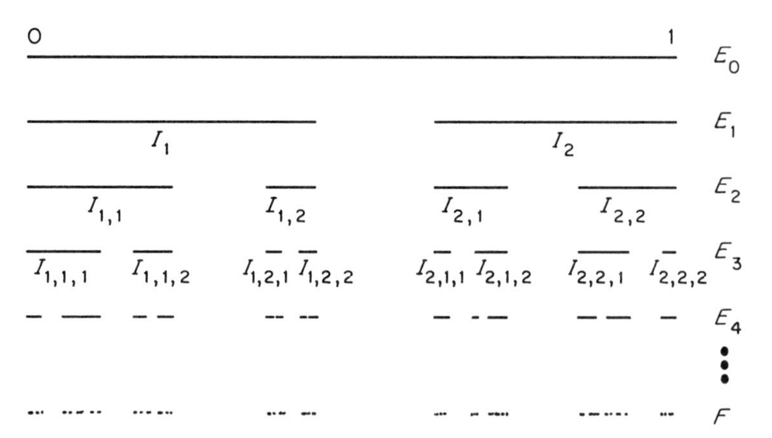

Figure 15.3 Construction of the random Cantor set analysed in Theorem 15.1. The length ratios $|I_{i_1,\ldots,i_k,1}|/|I_{i_1,\ldots,i_k}|$ have the same statistical distribution for each $i_1, \ldots, i_k$, and similarly for $|I_{i_1,\ldots,i_k,2}|/|I_{i_1,\ldots,i_k}|$

We take Ω to be our sample space, and assume that a probability measure P is defined on a suitably large family $\mathscr{F}$ of subsets of Ω, such that the ratios $C_{i_1,\ldots,i_k}$ are random variables. We impose statistical self-similarity on the construction by requiring $C_{i_1,\ldots,i_k,1}$ to have the same distribution as $C_1=|I_1|$, and $C_{i_1,\ldots,i_k,2}$ to have the same distribution as $C_2=|I_2|$ for every sequence $i_1,\ldots,i_k$. We assume that the $C_{i_1,\ldots,i_k}$ are independent random variables, except that for each sequence $i_1,\ldots,i_k$ we do not require $C_{i_1,\ldots,i_k,1}$ and $C_{i_1,\ldots,i_k,2}$ to be independent. It may be shown that $\dim_H F$ is a random variable which may be expressed in terms of the $C_{i_1,\ldots,i_k}$.

The following result is a random version of Theorem 9.3.

Theorem 15.1

With probability 1, *the random Cantor set F described above has* $\dim_H F=s$, *where s is the solution of the expectation equation*

$$\mathsf{E}(C_1^s+C_2^s)=1. \tag{15.1}$$

* ***Proof.*** It is easy to see that $\mathsf{E}(C_1^s+C_2^s)$ is strictly decreasing in s, so that (15.1) has a unique solution.

By slight abuse of notation, we write $I\in E_k$ to mean that the interval I is one of the basic intervals $I_{i_1,\ldots,i_k}$ of E_k. For such an interval I we write I_L and I_R for $I_{i_1,\ldots,i_k,1}$ and $I_{i_1,\ldots,i_k,2}$ respectively. We write $\mathsf{E}(X|\mathscr{F}_k)$ for the conditional expectation of a random variable X given a knowledge of the $C_{i_1,\ldots,i_j}$ for all sequences $i_1,\ldots,i_j$ with $j\leqslant k$. (Intuitively, we imagine that $E_0,\ldots,E_k$ have been constructed, and we are analysing what happens thereafter.) Let $I_{i_1,\ldots,i_k}$ be an interval in E_k. Then

$$\begin{aligned}\mathsf{E}((|I_{i_1,\ldots,i_k,1}|^s+|I_{i_1,\ldots,i_k,2}|^s)|\mathscr{F}_k)&=\mathsf{E}(C^s_{i_1,\ldots,i_k,1}+C^s_{i_1,\ldots,i_k,2})|I_{i_1,\ldots,i_k}|^s\\&=\mathsf{E}(C_1^s+C_2^s)|I_{i_1,\ldots,i_k}|^s\end{aligned}$$

by virtue of the identical distribution. Summing over all the intervals in E_k,

$$\mathsf{E}\Bigg(\sum_{I\in E_{k+1}}|I|^s|\mathscr{F}_k\Bigg)=\sum_{I\in E_k}|I|^s\mathsf{E}(C_1^s+C_2^s). \tag{15.2}$$

It follows that the unconditional expectation satisfies

$$\mathsf{E}\Bigg(\sum_{I\in E_{k+1}}|I|^s\Bigg)=\mathsf{E}\Bigg(\sum_{I\in E_k}|I|^s\Bigg)\mathsf{E}(C_1^s+C_2^s). \tag{15.3}$$

If s is the solution of (15.1), equation (15.2) becomes

$$\mathsf{E}\Bigg(\sum_{I\in E_{k+1}}|I|^s|\mathscr{F}_k\Bigg)=\sum_{I\in E_k}|I|^s. \tag{15.4}$$

Readers familiar with probability theory will recognise (15.4) as saying that the sequence of random variables

$$X_k=\sum_{I\in E_k}|I|^s \tag{15.5}$$

is a martingale with respect to $\mathscr{F}_k$. The crucial fact for our purposes, which we ask other readers to take on trust, is that, in this situation, X_k converges with probability 1 as $k \to \infty$ to a random variable X such that $\mathsf{E}(X) = \mathsf{E}(X_0) = \mathsf{E}(1^s) = 1$. In particular $0 \leqslant X < \infty$ with probability 1, and $X = 0$ with probability $q < 1$. But $X = 0$ if and only if $\sum_{I \in E_k \cap I_1} |I|^s$ and $\sum_{I \in E_k \cap I_2} |I|^s$ both converge to 0 as $k \to \infty$, where I_1 and I_2 are the intervals of E_1, and this happens with probability q^2, by virtue of the self-similarity of the construction. Hence $q = q^2$, so $q = 0$, and we conclude that $0 < X < \infty$ with probability 1. In particular, this implies that with probability 1 there are (random) numbers M_1, M_2 such that

$$0 < M_1 \leqslant X_k = \sum_{I \in E_k} |I|^s \leqslant M_2 < \infty \tag{15.6}$$

for all k. We have $|I| \leqslant 2^{-k}$ for all $I \in E_k$, so, $\mathscr{H}^s_\delta(F) \leqslant \sum_{I \in E_k} |I|^s \leqslant M_2$ if $k \geqslant -\log \delta / \log 2$, giving $\mathscr{H}^s(F) \leqslant M_2$. Thus $\dim_{\mathrm{H}} F \leqslant s$ with probability 1.

We use the potential theoretic method of Section 4.3 to derive the almost sure lower bound. To do this we introduce a random mass distribution μ on the random set F. Let s satisfy (15.1). For $I \in E_k$ let $\mu(I)$ be the random variable

$$\mu(I) = \lim_{j \to \infty} \{\Sigma |J|^s : J \in E_j \text{ and } J \subset I\}.$$

As with (15.5) this limit exists, and is positive and finite with probability 1. Furthermore, if $I \in E_k$,

$$\mathsf{E}(\mu(I) | \mathscr{F}_k) = |I|^s. \tag{15.7}$$

If $I \in E_k$ then $\mu(I) = \mu(I_{\mathrm{L}}) + \mu(I_{\mathrm{R}})$ and μ is a mass distribution with support contained in $\bigcap_{k=0}^\infty E_k = F$; see Proposition 1.7. (We ignore measure theoretic questions connected with the definition of μ.)

We fix $0 < t < s$ and estimate the expectation of the t-energy of μ. If $x, y \in F$, there is a greatest integer k such that x and y belong to a common interval of E_k; denote this interval by $x \wedge y$. If I is an interval of E_k, the subintervals I_{L} and I_{R} of I in E_{k+1} are separated by a gap of at least $d|I|$, where $d = 1 - 2b$. Thus

$$\begin{aligned}\iint_{x \wedge y = I} |x - y|^{-t} \mathrm{d}\mu(x)\, \mathrm{d}\mu(y) &= 2 \int_{x \in I_L} \int_{y \in I_R} |x - y|^{-t} \mathrm{d}\mu(x)\, \mathrm{d}\mu(y) \\ &\leqslant 2d^{-t} |I|^{-t} \mu(I_{\mathrm{L}}) \mu(I_{\mathrm{R}}).\end{aligned}$$

If $I \in E_k$,

$$\begin{aligned}\mathsf{E}\left(\iint_{x \wedge y = I} |x - y|^{-t} \mathrm{d}\mu(x) \mathrm{d}\mu(y) | \mathscr{F}_{k+1}\right) &\leqslant 2d^{-t} |I|^{-t} \mathsf{E}(\mu(I_{\mathrm{L}}) | \mathscr{F}_{k+1}) \mathsf{E}(\mu(I_{\mathrm{R}}) | \mathscr{F}_{k+1}) \\ &\leqslant 2d^{-t} |I|^{-t} |I_{\mathrm{L}}|^s |I_{\mathrm{R}}|^s \\ &\leqslant 2d^{-t} |I|^{2s - t}\end{aligned}$$

using (15.7). Using a variation of (1.21) this gives an inequality for the unconditional expectation

$$\mathsf{E}\left(\iint_{x \wedge y = I} |x-y|^{-t} \,\mathrm{d}\mu(x)\,\mathrm{d}\mu(y)\right) \leqslant 2d^{-t}\mathsf{E}(|I|^{2s-t}).$$

Summing over $I \in E_k$,

$$\mathsf{E}\left(\sum_{I\in E_k}\iint_{x \wedge y = I} |x-y|^{-t} \,\mathrm{d}\mu(x)\,\mathrm{d}\mu(y)\right) \leqslant 2d^{-t}\mathsf{E}\left(\sum_{I\in E_k} |I|^{2s-t}\right) = 2d^{-t}\lambda^k$$

where $\lambda = \mathsf{E}(C_1^{2s-t} + C_2^{2s-t}) < 1$, using (15.3) repeatedly. Then

$$\mathsf{E}\left(\int_F\int_F |x-y|^{-t} \,\mathrm{d}\mu(x)\,\mathrm{d}\mu(y)\right) = \mathsf{E}\left(\sum_{k=0}^{\infty}\sum_{I\in E_k}\iint_{x \wedge y = I} |x-y|^{-t} \,\mathrm{d}\mu(x)\,\mathrm{d}\mu(y)\right)$$

$$\leqslant 2d^{-t}\sum_0^{\infty}\lambda^k < \infty,$$

so that the t-energy of μ is finite, with probability 1. With probability 1, $0 < \mu(F) < \infty$, using (15.6), and $\dim_{\mathrm{H}} F \geqslant t$ by Theorem 4.13(a) □

This theorem and proof generalize in many directions. Each interval in E_k might give rise to a random number of intervals of random lengths in E_{k+1}. Of course, the construction generalizes to $\mathbb{R}^n$, and the separation condition between different component intervals can be relaxed, provided some sort of 'open set condition' (see (9.11)) is satisfied. The following construction is a full random analogue of the sets discussed in Section 9.2.

Let V be an open subset of $\mathbb{R}^n$ with closure $\bar{V}$, let $m \geqslant 2$ be an integer, and let $0 < b < 1$. We take Ω to be the class of all decreasing sequences $\bar{V} = E_0 \supset E_1 \supset E_2 \supset \cdots$ of closed sets satisfying the following conditions. The set E_k is a union of the m^k closed sets $\bar{V}_{i_1,\ldots,i_k}$ where $i_j = 1,\ldots,m$ $(1 \leqslant j \leqslant k)$ and $V_{i_1,\ldots,i_k}$ is either similar to V or is the empty set.

We assume that, for each $i_1,\ldots,i_k$, the set $V_{i_1,\ldots,i_k}$ contains $V_{i_1,\ldots,i_k,i}$ $(1 \leqslant i \leqslant m)$ and that these sets are disjoint; this is, essentially, equivalent to the open set condition. If $V_{i_1,\ldots,i_k}$ is non-empty, we write $C_{i_1,\ldots,i_k} = |V_{i_1,\ldots,i_k}|/|V_{i_1,\ldots,i_{k-1}}|$ for the similarity ratio between successive sets and we take $C_{i_1,\ldots,i_k} = 0$ if $V_{i_1,\ldots,i_k}$ is the empty set. We write $F = \bigcap_{k=0}^{\infty} E_k$.

Let P be a probability measure on a family of subsets of Ω such that the $C_{i_1,\ldots,i_k}$ are random variables. Suppose that given $C_{i_1,\ldots,i_k} > 0$, i.e. given that $V_{i_1,\ldots,i_k}$ is non-empty, $C_{i_1,\ldots,i_k,i}$ has identical distribution to C_i for each sequence $i_1,\ldots,i_k$ and for $1 \leqslant i \leqslant m$. We assume that the $C_{i_1,\ldots,i_k}$ are independent, except that, for each sequence $i_1,\ldots,i_k$, the random variables $C_{i_1,\ldots,i_k,1},\ldots,C_{i_1,\ldots,i_k,m}$ need not be independent. This defines a self-similar probability distribution on the constructions in Ω. We write N for the (random) number of the $C_1,\ldots,C_k$ that are positive; that is, the number of the sets $V_1,\ldots,V_k$ that are non-empty.

Theorem 15.2

The set F described above has probability q of being empty, where q is the smaller non-negative root of the polynomial equation

$$f(t) \equiv \sum_{j=0}^{m} \mathsf{P}(N=j)t^j = t. \tag{15.8}$$

With probability $1-q$ the set F has Hausdorff and box dimensions given by the solution s of

$$\mathsf{E}\left(\sum_{j=0}^{m} C_i^s\right) = 1. \tag{15.9}$$

* *Note on proof.* Basically, this is a combination of the probabilistic argument of Theorem 15.1 and the geometric argument of Theorem 9.3. Note that, if there is a positive probability that $N=0$, then there is a positive probability that $E_1 = \emptyset$ and therefore that $F = \emptyset$. This 'extinction' occurs if each of the basic sets in E_1 becomes extinct. By the self-similarity of the process, the probability q_0 of this happening is $f(q_0)$, so $q_0 = f(q_0)$. If q is the least non-negative root of f, then, using that f is increasing an inductive argument shows that $\mathsf{P}(E_k = \emptyset) = f(\mathsf{P}(E_{k-1} = \emptyset)) \leqslant f(q) = q$ for all k, so that $q_0 \leqslant q$. Thus $q_0 = q$.

Observe that F has probability 0 of being empty, i.e. $q=0$, if and only if $N \geqslant 1$ with probability 1. It is also not hard to show that F is empty with probability 1, i.e. $q=1$, if and only if either $\mathsf{E}(N) < 1$ or $\mathsf{E}(N) = 1$ and $\mathsf{P}(N=1) < 1$. (These extinction probabilities are closely related to the theory of branching processes.) □

Example 15.3. Random von Koch curve

Let C be a random variable with uniform distribution on the interval $(0, \frac{1}{3})$. Let E_0 be a unit line segment in $\mathbb{R}^2$. We form E_1 by removing a proportion C from the middle of E_0 and replacing it by the other two sides of an equilateral triangle based on the removed interval. We repeat this for each of the four segments in E_1 independently and continue in this way to get a limiting curve F. Then with probability 1, $\dim_H F = \dim_B F = 1.144$.

Calculation. This is a special case of Theorem 15.2. The set V may be taken as the isosceles triangle based on E_0 and of height $\frac{1}{6}\sqrt{3}$. At each stage, a segment of length L is replaced by four segments of lengths $\frac{1}{2}(1-C)L, CL, CL$ and $\frac{1}{2}(1-C)L$, so we have $m=4$ and $C_1 = C_4 = \frac{1}{2}(1-C)$ and $C_2 = C_3 = C$. Since C is uniformly distributed on $(0, \frac{1}{3})$, expression (15.9) becomes

$$1 = \mathsf{E}(2(\tfrac{1}{2}(1-C))^s + 2C^s) = \int_0^{1/3} 3 \times 2[(\tfrac{1}{2}(1-c))^s + c^s]\,\mathrm{d}c$$

or

$$s+1=12\times 2^{-(s+1)}-6\times 3^{-(s+1)}$$

giving the dimension stated. □

15.2 Fractal percolation

Our discussion of percolation centres around certain random fractals of the type discussed in the previous section.

Let p be a number with $0<p<1$. We divide the unit square E_0 into 9 squares of side $\frac{1}{3}$ in the obvious way. We select a subset of these squares to form E_1 in such a way that each square has independent probability p of being selected. Similarly, each square of E_1 is divided into 9 squares of side $\frac{1}{9}$, and each of these has independent probability p of being chosen to be a square of E_2. We continue in this way, so that E_k is a random collection of squares of side 3^{-k}. This procedure, which depends on the parameter p, defines a random fractal $F_p=\bigcap_{k=0}^{\infty}E_k$; see figures 15.4 and 15.5. (It is not difficult to describe this construction in precise probabilistic terms; for example, by taking the possible nested sequences of squares E_k as the sample space.)

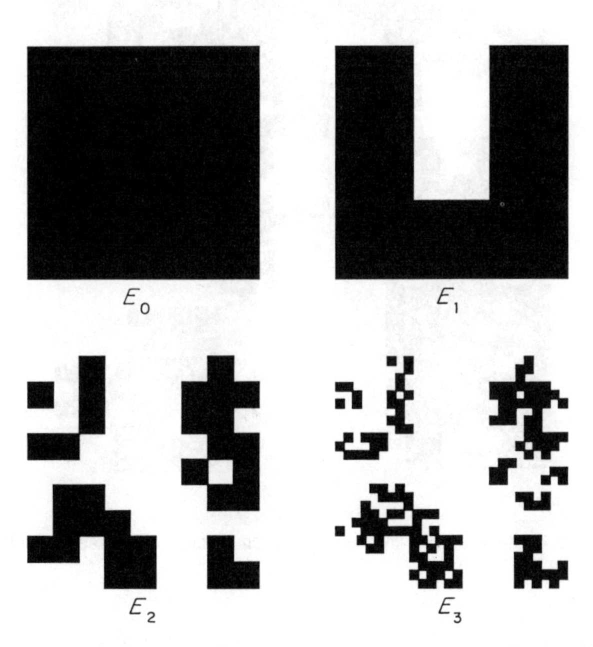

Figure 15.4 Steps in the construction of the random fractal discussed in Section 15.2 with $p=0.6$. The fractal obtained is shown in figure 15.5(a)

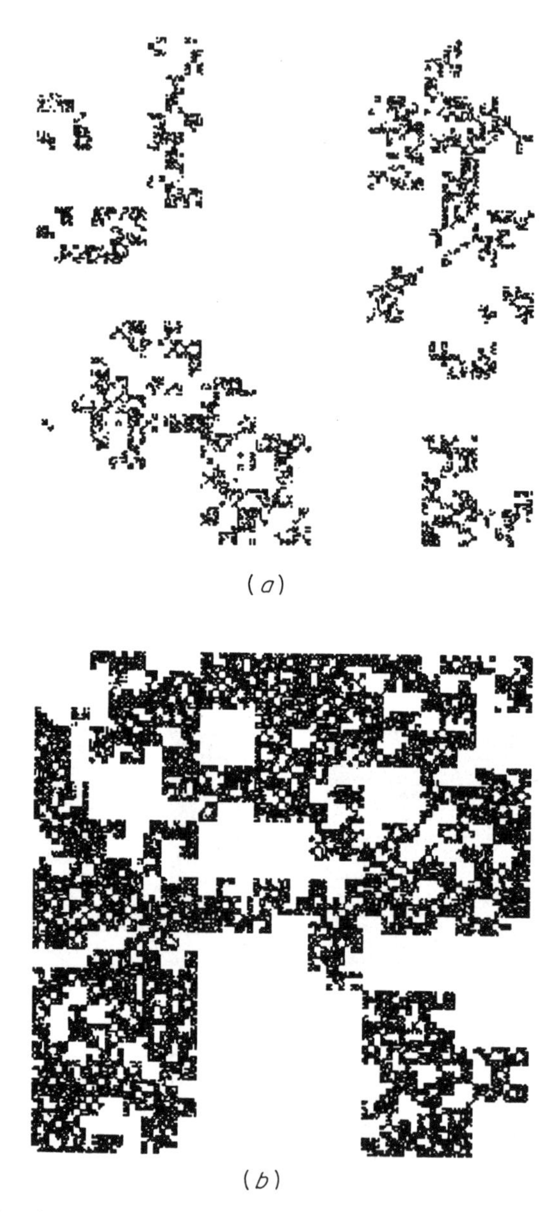

Figure 15.5 Random fractals realized by the percolation process discussed in Section 15.2 with (*a*) $p = 0.6$ and (*b*) $p = 0.8$

Proposition 15.4

Given p, let $t = q$ be the least positive solution of the equation

$$t = (pt + 1 - p)^9.$$

Then F_p is empty with probability q. If $p \leqslant \frac{1}{9}$ then $q = 1$. If $\frac{1}{9} < p \leqslant 1$ then $0 < q < 1$ and, with probability $1 - q$, $\dim_H F = \dim_B F = \log 9p/\log 3$.

Proof. Let N be the (random) number of squares in E_1. Then $P(N = j) = \binom{9}{j} p^j (1-p)^{9-j}$, where $\binom{n}{r} = n!/r!(n-r)!$ is the binomial coefficient, so the probability that $F_p = \varnothing$ is, by (15.8), the smallest positive root of

$$t = \sum_{j=0}^{9} \binom{9}{j} p^j (1-p)^{9-j} t^j = (pt + 1 - p)^9.$$

Each square of E_1 has side $\frac{1}{3}$, so (15.9) becomes

$$1 = \mathsf{E}\left(\sum_{j=0}^{N} C_i^s \right) = \mathsf{E}\left(\sum_{j=0}^{N} 3^{-s} \right) = 3^{-s} \mathsf{E}(N) = 3^{-s} 9p$$

(each of the nine squares of side $\frac{1}{3}$ is selected with probability p, so the expected number chosen is $9p$). Thus by Theorem 15.2, the almost sure dimension of F_p, given it is non-empty, is $\log 9p/\log 3$. □

In this section we discuss briefly the qualitative way in which the random set F_p changes as p increases from 0 to 1. We have already noted that F_p is almost surely empty if $0 < p \leqslant \frac{1}{9}$. If $\frac{1}{9} < p < \frac{1}{3}$ we have, with probability 1, that either $F_p = \varnothing$ or $\dim_H F_p = \log 9p/\log 3 < 1$, so by Proposition 2.5 F_p is totally disconnected. At the other extreme, if p is close to 1, it is plausible that such a high proportion of the squares are retained at each stage of the construction that F_p will connect the left and right sides of the square E_0; when this happens we say that *percolation* occurs between the sides. We show that this is the case at least if p is very close to 1; the ridiculous bound 0.999 obtained can certainly be reduced considerably.

Proposition 15.5

Suppose that $0.999 < p < 1$. Then there is a positive probability (in fact bigger than 0.9999) that the random fractal F_p joins the left and right sides of E_0.

∗*Proof.* The proof depends on the observation that if I_1 and I_2 are abutting squares in E_k and both I_1 and I_2 contain either 8 or 9 subsquares of E_{k+1}, then there is a subsquare in I_1 and one in I_2 that abut, with the squares of E_{k+1} in I_1 and I_2 forming a connected unit.

We say that a square of E_k is *full* if it contains either 8 or 9 squares of E_{k+1}.

We say that a square of E_k is *2-full* if it contains 8 or 9 full squares of E_{k+1}, and, inductively, that a square of E_k is *m-full* if it contains either 8 or 9 squares of E_{k+1} that are $(m-1)$-full. By the above remark, if E_0 is m-full, then opposite sides of E_0 are joined by a sequence of abutting squares of E_m.

The square E_0 is m-full ($m \geqslant 1$) if either

(a) E_1 contains 9 squares all of which are $(m-1)$-full, or
(b) E_1 contains 9 squares of which 8 are $(m-1)$-full, or
(c) E_1 contains 8 squares all of which are $(m-1)$-full.

Thus, if p_m is the probability that E_0 is m-full, we get, summing the probabilities of these three alternatives using (1.16), and using the self-similarity of the process,

$$p_m = p^9 p_{m-1}^9 + p^9 9 p_{m-1}^8 (1 - p_{m-1}) + 9p^8(1-p)p_{m-1}^8 = 9p^8 p_{m-1}^8 - 8p^9 p_{m-1}^9 \tag{15.10}$$

if $m \geqslant 2$. Furthermore, $p_1 = p^9 + 9p^8(1-p) = 9p^8 - 8p^9$, so we have an iterative scheme $p_m = f(p_{m-1})$ for $m \geqslant 1$, where $p_0 = 1$ and

$$f(t) = 9p^8 t^8 - 8p^9 t^9. \tag{15.11}$$

Suppose that $p = 0.999$. Then (15.11) becomes

$$f(t) = 8.928\,2515\,t^8 - 7.928\,2874\,t^9$$

and a little calculation shows that $t_0 = 0.999\,9613$ is a fixed point of f which is stable in the sense that $0 < f(t) - t_0 \leqslant \frac{1}{2}(t - t_0)$ if $t_0 < t \leqslant 1$. It follows that p_m is decreasing and converges to t_0 as $m \to \infty$, so there is a probability $t_0 > 0$ that E_0 is m-full for all m. When this happens, opposite sides of E_0 are joined by a sequence of squares in E_m for each m, so the intersection $F = \bigcap_{k=0}^{\infty} E_k$ joins opposite sides of E_0. Thus, there is a positive probability of percolation occurring if $p = 0.999$, and consequently for larger values of p. $\square$

We have seen that if $0 < p < \frac{1}{3}$ then, with probability 1, F_p is empty or totally disconnected. On the other hand, if $p > 0.999$ then there is a high probability of percolation. The next theorem states that one or other of these situations pertains for each value of p.

Theorem 15.6

There is a critical number p_c with $0.333 < p_c < 0.999$ such that if $0 < p < p_c$ then F_p is totally disconnected with probability 1, but if $p_c < p < 1$ then there is positive probability that F_p connects the left and right sides of E_0.

Idea of proof. Suppose p is such that there is a positive probability of F_p not being totally disconnected. Then there is positive probability of some two distinct points of F being joined by a path in F. This implies that there is a positive probability of the path passsing through opposite sides of one of the squares in E_k for some k; by virtue of the statistical self-similarity of the construction, there is a positive probability of a path crossing E_0 from left to right. Clearly, if F_p has probability 1 of being totally disconnected, the same is true of $F_{p'}$ if $p' < p$. Thus the critical probability p_c exists with the properties stated. □

Experiment suggests that $0.7 < p_c < 0.8$.

The change in form of F_p as p increases through p_c is even more dramatic than Theorem 15.6 suggests. Let F'_p be a random set obtained by tiling the plane with independent random copies of F_p. If $p < p_c$ then, almost surely, F'_p is totally disconnected. However, if $p \geqslant p_c$ then, with probability 1, F'_p has a single unbounded connected component. Thus as p increases through p_c there is a 'phase transition' as the isolated points of F'_p suddenly coalesce to form what is basically a single unit. The idea underlying the proof of this is that, if $p > p_c$, then given that parts of F'_p lie in two disjoint discs of unit radius, there is a positive probability of them being joined by a path in F'_p. There are infinitely many such discs in an unbounded component of F'_p, so if F'_p had two unbounded components, there would be probability 1 of their being joined.

15.3 Notes and references

The main references on random fractals of the type discussed in Section 15.1 are Kahane (1974), Peyrière (1976), Falconer (1986b, 1987), Mauldin and Williams (1986a), Graf (1987) and Graf, Mauldin and Williams (1988). An interesting construction for fractals by random deletion is analysed by Zähle (1984). The fractal percolation model was suggested by Mandelbrot (1974), and detailed mathematical analysis was given by Chayes, Chayes and Durrett (1988). Much has been written on 'discrete' percolation, where squares are selected at random from a large square mesh (see Kesten (1982) or Grimmett (1989)) and there are many parallels between this and the fractal case.

Exercises

15.1 Find the almost sure Hausdorff dimension of the random Cantor set constructed by removing the middle third of each interval with probability $\frac{1}{2}$ and the middle two-thirds of the interval with probability $\frac{1}{2}$ at each step of the construction.

15.2 Consider the following random version of the von Koch construction. We start with a unit segment. With probability $\frac{1}{2}$ we replace the middle third of the segment by the other two sides of the (upwards pointing) equilateral triangle, and with probability $\frac{1}{2}$ we remove the middle third altogether. We repeat this procedure with the segments that remain, in the usual way. Show that, with probability 1, this random fractal has Hausdorff dimension 1.

15.3 Show that the random von Koch curve depicted in figure 15.1 *always* has Hausdorff dimension $s = \log 4/\log 3$ and, indeed, is an s-set. (This is not typical of random constructions.)

15.4 Let $0 < p < 1$. We may randomize the construction of the Sierpiński gasket (figure 0.3) by selecting each of the three equilateral subtriangles independently with probability p at each step. (Thus we have a percolation process based on the Sierpiński gasket.) Show that if $p \leqslant \frac{2}{3}$ then the limiting set F is empty with probability 1, but if $\frac{2}{3} < p < 1$ then there is a positive probability that F is non-empty. Find an expression for this probability, and show that, given F is non-empty, $\dim_H = \log 3p / \log 2$ with probability 1.

15.5 For the random Sierpiński gasket described in Exercise 15.4 show that F is totally disconnected with probability 1 for any $p < 1$. (We regard two triangles as being joined if they touch at a vertex.)

15.6 Consider the random Cantor set analysed in Theorem 15.1. With $\mathscr{H}^s_\infty(F)$ denoting the infimum of the sums in (2.1) over arbitrary coverings of F, show that

$$\mathscr{H}^s_\infty(F) = \min\{1, \mathscr{H}^s_\infty(F \cap I_1) + \mathscr{H}^s_\infty(F \cap I_2)\}$$

where s is the solution of (15.1). Use statistical self-similarity to deduce that, unless $\mathsf{P}(C_1^s + C_2^s = 1) = 1$, then, almost surely, $\mathscr{H}^s_\infty(F) = 0$, and thus $\mathscr{H}^s(F) = 0$.

Chapter 16 Brownian motion and Brownian surfaces

In 1827 the botanist R. Brown noticed that minute particles suspended in a liquid moved on highly irregular paths. This, and a similar phenomenon for smoke particles in air, was explained ultimately as resulting from molecular bombardment of the particles. Einstein published a mathematical study of this motion, which eventually led to Perrin's Nobel Prize-winning calculation of Avogadro's number.

In 1923 Wiener proposed a rigorous mathematical model that exhibited random behaviour similar to that observed in Brownian motion. The paths described by this 'Wiener process' in 3-dimensional space are so irregular as to have Hausdorff dimension equal to 2. This is a good example of a natural phenomenon with a fractal appearance that can be explained by a simple mathematical model.

A path may be described by a function $f:\mathbb{R}\to\mathbb{R}^n$ where $f(t)$ is the position of a particle at time t. We can study f from two differing viewpoints. Either we can think of the *path* or *trail* $f([t_1,t_2])=\{f(t):t_1\leqslant t\leqslant t_2\}$ as a subset of $\mathbb{R}^n$ with t regarded merely as a parameter, or we can consider the *graph* of f, graph $f=\{(t,f(t)):t_1\leqslant t\leqslant t_2\}$, as a record of the variation of f with time. Brownian paths and their graphs are, in general, fractals.

In this chapter, our aim is to define a probability measure on a space of functions, such that the paths likely to occur resemble observed Brownian motion. We begin by investigating the fractal form of classical Brownian motion, and then we examine some variants that have been used to model a wide variety of phenomena, from polymer chains to topographical surfaces.

16.1 Brownian motion

We first define Brownian motion in one dimension, and then extend the definition to the higher-dimensional cases.

To motivate the definition, let us consider a particle performing a random walk on the real line. Suppose at small time intervals τ the particle jumps a small distance δ, randomly to the left or to the right. (This might be a reasonable description of a particle undergoing random molecular bombardment in one dimension.) Let $X_\tau(t)$ denote the position of the particle at time t. Then, given the position $X_\tau(k\tau)$ at time $k\tau$, $X_\tau((k+1)\tau)$ is equally likely to be $X_\tau(k\tau)+\delta$ or $X_\tau(k\tau)-\delta$. Assuming that the particle starts at the origin at time 0, then for $t>0$, the position at time t is described by the random variable

$$X_\tau(t)=\delta(Y_1+\cdots+Y_{[t/\tau]})$$

where $Y_1, Y_2, \ldots$ are independent random variables, each having probability $\frac{1}{2}$ of equalling 1 and probability $\frac{1}{2}$ of equalling -1. Here $[t/\tau]$ denotes the largest integer less than or equal to t/τ. We normalize the step length δ as $\sqrt{\tau}$ so that

$$X_\tau(t)=\sqrt{\tau}(Y_1+\cdots+Y_{[t/\tau]}). \tag{16.1}$$

The central limit theorem (see (1.26)) tells us that, for fixed t, if τ is small then the distribution of the random variable $X_\tau(t)$ is approximately normal with mean 0 and variance t, since the Y_i have mean 0 and variance 1. In the same way, if t and h are fixed, and τ is sufficiently small, then $X_\tau(t+h)-X_\tau(t)$ is approximately normal with mean 0 and variance h. We also note that, if $0\leqslant t_1\leqslant t_2\leqslant\cdots\leqslant t_{2m}$, then the increments $X_\tau(t_2)-X_\tau(t_1)$, $X_\tau(t_4)-X_\tau(t_3),\ldots$, $X_\tau(t_{2m})-X_\tau(t_{2m-1})$ are independent random variables. We define Brownian motion with the limit of the random walk $X_\tau(t)$ as $\tau\to 0$ in mind.

Let $(X, \mathscr{F}, \mathsf{P})$ be a probability space. For our purposes we call X a *random process* or *random function* from $[0,\infty)$ to $\mathbb{R}$ if $X(t)$ is a random variable for each t with $0\leqslant t<\infty$. Occasionally, we consider random functions on a finite interval $[t_1, t_2]$ instead, in which case the development is similar. (In the formal definition of a random process there is an additional 'measurability' condition, which need not concern us here.) Of course, we should think of X as defining a *sample function* $t\mapsto X(\omega,t)$ for each point ω in the sample space Ω. Thus we think of the points of Ω as parametrizing the functions $X:[0,\infty)\to\mathbb{R}$, and we think of P as a probability measure on this class of functions.

We define *Brownian motion* or the *Wiener process* to be a random process X such that:

(BM) (i) with probability 1, $X(0)=0$ (i.e. the process starts at the origin) and $X(t)$ is a continuous function of t;

(ii) for any $t\geqslant 0$ and $h>0$ the increment $X(t+h)-X(t)$ is normally distributed with mean 0 and variance h, thus

$$\mathsf{P}(X(t+h)-X(t)\leqslant x)=(2\pi h)^{-1/2}\int_{-\infty}^{x}\exp\left(\frac{-u^2}{2h}\right)du; \tag{16.2}$$

(iii) if $0\leqslant t_1\leqslant t_2\leqslant\cdots\leqslant t_{2m}$, the increments $X(t_2)-X(t_1),\ldots,X(t_{2m})-X(t_{2m-1})$ are independent.

(There is some overkill in this definition: (iii) may be deduced from (i) and (ii)).

Note that it is immediate from (i) and (ii) that $X(t)$ is itself normally distributed with mean 0 and variance t for each t. Observe that the increments of X are *stationary*; that is, $X(t+h)-X(t)$ has distribution independent of t.

(On a point of notation: we write $\mathsf{E}(X(t))$ to denote the expectation or mean value of $X(t)$; some readers may be used to seeing $\langle X(t) \rangle$, thought of as the average of $X(t)$ over the functions in the sample space.)

The first question that arises is whether there actually is a random function satisfying the conditions (BM). It is quite hard to show that Brownian motion does exist, and we do not do so here. The proof uses the special properties of the normal distribution. For example, given that $X(t_2) - X(t_1)$ and $X(t_3) - X(t_2)$ are independent and normal with means 0 and variances $t_2 - t_1$ and $t_3 - t_2$ respectively, the sum $X(t_3) - X(t_1)$ is necessarily normal with mean 0 and variance $t_3 - t_1$; see (1.24) *et seq*. This is essential for the definition (BM) to be self-consistent. It should at least seem plausible that a process $X(t)$ satisfying (BM) exists, if only as a limit of the random walks $X_\tau(t)$ as $\tau \to 0$.

Instead of proving existence, we mention two methods of constructing Brownian sample functions, for example, with a computer. Both methods can, in fact, be used as a basis for existence proofs. The first method uses the random walk approximation (16.1). Values of 1 or -1 are assigned by 'coin tossing' to Y_i for $1 \leqslant i \leqslant m$, where m is large, and $X_\tau(t)$ is plotted accordingly. If τ is small compared with t, then this should give a good approximation to a Brownian sample function.

Alternatively, the 'random midpoint displacement' method may be used to obtain a sample function $X:[0,1] \to \mathbb{R}$. We define the values of $X(k2^{-j})$ where $0 \leqslant k \leqslant 2^j$ by induction on j. We set $X(0) = 0$ and choose $X(1)$ at random from a normal distribution with mean 0 and variance 1. Next we select $X(\frac{1}{2})$ from a normal distribution with mean $\frac{1}{2}(X(0) + X(1))$ and variance $\frac{1}{2}$. At the next step $X(\frac{1}{4})$ and $X(\frac{3}{4})$ are chosen, and so on. At the jth stage the values $X(k2^{-j})$ for odd k are chosen independently from a normal distribution with mean $\frac{1}{2}(X((k-1)2^{-j}) + X((k+1)2^{-j}))$ and variance 2^{-j}. This procedure determines $X(t)$ at all binary points $t = k2^{-j}$. Assuming that X is continuous, then X is completely determined. It may be shown, using properties of normal distributions, that the functions thus generated have the distribution given by (BM).

The graph of a Brownian sample function is shown in figure 16.1.

It is easy to extend the definition of Brownian motion from $\mathbb{R}$ to $\mathbb{R}^n$: we just define Brownian motion on $\mathbb{R}^n$ so that the coordinate components are independent 1-dimensional Brownian motions. Thus $X:[0,\infty) \to \mathbb{R}^n$ given by $X(t) = (X_1(t), \ldots, X_n(t))$ is an *n-dimensional Brownian motion* on some probability space if the random process $X_i(t)$ is a 1-dimensional Brownian motion for each i, and $X_1(t_1), \ldots, X_n(t_n)$ are independent for any set of times $t_1, \ldots, t_n$. A sample path of Brownian motion in $\mathbb{R}^2$ is shown in figure 16.2.

By definition, the projection of $X(t)$ onto each of the coordinate axes is a 1-dimensional Brownian motion. However, the coordinate axes are not special in this respect: n-dimensional Brownian motion is *isotropic*; that is, it has the same characteristics in every direction. To see this, consider, for convenience, the case of 2-dimensional Brownian motion $X(t) = (X_1(t), X_2(t))$. The projection of $X(t)$ onto the line L_θ at angle θ through the origin is $X_1(t)\cos\theta + X_2(t)\sin\theta$. For $t \geqslant 0$ and $h > 0$ the random variables $X_1(t+h) - X_1(t)$ and $X_2(t+h) - X_2(t)$

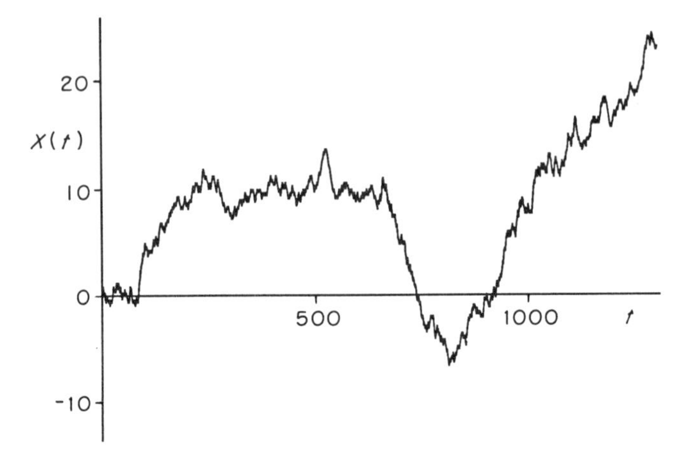

Figure 16.1 Graph of a Brownian sample functon

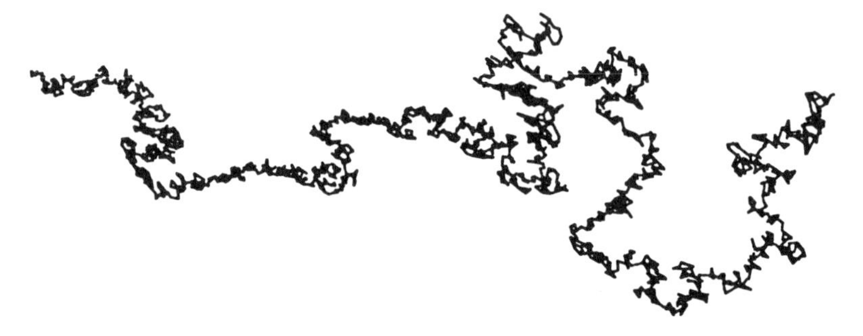

Figure 16.2 A simulation of a Brownian path in $\mathbb{R}^2$

are independent and normally distributed with means 0 and variances h. Thus the increments of the projection onto L_θ, given by

$$(X_1(t+h) - X_1(t))\cos\theta + (X_2(t+h) - X_2(t))\sin\theta,$$

are normally distributed with mean 0 and variance $h\cos^2\theta + h\sin^2\theta = h$; see (1.24) *et seq*. In a similar way, the increments of the projection are independent, so the projection of $X(t)$ onto L_θ is 1-dimensional Brownian motion, for all angles θ.

If $\gamma > 0$, replacing h by γh and x by $\gamma^{1/2}x$ does not alter the value of the right-hand side of (16.2) (by substituting $u_1 = u\gamma^{-1/2}$ in the integral). Thus

$$\mathsf{P}(X_i(t+h) - X_i(t) \leqslant x_i) = \mathsf{P}(X_i(\gamma t + \gamma h) - X_i(\gamma t) \leqslant \gamma^{1/2}x_i)$$

for all x_i. It follows that $X(t)$ and $\gamma^{-1/2}X(\gamma t)$ have the same distribution, changing the temporal scale by a factor γ and the spatial scale by a factor $\gamma^{1/2}$ gives a process indistinguishable from the original. Thus the Brownian paths are *statistically self-similar*, in that the paths $X(t)$ and $X(\gamma t)$ $(0 \leqslant t < \infty)$ are indistinguishable, and the graphs are *statistically self-affine*, in that the scaling factor is different in the t and x directions.

Suppose that $X(t)=(X_1(t),\ldots,X_n(t))$ is n-dimensional Brownian motion. Since $X_i(t+h)-X_i(t)$ has independent normal distribution for each i, it follows from (16.2) that if $[a_i, b_i]$ are intervals, then

$$\mathsf{P}(X_i(t+h)-X_i(t)\in[a_i,b_i])=(2\pi h)^{-1/2}\int_{a_i}^{b_i}\exp\left(-\frac{x_i^2}{2h}\right)\mathrm{d}x_i.$$

Hence if E is the parallelepiped $[a_1,b_1]\times\cdots\times[a_n,b_n]$

$$\begin{aligned}\mathsf{P}(X(t+h)-X(t)\in E)&=\prod_{i=1}^{n}\left[(2\pi h)^{-1/2}\int_{a_i}^{b_i}\exp\left(-\frac{x_i^2}{2h}\right)\mathrm{d}x_i\right]\\&=(2\pi h)^{-n/2}\int_E\exp\left(-\frac{|x|^2}{2h}\right)\mathrm{d}x \qquad (16.3)\end{aligned}$$

where $x=(x_1,\ldots,x_n)$. By approximating sets by unions of such parallelepipeds, it follows that (16.3) holds for any Borel set E. (We sometimes say that $X(t+h)-X(t)$ has *multidimensional normal* distribution.) Thus, taking E as the ball $B_\rho(0)$, and converting into polar coordinates,

$$\mathsf{P}(|X(t+h)-X(t)|\leqslant\rho)=ch^{-n/2}\int_{r=0}^{\rho}r^{n-1}\exp\left(-\frac{r^2}{2h}\right)\mathrm{d}r \qquad (16.4)$$

when c is a constant depending only on n.

A fundamental property of a Brownian motion is that, with probability 1, the sample functions satisfy a Hölder condition of exponent λ for each $\lambda<\frac{1}{2}$.

Proposition 16.1

Suppose $0<\lambda<\frac{1}{2}$. *With probability* 1 *the Brownian sample function* $X:[0,1]\to\mathbb{R}^n$ *satisfies*

$$|X(t+h)-X(t)|\leqslant b|h|^\lambda \qquad (|h|<H_0) \qquad (16.5)$$

for some $H_0>0$, *where* b *depends only on* λ.

Proof. If $h>0$ we have, by (16.4),

$$\begin{aligned}\mathsf{P}(|X(t+h)-X(t)|>h^\lambda)&=ch^{-n/2}\int_{h^\lambda}^{\infty}r^{n-1}\exp\left(\frac{-r^2}{2h}\right)\mathrm{d}r\\&=c\int_{h^{\lambda-1/2}}^{\infty}u^{n-1}\exp\left(\frac{-u^2}{2}\right)\mathrm{d}u\\&\leqslant c_1\int_{h^{\lambda-1/2}}^{\infty}\exp(-u)\,\mathrm{d}u\\&=c_1\exp(-h^{\lambda-1/2})\\&\leqslant c_2h^{-2} \qquad (16.6)\end{aligned}$$

after a substitution $u = rh^{-1/2}$ and some sweeping estimates, where c_1 and c_2 do not depend on h or t. Taking $[t, t+h]$ as the binary intervals $[(m-1)2^{-j}, m2^{-j}]$ we have

$$\begin{aligned}\mathsf{P}(|X((m-1)2^{-j}) - X(m2^{-j})| > 2^{-j\lambda} \quad \text{for some} \quad j \geqslant k \quad \text{and} \quad 1 \leqslant m \leqslant 2^j) \\ \leqslant c_2 \sum_{j=k}^{\infty} 2^j 2^{-2j} \\ = c_2 2^{-k+1}.\end{aligned}$$

Thus with probability 1 there is an integer K such that

$$|X((m-1)2^{-j}) - X(m2^{-j})| \leqslant 2^{-j\lambda} \tag{16.7}$$

for all $j > K$ and $1 \leqslant m \leqslant 2^j$. If $h < H_0 = 2^{-K}$ the interval $[t, t+h]$ may, except possibly for the endpoints, be expressed as a countable union of contiguous binary intervals of the form $[(m-1)2^{-j}, m2^{-j}]$ with $2^{-j} \leqslant h$ and with no more than two intervals of any one length. (Take all the binary intervals in $[t, t+h]$ not contained in any other such intervals.) Then, using the continuity of X, if k is the least integer with $2^{-k} \leqslant h$,

$$|X(t) - X(t+h)| \leqslant 2 \sum_{j=k}^{\infty} 2^{-j\lambda} = \frac{2^{-k\lambda} 2}{(1 - 2^{-\lambda})} < \frac{2h^{\lambda}}{(1 - 2^{-\lambda})}. \qquad \square$$

Theorem 16.2

With probability 1, *a Brownian sample path in* $\mathbb{R}^n$ ($n \geqslant 2$) *has Hausdorff dimension and box dimension equal to* 2.

Proof. For each $\lambda < \frac{1}{2}$, $X:[0,1] \to \mathbb{R}^n$ satisfies a Hölder condition (16.5) with probability 1, so by Proposition 2.3, $\dim_H X([0,1]) \leqslant (1/\lambda) \dim_H [0,1] < 1/\lambda$, with a similar inequality for box dimensions. Thus, almost surely, Brownian paths have dimension at most 2.

For the lower bound we use the potential theoretic method. Take $1 < s < 2$. For given t and h let $p(\rho)$ denote the expression in (16.4). Averaging over all functions, it follows that

$$\begin{aligned}\mathsf{E}(|X(t+h) - X(t)|^{-s}) &= \int_0^{\infty} r^{-s} \mathrm{d}p(r) \\ &= ch^{-n/2} \int_0^{\infty} r^{-s+n-1} \exp\left(\frac{-r^2}{2h}\right) \mathrm{d}r \\ &= \tfrac{1}{2} ch^{-s/2} \int_0^{\infty} w^{(n-s-2)/2} \exp\left(\frac{-w}{2}\right) \mathrm{d}w \\ &= c_1 h^{-s/2} \qquad (16.8)\end{aligned}$$

after substituting $w = r^2/h$, where c_1 is independent of h and t. Then

$$\mathsf{E}\left(\int_0^1\int_0^1 |X(t)-X(u)|^{-s}\,\mathrm{d}t\,\mathrm{d}u\right) = \int_0^1\int_0^1 \mathsf{E}|X(t)-X(u)|^{-s}\,\mathrm{d}t\,\mathrm{d}u$$

$$= \int_0^1\int_0^1 c_1|t-u|^{-s/2}\,\mathrm{d}t\,\mathrm{d}u$$

$$< \infty \tag{16.9}$$

since $s<2$. There is a natural way of defining a mass distribution μ_f on a path f, with the mass of a set equal to the time the path spends in the set, i.e. $\mu_f(A) = \mathscr{L}\{t: 0 \leqslant t \leqslant 1 \text{ and } f(t)\in A\}$ where $\mathscr{L}$ is Lebesgue measure. Then $\int g(x)\,\mathrm{d}\mu_f(x) = \int_0^1 g(f(t))\,\mathrm{d}t$ for any function g, so (16.9) becomes

$$\mathsf{E}\left(\int\int |x-y|^{-s}\,\mathrm{d}\mu_X(x)\,\mathrm{d}\mu_X(y)\right) < \infty.$$

Hence if $s<2$ then $\int\int |x-y|^{-s}\,\mathrm{d}\mu_X(x)\,\mathrm{d}\mu_X(y) < \infty$ almost surely, where μ_X is a mass distribution on $X(t)$, so $\dim_{\mathrm{H}} X([0,1]) \geqslant s$ by Theorem 4.13(*a*). □

In fact, with probability 1, Brownian paths in $\mathbb{R}^n$ ($n \geqslant 2$) have 2-dimensional Hausdorff measure 0. More delicate arguments involving the finer definitions of dimension given in Section 2.5 show that, with probability 1, the paths $X([0,1])$ have positive finite measure with respect to the function $h(t) = t^2 \log(1/t)\log\log\log(1/t)$, if $n=2$, and with respect to $h(t) = t^2\log\log(1/t)$, if $n \geqslant 3$. In this sense, Brownian paths have a dimension that is 'logarithmically smaller' than 2.

An obvious qualitative question about Brownian paths is whether they are simple curves, or whether they are self-intersecting. Given a function f, we say that x is a point of *multiplicity* k if $f(t) = x$ for k distinct values of t. Dimensional methods may be used to determine whether Brownian functions have multiple points.

Theorem 16.3

With probability 1, *a Brownian sample function* $B:[0,\infty)\to\mathbb{R}^n$ *has multiple points as follows:*
$n=2$: *there are points of multiplicity* k *for every positive integer* k;
$n=3$: *there are double points but no triple points;*
$n\geqslant 4$: *there are no multiple points.*

Idea of proof. One approach is to use the intersection theorems of Chapter 8. For the case $n=3$, suppose that $\dim_{\mathrm{H}}(X([0,1])\cap X([2,3])) < 1$ with probability 1. Using isotropy and scaling of Brownian motion it is not difficult to see that this implies that $\dim_{\mathrm{H}}(X([0,1])\cap\sigma(X([2,3]))) < 1$ with probability 1 for any similarity transformation σ. It follows that, with probability 1, $\dim_{\mathrm{H}}(X([0,1])\cap \sigma(X([2,3]))) < 1$ for almost all similarities σ. Since, by Theorem 16.2,

$\dim_H X([0,1]) = \dim_H X([0,1]) = 2$ with probability 1, this contradicts Theorem 8.2(*a*), and we conclude that $\dim_H(X([0,1]) \cap X([2,3])) = 1$ with positive probability, p say. Using the statistical self-similarity of $X(t)$ it follows that $\dim_H(X([t, t+\delta]) \cap X([t+2\delta, t+3\delta])) = 1$ with probability p for any t and δ, so, since the increments are independent, the set of double points of F has Hausdorff dimension 1 with probability 1.

Similar techniques may be used to prove the other results. □

The derivation of the almost sure dimension of Brownian graphs is similar to that for Brownian paths.

Theorem 16.4

With probability 1, *the graph of a Brownian sample function* $X:[0,1] \to \mathbb{R}$ *has Hausdorff and box dimension* $1\frac{1}{2}$.

Proof. From the Hölder condition (16.5) and Corollary 11.2(*a*) it is clear that, with probability 1, graph X has Hausdorff dimension and upper box dimension at most $2 - \lambda$ for every $\lambda < \frac{1}{2}$, so has dimensions at most $1\frac{1}{2}$. For the lower estimate, as in the proof of Theorem 16.2,

$$
\begin{aligned}
\mathsf{E}((|X(t+h) - X(t)|^2 + h^2)^{-s/2}) &= \int_0^\infty (r^2 + h^2)^{-s/2}\,\mathrm{d}p(r) \\
&= ch^{-1/2} \int_0^\infty (r^2 + h^2)^{-s/2} \exp\left(\frac{-r^2}{2h}\right) \mathrm{d}r \\
&= \tfrac{1}{2}c \int_0^\infty (uh + h^2)^{-s/2} u^{-1/2} \exp\left(\frac{-u}{2}\right) \mathrm{d}u \\
&\leqslant \tfrac{1}{2}c \int_0^h (h^2)^{-s/2} u^{-1/2}\,\mathrm{d}u \\
&\quad + \tfrac{1}{2}c \int_h^\infty (uh)^{-s/2} u^{-1/2}\,\mathrm{d}u \\
&\leqslant c_1 h^{1/2 - s}
\end{aligned}
$$

on splitting the range of integration and estimating the integral in two ways. We may lift Lebesgue measure from the t axis to get a mass distribution μ_f on the graph of a function f given by $\mu_f(A) = \mathcal{L}\{t : 0 \leqslant t \leqslant 1 \text{ and } (t, f(t)) \in A\}$. Using Pythagoras's Theorem,

$$
\begin{aligned}
\mathsf{E}\left(\int\int |x - y|^{-s}\,\mathrm{d}\mu_X(x)\,\mathrm{d}\mu_X(y)\right) &= \int_0^1 \int_0^1 \mathsf{E}(|X(t) - X(u)|^2 + |t - u|^2)^{-s/2})\,\mathrm{d}t\,\mathrm{d}u \\
&\leqslant \int_0^1 \int_0^1 c_1 |t - u|^{1/2 - s}\,\mathrm{d}t\,\mathrm{d}u \\
&< \infty
\end{aligned}
$$

if $s < 1\frac{1}{2}$. With probability 1, the mass distribution μ_X on graph X is positive and finite and has finite s-energy, so Theorem 4.13(a) gives $\dim_H \text{graph}\, X \geqslant 1\frac{1}{2}$. □

Since, with probability 1, the graph of X over any interval has dimension $1\frac{1}{2}$, it is immediate that Brownian functions, though continuous, are not continuously differentiable. In fact, with probability 1, a Brownian function is nowhere differentiable.

As with Brownian paths, Brownian graphs have dimension logarithmically smaller than $1\frac{1}{2}$: with probability 1, the graph of X over the range $[0, 1]$ has positive finite measure with respect to the function $h(t) = t^{3/2} \log\log 1(/t)$.

The sets of times at which a Brownian sample function takes particular values are often of interest. If $f:[0, 1] \to \mathbb{R}$ is a function, we define the *level sets* $f^{-1}(c) = \{t: f(t) = c\}$ for each value of c. The level sets are, essentially, the intersections of the graph of f with lines parallel to the t-axis.

Proposition 16.5

With probability 1, *a Brownian sample function* $X:[0, 1] \to \mathbb{R}$ *satisfies* $\dim_H X^{-1}(c) \leqslant \frac{1}{2}$ *for almost all* c *(in the sense of* 1*-dimensional Lebesgue measure). Moreover, for any given* c, $\dim_H X^{-1}(c) = \frac{1}{2}$ *with positive probability.*

Note on proof. With probability 1, $\dim_H \text{graph}\ X = 1\frac{1}{2}$, by Theorem 16.4. Thus $\dim_H((\text{graph}\, X) \cap L_c) \leqslant \frac{1}{2}$ for almost all c, where L_c is the line $y = c$; otherwise Corollary 7.10 would imply that $\dim_H \text{graph}\, X > 1\frac{1}{2}$.

It is much harder to show that $\dim_H X^{-1}(c) = \frac{1}{2}$ with positive probability. The argument is not unlike that indicated for the proof of Theorem 8.2. □

16.2 Fractional Brownian motion

Brownian motion, although of central theoretical importance, is, for many purposes, too restrictive. A Brownian sample function is often regarded as a 'typical' random function, although its graph has dimension $1\frac{1}{2}$ almost surely. However, random functions with graphs of other dimensions are required for a variety of modelling purposes.

It may be shown that the Brownian process is the unique probability distribution of functions, which has independent increments that are stationary and of finite variance. To obtain sample functions with different characteristics it is necessary to relax one or more of these conditions.

There are two usual variations. *Fractional Brownian motion* has increments which are normally distributed but no longer independent. *Stable processes*, on the other hand, dispense with the finite-variance condition and this can lead to discontinuous functions. For simplicity, we just discuss the graphs of these processes in the 1-dimensional case; analogous processes may be defined in n-dimensions.

Fractional Brownian motion of index-α $(0<\alpha<1)$ is defined to be a random process $X:[0,\infty)\to\mathbb{R}$ on some probability space such that:

(FBM) (i) with probability 1, $X(t)$ is continuous and $X(0)=0$;
(ii) for any $t\geqslant 0$ and $h>0$ the increment $X(t+h)-X(t)$ has the normal distribution with mean zero and variance $h^{2\alpha}$, so that

$$\mathsf{P}(X(t+h)-X(t)\leqslant x)=(2\pi)^{-1/2}h^{-\alpha}\int_{-\infty}^{x}\exp(-u^2/2h^{2\alpha})\,du. \tag{16.10}$$

It may be shown that, for $0<\alpha<1$, a process satisfying (FBM) exists.

It is implicit in the above definition that the increments $X(t+h)-X(t)$ are stationary; that is, they have probability distribution independent of t. However, the distribution of functions specified by (FBM) cannot have independent increments except in the Brownian case of $\alpha=\frac{1}{2}$. By condition (ii), $\mathsf{E}((X(t+h)-X(t))^2)=h^{2\alpha}$, from which it may be shown that

$$\mathsf{E}(X(t)(X(t+h)-X(t)))=\tfrac{1}{2}[(t+h)^{2\alpha}-t^{2\alpha}-h^{2\alpha}] \tag{16.11}$$

which is non-zero if $\alpha\neq\frac{1}{2}$. Hence $\mathsf{E}((X(t)-X(0))(X(t+h)-X(t)))$ is positive or negative according to whether $\alpha>\frac{1}{2}$ or $\alpha<\frac{1}{2}$. Thus the increments are not independent—if $\alpha>\frac{1}{2}$ then $X(t)-X(0)$ and $X(t+h)-X(t)$ tend to be of the same sign, so that $X(t)$ tends to increase in the future if it has had an increasing tendency in the past. Similarly, if $\alpha<\frac{1}{2}$ then $X(t)-X(0)$ and $X(t+h)-X(t)$ tend to be of opposite sign. Note also that it may be deduced from condition (16.10) that the scaled paths $\gamma^{-\alpha}X(\gamma t)$ have the same statistical distribution as $X(t)$ for $\gamma>0$.

The almost sure dimension of fractional Brownian graphs may be determined in a similar way to the strict Brownian case.

Proposition 16.6

Suppose $0<\lambda<\alpha$. *With probability* 1, *an index-α Brownian sample function* $X:[0,1]\to\mathbb{R}$ *satisfies*

$$|X(t+h)-X(t)|\leqslant b|h|^{\lambda} \tag{16.12}$$

if $|h|<H_0$, *for some* $H_0>0$ *and* $b>0$.

Note on proof. Provided that $\lambda<\frac{1}{2}$, the proof goes through as in Proposition 16.1, using (16.10) instead of (16.4). However, if $\alpha>\lambda\geqslant\frac{1}{2}$ this leads to an estimate $c_2h^{1/2-\alpha}$ in place of (16.6) and rather more sophisticated techniques from probability theory are required to show that the Hölder condition (16.12) is valid uniformly for all t. □

Theorem 16.7

With probability 1 *an index-α Brownian sample function* $X:[0,1]\to\mathbb{R}$ *has graph with Hausdorff and box dimensions* $2-\alpha$.

Proof. Corollary 11.2(*a*) together with the Hölder condition (16.12) show that the dimension is almost surely at most $2-\alpha$. The lower bound is obtained as in Theorem 16.4 using the probability distribution (16.10). □

The autocorrelation theory discussed in Section 11.2, may be applied to fractional Brownian functions. It is convenient to assume that X is defined for all time, i.e. $X:(-\infty,\infty)\to\mathbb{R}$. This requires only trivial modification to the definition (FBM). Since the variance $\mathsf{E}(|X(t+h)-X(t)|^2)$ tends to infinity with h, we have

$$\lim_{T\to\infty}\mathsf{E}\left(\frac{1}{2T}\int_{-T}^{T}X(t)^2\,\mathrm{d}t\right)=\infty$$

so the sample functions tend to have an infinite mean square. Nevertheless,

$$\mathsf{E}\left(\frac{1}{2T}\int_{-T}^{T}(X(t+h)-X(t))^2\,\mathrm{d}t\right)=\frac{1}{2T}\int_{-T}^{T}\mathsf{E}(X(t+h)-X(t))^2\,\mathrm{d}t=h^{2\alpha}$$

It may be deduced that 'on average' the sample functions satisfy

$$\frac{1}{2T}\int_{-T}^{T}(X(t+h)-X(t))^2\,\mathrm{d}t\simeq ch^{2\alpha}$$

and, according to (11.18) and (11.19), this does indeed correspond to a graph of dimension $2-\alpha$. Taking this parallel further, we might expect $X(t)$ to have a power spectrum (11.15) approximately $1/\omega^{1+2\alpha}$.

Because of the correlations between the increments, simulating index-α fractional Brownian motion can be awkward. The random midpoint displacement method for constructing graphs of Brownian motion does not generalize to the fractional case. If we take $X(k2^{-j})$ to have the normal distribution of variance $2^{-2\alpha j}$ for k odd and independent for each k and j, the resulting function fails to have stationary increments. It is possible to approximate X by a 'random walk' using certain sums of normal random variables, but the formula is quite complicated.

An alternative method of constructing functions with characteristics similar to index-α Brownian functions is to randomize the Weierstrass function (11.4). Consider the random function

$$X(t)=\sum_{k=1}^{\infty}C_k\lambda^{-\alpha k}\sin(\lambda^k t+A_k) \tag{16.13}$$

where $\lambda>1$ and where the C_k are independent random variables with the normal distribution of zero mean and variance one, and the 'phases' A_k are independent, each having the uniform distribution on $[0,2\pi)$. Clearly $\mathsf{E}(X(t+h)-X(t))=0$. Furthermore

$$\mathsf{E}(X(t+h)-X(t))^2=\mathsf{E}\left(\sum_{k=1}^{\infty}C_k\lambda^{-\alpha k}2\sin(\tfrac{1}{2}\lambda^k h)\cos(\lambda^k(t+\tfrac{1}{2}h)+A_k)\right)^2$$

$$=2\sum_{k=1}^{\infty}\lambda^{-2\alpha k}\sin^2(\tfrac{1}{2}\lambda^k h)$$

using that $\mathsf{E}(C_k C_j) = 1$ or 0 according as to whether $k = j$ or not, and that the mean of $\cos^2(a + A_k)$ is $\frac{1}{2}$. Choosing N so that $\lambda^{-(N+1)} \leqslant h < \lambda^{-N}$, gives

$$\begin{aligned}\mathsf{E}(X(t+h) - X(t))^2 &\simeq \tfrac{1}{2} \sum_{k=1}^{N} \lambda^{-2\alpha k} \lambda^{2k} h^2 + 2 \sum_{k=N+1}^{\infty} \lambda^{-2\alpha k} \\ &\simeq c\lambda^{-2\alpha N} \\ &\simeq c h^{2\alpha}\end{aligned}$$

in the sense that $0 < c_1 \leqslant \mathsf{E}(X(t+h) - X(t))^2/h^{2\alpha} \leqslant c_2 < \infty$ for $h < 1$. Thus (16.13) has certain statistical characteristics resembling index-α fractional Brownian motion, and provides a usable method for drawing random graphs of various dimensions. Such functions are often used in fractal modelling. A value of $\alpha = 0.8$, corresponding to a graph of dimension 1.2, is about right for a 'mountain skyline'.

As might be expected, the level sets of index-α Brownian sample functions are typically of dimension $1 - \alpha$. Proposition 16.5 generalizes to give that, with probability 1, $\dim_{\mathrm{H}} X^{-1}(c) \leqslant 1 - \alpha$ for almost all c, and that, for given c, $\dim_{\mathrm{H}} X^{-1}(c) = 1 - \alpha$ with positive probability.

16.3 Stable processes

An alternative generalization of Brownian motion gives the *stable processes* introduced by Lévy. A stable process is a random function $X:[0, \infty) \to \mathbb{R}$ such that the increments $X(t+h) - X(t)$ are stationary, i.e. with distribution depending only on h, and independent, i.e. with $X(t_2) - X(t_1), \ldots, X(t_{2m}) - X(t_{2m-1})$ independent if $0 \leqslant t_1 \leqslant t_2 \cdots < t_{2m}$. However, except in very special cases such as Brownian motion, stable processes have infinite variance and are discontinuous with probability 1.

It is not, in general, possible to specify the probability distribution of stable processes directly. Fourier transforms are usually used to define such distributions, and analysis of the dimensions of graphs and paths of stable processes requires Fourier transform methods.

The probability distribution of a random variable Y may be specified by its *characteristic function*, i.e. the Fourier transform $E(\exp(\mathrm{i}uY))$ for $u \in \mathbb{R}$. To define a stable process, we take a suitable function $\psi:\mathbb{R} \to \mathbb{R}$ and require that the increments $X(t+h) - X(t)$ satisfy

$$\mathsf{E}(\exp(\mathrm{i}u(X(t+h) - X(t)))) = \exp(-h\psi(u)) \tag{16.14}$$

with $X(t_2) - X(t_1), \ldots, X(t_{2m}) - X(t_{2m-1})$ independent if $0 \leqslant t_1 \leqslant t_2 \leqslant \cdots \leqslant t_{2m}$. Clearly the increments are stationary. This definition is, at least, consistent in the following sense. If $t_1 < t_2 < t_3$, then, averaging over all paths, and using indpendence,

$$\begin{aligned}\mathsf{E}(\exp(iu(X(t_3)-X(t_1)))) &= \mathsf{E}(\exp iu((X(t_3)-X(t_2))+(X(t_2)-X(t_1))))\\ &= \mathsf{E}(\exp iu(X(t_3)-X(t_2)))\mathsf{E}(\exp iu(X(t_2)-X(t_1)))\\ &= \exp(-(t_3-t_2)\psi(u))\exp(-(t_2-t_1)\psi(u))\\ &= \exp(-(t_3-t_1)\psi(u)).\end{aligned}$$

It may be shown that, for suitable ψ, stable processes do exist.

Taking $\psi(u)=c|u|^{\alpha}$ with $0<\alpha\leqslant 2$, gives the *stable symmetric process of index-*α. Replacing h by γh and u by $\gamma^{-1/\alpha}u$ leaves the right-hand side of (16.14) unaltered, so it follows that $\gamma^{-1/\alpha}X(\gamma t)$ has the same statistical distribution as $X(t)$. The case $\alpha=2$ is standard Brownian motion.

Theorem 16.8

With probability 1, *the graph of the stable symmetric process of index-*α *has Hausdorff and box dimension* $\max\{1,2-1/\alpha\}$.

Partial proof. We show that $\dim_{\mathrm{H}}$ graph $X\leqslant\overline{\dim}_B$ graph $X\leqslant\max\{1,2-1/\alpha\}$. Write $R_f[t_1,t_2]=\sup\{|f(t)-f(u)|:t_1\leqslant t,u\leqslant t_2\}$ for the maximum range of a function f over the interval $[t_1,t_2]$. By virtue of the statistical self-similarity of the process X, for any t and $\delta>0$

$$\mathsf{E}(R_X[t,t+\delta])=\delta^{1/\alpha}\mathsf{E}(R_X[0,1]).$$

If N_δ squares of the δ-coordinate mesh are intersected by graph X, it follows from Proposition 11.1 that

$$\begin{aligned}\mathsf{E}(N_\delta) &\leqslant 2m+\delta^{-1}\sum_{i=0}^{m-1}\mathsf{E}(R_X[i\delta,(i+1)\delta])\\ &= 2m+m\delta^{-1}\delta^{1/\alpha}\mathsf{E}(R_X[0,1])\end{aligned}$$

where m is the least integer greater than or equal to $1/\delta$. It may be shown, and is at least plausible, that $\mathsf{E}(R_X[0,1])<\infty$, so there is a constant c such that $\mathsf{E}(N_\delta\delta^{\beta})\leqslant c$ for all small δ, where $\beta=\max\{1,2-1/\alpha\}$. Then $\mathsf{E}(N_\delta\delta^{\beta+\varepsilon})\leqslant c\delta^{\varepsilon}$ if $\varepsilon>0$, so that

$$\mathsf{E}\left(\sum_{k=1}^{\infty}N_{2^{-k}}(2^{-k})^{\beta+\varepsilon}\right)\leqslant c\sum_{k=1}^{\infty}(2^{-k})^{\varepsilon}<\infty.$$

It follows that, with probability 1, $\sum_{k=1}^{\infty}N_{2^{-k}}(2^{-k})^{\beta-\varepsilon}<\infty$. In particular, $N_\delta\delta^{\beta-\varepsilon}\to 0$ as $\delta=2^{-k}\to 0$, so $\overline{\dim}_B$ graph $X\leqslant\beta$ with probability 1. □

If $\alpha<1$ then almost surely $\dim_{\mathrm{H}}$ graph $X=1$, the smallest dimension possible for the graph of any function on $[0,1]$. This reflects the fact that the sample paths are constant except between certain jump discontinuities. The image of X, that is $\{X(t):0\leqslant t\leqslant 1\}$, has dimension α with probability 1, which is indicative of the distribution of the jumps. It may be shown that the probability of there being k jumps of absolute value at least a in the interval $[t, t+h]$ is $(ha^{-\alpha})^k \exp(-ha^{-\alpha})/k!$, corresponding to a Poisson distribution of mean $ha^{-\alpha}$.

If $1 < \alpha < 2$ the stable symmetric process combines a 'continuous' component and a 'jump' component.

16.4 Brownian surfaces

We end this chapter with a brief discussion of fractional Brownian surfaces which have been used so effectively in creating computer-generated landscapes.

We replace the time variable t by coordinates (x, y) so the random variable $X(x, y)$ may be thought of as the height of a surface at the point (x, y).

For $0 < \alpha < 1$ we define an *index-α Brownian function* $X: \mathbb{R}^2 \to \mathbb{R}$ to be a random function such that:

(FBS) (i) with probability 1, $X(0,0) = 0$, and $X(x, y)$ is a continuous function of (x, y);

(ii) for (x, y), $(h, k) \in \mathbb{R}^2$ the height increments $X(x+h, y+k) - X(x, y)$ have the normal distribution with mean zero and variance $(h^2 + k^2)^\alpha$, thus

$$\mathsf{P}(X(x+h, y+k) - X(x, y) \leqslant z)$$
$$= (2\pi)^{-1/2}(h^2 + k^2)^{-\alpha/2} \int_{-\infty}^{z} \exp\left(\frac{-r^2}{2(h^2+k^2)^\alpha}\right) dr. \qquad (16.15)$$

Some effort is required to demonstrate the existence of a process satisfying these conditions. The correlations between the random variables $X(x, y)$ at different points are quite involved.

We term $\{(x, y, X(x, y)) : (x, y) \in \mathbb{R}^2\}$ an *index-α Brownian surface*. Some sample surfaces are depicted in figure 16.3.

Comparing (16.15) with the distribution (16.10) we see that the graph obtained by intersecting $X(x, y)$ with any vertical plane is that of a 1-dimensional index-α Brownian function (after adding a constant to ensure $X(0) = 0$). We can often gain information about surfaces by considering such vertical sections.

Theorem 16.9

With probability 1, *an index-α Brownian sample surface has Hausdorff and box dimensions equal to* $3 - \alpha$.

Proof. It may be shown that if $\lambda < \alpha$ then, with probability 1, an index-α Brownian function $X: [0,1] \times [0,1] \to \mathbb{R}$ satisfies

$$|X(x+h, y+k) - X(x, y)| \leqslant b(h^2 + k^2)^{\lambda/2} = b|(h, k)|^\lambda$$

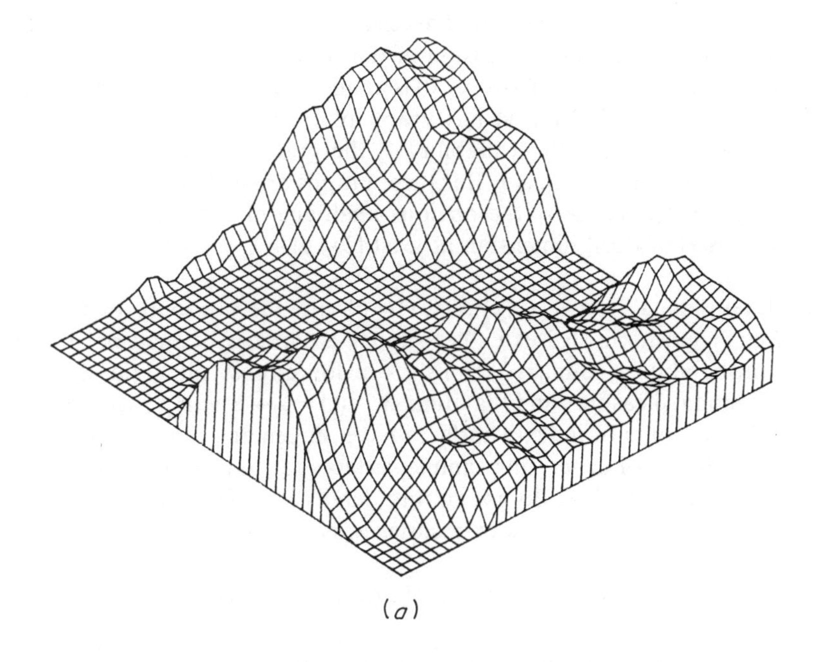

(a)

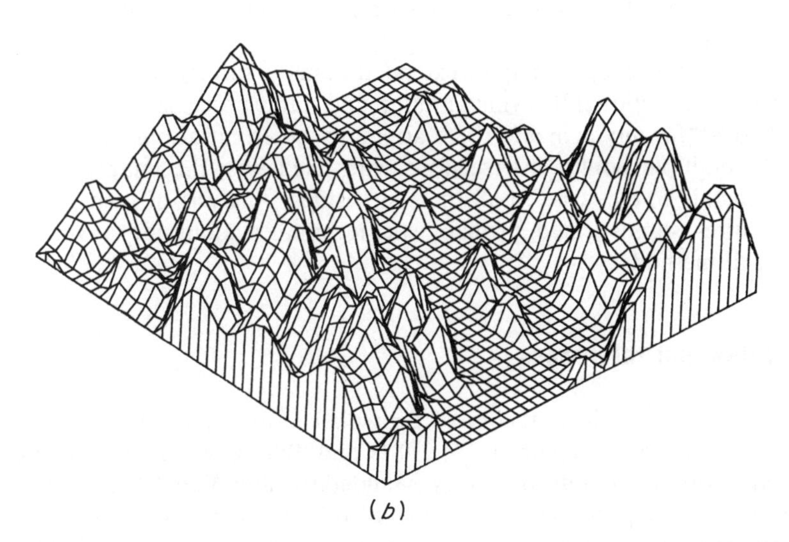

(b)

Figure 16.3 Index-α fractional Brownian surfaces: (*a*) $\alpha = 0.95$, dimension $= 2.05$; (*b*) $\alpha = 0.8$, dimension $= 2.20$

provided that (h, k) is sufficiently small. The analogue of Corollary 11.2(*a*) for a function of two variables (see Exercise 11.9) then gives $3-\lambda$ as an upper bound for the upper box dimension of the surface.

If we fix x_0, then $X(x_0, y) - X(x_0, 0)$ is an index-α Brownian function on $[0, 1]$, so by Theorem 16.7 $X(x_0, y)$ has graph of Hausdorff dimension $2-\alpha$ with probability 1. Thus, with probability 1, the graph of $X(x_0, y)$ has dimension $2-\alpha$ for almost all $0 \leqslant x_0 \leqslant 1$. But these graphs are just parallel sections of the surface given by X, so by the obvious analogue of Corollary 7.10 in $\mathbb{R}^3$ the surface has Hausdorff dimension at least $(2-\alpha)+1$. □

The level sets $X^{-1}(c) = \{(x, y): X(x, y) = c\}$ are the contours of the random surface. Proposition 16.5 extends to index-α surfaces. It may be shown that, with probability 1, $\dim_H X^{-1}(c) \leqslant 2-\alpha$ for almost all c (in the sense of 1-dimensional measure), and that $\dim_H X^{-1}(c) = 2-\alpha$ with positive probability. Thus the contours of index-α surfaces have, in general, dimension $2-\alpha$.

The problems of generating sample surfaces for index-α Brownian functions are considerable, and we do not go into details here. However, we remark that an analogue of (16.13) for index-α surfaces, is

$$X(x, y) = \sum_{k=1}^{\infty} C_k \lambda^{-\alpha k} \sin(\lambda^k (x \cos B_k + y \sin B_k) + A_k)$$

where the C_k are independent having the normal distribution of mean zero and variance 1, and the A_k and B_k are independent with the uniform distribution on $[0, 2\pi)$. Such functions provide one possible approach to computer generation of random surfaces.

The ideas in this chapter may be extended in many directions and combined in many ways. Fractional Brownian motion and stable processes may be defined from $\mathbb{R}^n$ to $\mathbb{R}^m$ for any n, m and there are many other variations. Questions of level sets, multiple points, intersections with fixed sets, the images $X(F)$ for various fractals F, etc, arise in all these situations. Analysis of such problems often requires sophisticated probabilistic techniques alongside a variety of geometrical methods.

16.5 Notes and references

Details of the probabilistic theory of Brownian motion may be found in the books by Carlin and Taylor (1975, 1981) and Billingsley (1979). Fractional Brownian motion was introduced by Mandelbrot and Van Ness (1968), and properties of stable processes were first studied by Lévy. The surveys by Taylor (1973, 1986) mention many dimensional properties of such processes. The books by Adler (1981) and Kahane (1985) are basic references for the mathematics of fractional Brownian functions and surfaces and their dimensions. Computational methods for generating Brownian paths and surfaces are discussed in Voss (1985), Feder (1988) and Peitgen and Saupe (1988).

Exercises

16.1 Use the statistical self-similarity of Brownian motion to show that, with probability 1, a Brownian sample path in $\mathbb{R}^3$ has box dimension of at most 2.

16.2 Let $X:[0, \infty) \to \mathbb{R}^3$ be usual Brownian motion. Show that, with probability 1, the image $X(F)$ of the middle third Cantor set F has Hausdorff dimension at most $\log 4/\log 3$. (Harder: show that it is almost surely equal to $\log 4/\log 3$.) What is the analogous result for index-α Brownian motion?

16.3 Let $X:[0, \infty) \to \mathbb{R}^3$ be usual Brownian motion, and let F be a compact subset of $\mathbb{R}^3$. Use Theorem 8.2 to show that if $\dim_H F > 1$ then there is a positive probability of the Brownian path $X(t)$ hitting F.

16.4 Show that, with probability 1, the Brownian sample function X: $[0,\infty) \to \mathbb{R}$ is not monotonic on any interval $[t, u]$.

16.5 Derive (16.11) from the definition of fractional Brownian motion.

16.6 Take $\frac{1}{2} \leqslant \alpha_1 \leqslant \alpha_2 < 1$ and let $X_1(t)$ and $X_2(t)$ be independent Brownian functions from $[0, 1]$ to $\mathbb{R}$ of indices α_1 and α_2 respectively. Show that, with probability 1, the path in $\mathbb{R}^2$ given by $\{(X_1(t), X_2(t)): 0 \leqslant t \leqslant 1\}$ has Hausdorff and box dimensions of $(1 + \alpha_2 - \alpha_1)/\alpha_2$.

Chapter 17 Multifractal measures

A mass distribution μ may be spread over a region in such a way that the concentration of mass varies widely. If often happens that the sets where the mass concentration has a given density, say where $\mu(B_r(x)) \simeq r^\alpha$ for small r, display fractal-like features, with different sets corresponding to different α. A mass distribution or measure μ with this sort of property is called a *multifractal measure*. As with fractals, an exact definition of multifractal measures tends to be avoided.

An important class of multifractal occurs in connection with attractors in dynamical systems, see Section 13.7. If $f: D \to D$ is a mapping on a domain D, we can define a measure by letting

$$\mu(A) = \lim_{m \to \infty} \frac{1}{m} \#\{k: 1 \leqslant k \leqslant m \text{ and } f^k(x) \in A\}$$

for subsets A of D, where x is some given initial point of D. Then $\mu(A)$ represents the 'proportion of time' that the iterates of x spend in A. We have seen that the support of μ is often an attractor of f and may be a fractal. However the non-uniformity of the distribution of μ may highlight further subsets of the attractor. The irregularity of distribution of μ contains much information about the system which can conveniently be recorded and analysed using multifractal theory.

Multifractals represent a move from the geometry of sets as such to geometric properties of measures. However, in view of the widespread recent interest in multifractals, and the 'fractal ideas' that are involved, it is appropriate to include a note on multifractals in this book. It could reasonably be argued that this chapter belongs to Part I of the book, on Foundations, since the theory may be applied in many different areas of mathematics. However, it has been left until now for reasons of exposition and pedagogy. Familiarity with a variety of fractals, many of which have multifractal counterparts, should make multifractal theory appear more natural. Moreover, multifractal measures are sometimes regarded as a device to generate a spectrum of fractals, though, as we shall see, this approach requires considerable care.

A number of approaches to multifractal measures have been presented; none have been entirely satisfactory from the mathematical point of view, in that they tend to lack rigour and to apply to restricted classes of measures. Our treatment brings out some of the main ideas of the subject, but it should be emphasized that it is not the only approach.

17.1 A multifractal formalism

Let μ be a measure supported by a bounded region of $\mathbb{R}^n$, with total mass $\mu(\mathbb{R}^n) = 1$. The support of μ itself may or may not be a fractal.

For each $0 < \delta < 1$, let $\{B_i\}$ be the cubes of the δ-coordinate mesh that intersect the support of μ. We count the number of these δ-mesh cubes where the measure is 'reasonably large'. For $-\infty < \alpha < \infty$ let

$$N_\delta(\alpha) = \#\{i : \mu(B_i) \geqslant \delta^\alpha\}. \tag{17.1}$$

(Note that our notation differs slightly from that of some other authors.) Although $N_\delta(\alpha)$ is obtained by 'box counting' and it is natural to examine the behaviour of $\log N_\delta(\alpha)/-\log\delta$ as $\delta \to 0$, caution is required in interpreting this limit—it need not be the box dimension of any particular set, since the boxes counted when δ is small need not be contained in those counted when δ is much larger.

We also define the sums over the δ-mesh cubes

$$S_\delta(q) = \sum_i \mu(B_i)^q \tag{17.2}$$

for $-\infty < q < \infty$. Since $S_\delta(0) = N_\delta(\text{support } \mu)$ i the number of δ-mesh cubes required to cover the support of μ, in one sense the limits $\lim_{\delta\to\infty} \log S_\delta(q)/-\log\delta$ generalize the idea of box dimension. It is worth noting that, just as with box dimensions (see (3.14)), the limiting behaviour of $\log S_\delta(q)/-\log\delta$ as $\delta \to 0$ is determined by that of $\log S_{\delta_k}(q)/-\log\delta_k$ for any sequence $\delta_k \to 0$ such that $\delta_{k+1}/\delta_k \geqslant c$ for some $c > 0$.

Observe that, for each δ, we have that $N_\delta(\alpha)$ increases as α increases, and $S_\delta(q)$ decreases as q increases.

The theory of multifractals depends on the fact that, in many cases, $N_\delta(\alpha)$ and $S_\delta(q)$ obey power laws as $\delta \to 0$; moreover there are fundamental relationships between the power-law exponents.

The following example of a 'self-similar' measure has features that are typical of multifractal measures.

Example 17.1

Let $0 < p < 1$ be given. We construct a measure μ on the middle third Cantor set $F = \bigcap_{k=0}^{\infty} E_k$ as follows. (As usual, E_k contains 2^k intervals of length 3^{-k}.) Split a unit mass so that the left interval of E_1 has mass p and the right interval has mass $1-p$. Divide the mass on each interval of E_1 between the two subintervals of E_2 in the ratio $p:1-p$. Continue in this way, so that the mass on each interval of E_k is divided in the ratio $p:1-p$ between its two subintervals in E_{k+1}; see figure 17.1. This defines a mass distribution μ on F. Then $S_\delta(q)$ and $N_\delta(\alpha)$ satisfy power laws for small δ, for each $-\infty < q < \infty$ and $\alpha \geqslant 0$.

Calculation (first part). The set E_k is made up of 2^k intervals of length 3^{-k}, and, for each r, a number $\binom{k}{r}$ of these have mass $p^r(1-p)^{k-r}$, where $\binom{k}{r}$ is the usual

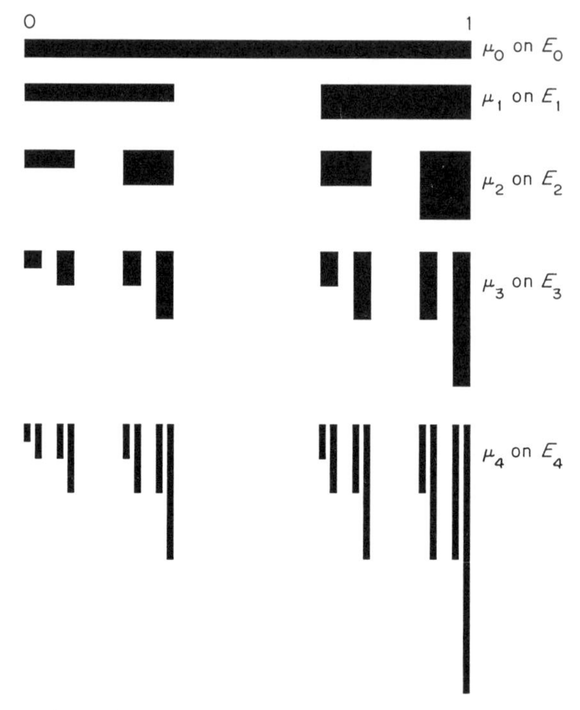

Figure 17.1 Construction of the self-similar measure analysed in Example 17.1. The mass on each interval of E_k in the construction of the Cantor set, indicated by the area of the rectangle, is divided in the ratio $p:1-p$ (in this case $\frac{1}{3}:\frac{2}{3}$) between the two subintervals of E_{k+1}. Continuing this process yields a mass distribution μ on the Cantor set F

binomial coefficient. By the binomial theorem,

$$S_{3^{-k}}(q) = \sum_{r=0}^{k} \binom{k}{r} p^{qr}(1-p)^{q(k-r)} = (p^q + (1-p)^q)^k$$

so

$$S_\delta(q) = \delta^{\log(p^q + (1-p)^q)/\log 3}$$

at least when $\delta = 3^{-k}$. Thus

$$\lim_{\delta \to 0} \frac{\log S_\delta(q)}{-\log \delta} = \frac{\log(p^q + (1-p)^q)}{\log 3}.$$

Direct evaluation of $N_\delta(\alpha)$ is rather harder. Assuming, without loss of generality, that $0 < p < \frac{1}{2}$, we have

$$N_{3^{-k}}(\alpha) = \sum_{r=0}^{m} \binom{k}{r} \tag{17.3}$$

where m is the largest integer such that

$$p^m(1-p)^{k-m} \geqslant 3^{-k\alpha}, \qquad \text{i.e.} \quad m \simeq \frac{k(\log(1-p) + \alpha \log 3)}{\log(1-p) - \log p},$$

assuming this lies between 0 and k.

It is now possible, but tedious, to use (17.3) to estimate $N_{3^{-k}}(\alpha)$ and thus examine its power-law exponent. We prefer to adjourn the calculation until we have developed some general theory enabling an indirect approach. □

We now assume that, for small ε, the number of δ-mesh cubes B_i with $\delta^{\alpha+\varepsilon} \leqslant \mu(B_i) < \delta^{\alpha}$ is roughly of the order $\delta^{-f(\alpha)}$ for small δ. (It is tempting to interpret this as meaning that a set such as

$$\left\{ x: \lim_{\delta \to 0} \frac{\log \mu(B(x))}{-\log \delta} = \alpha \text{ where } B(x) \text{ is the } \delta\text{-mesh cube containing } x \right\}$$

has box dimension $f(\alpha)$. However, this interpretation is misleading—such sets are often dense in support μ and so have box dimension equal to that of support μ). A crude estimate based on such an assumption gives

$$\begin{aligned} S_\delta(q) &\sim \int_0^\infty \delta^{q\alpha} \delta^{-f(\alpha)} \, \mathrm{d}\alpha \\ &= \int_0^\infty \delta^{q\alpha - f(\alpha)} \, \mathrm{d}\alpha. \end{aligned}$$

For small δ, the dominant contribution to this integral comes from the value of α for which $f(\alpha) - q\alpha$ is greatest. Writing $\tau(q)$ for this maximum value, gives a power law $S_\alpha(q) \sim \delta^{-\tau(q)}$.

This handwaving argument may be made precise under the following hypothesis on $N_\delta(\alpha)$: for $0 \leqslant \alpha < \infty$ the following double limit exists:

$$\lim_{\varepsilon \to 0} \lim_{\delta \to 0} \frac{\log(N_\delta(\alpha + \varepsilon) - N_\delta(\alpha - \varepsilon))}{-\log \delta} \equiv f(\alpha). \tag{17.4}$$

(We allow the possibility that $f(\alpha) = -\infty$.) This implies that, given $\eta > 0$, for small enough $\varepsilon > 0$ we have

$$\delta^{-f(\alpha)+\eta} \leqslant N_\delta(\alpha + \varepsilon) - N_\delta(\alpha - \varepsilon) \leqslant \delta^{-f(\alpha)-\eta} \tag{17.5}$$

for all sufficiently small δ.

Proposition 17.2

Assume that $N_\delta(\alpha)$ *satisfies* (17.4), *and define*

$$\tau(q) = \sup_{0 \leqslant \alpha < \infty} (f(\alpha) - q\alpha). \tag{17.6}$$

Then

$$\tau(q) = \lim_{\delta \to 0} \frac{\log S_\delta(q)}{-\log \delta} \tag{17.7}$$

and, in particular, this limit exists.

**Proof.* We give the proof where $q>0$. For each $\alpha \geqslant 0$, if $\eta>0$ there exists ε with $0<\varepsilon<\eta/q$ such that, if $0<\delta \leqslant \delta_0$,

$$\delta^{-f(\alpha)+\eta} \leqslant N_\delta(\alpha+\varepsilon) - N_\delta(\alpha-\varepsilon) \leqslant N_\delta(\alpha+\varepsilon)$$

Then

$$\begin{aligned} S_\delta(q) &\geqslant N_\delta(\alpha+\varepsilon)\delta^{q(\alpha+\varepsilon)} \\ &\geqslant \delta^{-f(\alpha)+2\eta+q\alpha}. \end{aligned}$$

It follows that

$$\begin{aligned} \lim_{\delta\to 0} \frac{\log S_\delta(q)}{-\log\delta} &\geqslant f(\alpha) - q\alpha - 2\eta \\ &\geqslant \tau(q) - 2\eta. \end{aligned} \tag{17.8}$$

For the opposite inequality, let $\eta>0$ be given. Choose β large enough to ensure that $n - q\beta \leqslant \tau(q)$ (n is just the dimension of the space $\mathbb{R}^n$). Using (17.5), a simple compactness argument gives a sequence $\alpha_1 < \cdots < \alpha_m$ and $\varepsilon < \eta/q$ such that $[0,\beta] \subset \bigcup_{k=1}^m (\alpha_k - \varepsilon, \alpha_k + \varepsilon)$ and such that for each k $(1 \leqslant k \leqslant m)$

$$N_\delta(\alpha_k+\varepsilon) - N_\delta(\alpha-\varepsilon) \leqslant \delta^{-f(\alpha_k)-\eta} \tag{17.9}$$

if δ is sufficiently small. Thus, since the number of δ-mesh cubes is at most $c\delta^{-n}$ for some constant c,

$$\begin{aligned} S_\delta(q) &\leqslant \sum_{k=1}^m (N_\delta(\alpha_k+\varepsilon) - N_\delta(\alpha_k-\varepsilon))\delta^{q(\alpha_k-\varepsilon)} + c\delta^{-n}\delta^{q\beta} \\ &\leqslant \sum_{k=1}^m \delta^{-f(\alpha_k)+q\alpha_k-2\eta} + c\delta^{-n}\delta^{q\beta} \\ &\leqslant (m+c)\delta^{-\tau(q)-2\eta} \end{aligned}$$

for all small enough δ. Thus,

$$\overline{\lim_{\delta\to 0}} \frac{\log S_\delta(q)}{-\log\delta} \leqslant \tau(q) + 2\eta.$$

Inequality (17.8) also holds for all $\eta>0$, so we get (17.7). □

The assumption (17.4), that the number of δ-mesh cubes with $\mu(B_i)$ approximately equal to δ^α obeys a power law, is at least plausible, and may be shown to be valid in many cases of interest. In such cases (17.7) relates the power-law exponents

$$S_\delta(q) \sim \delta^{-\tau(q)} \quad \text{and} \quad N_\delta(\alpha+\varepsilon) - N_\delta(\alpha-\varepsilon) \sim \delta^{-f(\alpha)}.$$

Additional assumptions on the form of f enable further relationships between the various exponents to be deduced. We assume from now on what is often the case in practice: that f is a differentiable function of α where $f(\alpha)>0$, and is strictly convex, i.e. $f'(\alpha)$ is strictly decreasing with α.

Suppose that, for each q, the supremum in (17.6) is attained at $\alpha = \alpha(q) > 0$. Then at $\alpha(q)$

$$\frac{d}{d\alpha}(f(\alpha) - q\alpha) = 0 \tag{17.10}$$

giving

$$q = \frac{df}{d\alpha}(\alpha(q)). \tag{17.11}$$

Thus $\alpha(q)$ is the value of α at which the graph of f has slope q. From (17.6)

$$\tau(q) = f(\alpha(q)) - q\alpha(q) \tag{17.12}$$

so, if α is differentiable as a function of q,

$$\frac{d\tau}{dq} = \frac{df}{d\alpha}\frac{d\alpha}{dq} - \alpha - q\frac{d\alpha}{dq}.$$

On putting $\alpha = \alpha(q)$ we get, using (17.11), that

$$\frac{d\tau}{dq}(q) = -\alpha(q). \tag{17.13}$$

Equations (17.12) and (17.13) are, essentially, a Legendre transform pair relating the independent variables q and τ to the independent variables α and f. In mathematical examples, it is usually easiest to find $\tau(q)$ as q varies, and then $\alpha(q)$ and $f(\alpha(q))$ may be found using (17.13) and (17.12), enabling a graph of $f(\alpha)$ against α to be plotted via the parameter q. For practical estimations, this procedure can present difficulties, and other methods of finding the graph of $f(\alpha)$ are often better. This $f(\alpha)$ curve is sometimes referred to as the *multifractal spectrum* of the measure μ; see figure 17.2.

There are a number of notable values of the parameter q. When $q = 0$, we have $S_\delta(0) = N_\delta(\text{support } \mu)$, the number of δ-mesh cubes required to cover the support of μ, so $\tau(0) = \dim_B(\text{support } \mu) = f(\alpha(0))$, by (17.12). Since $df(\alpha(0))/d\alpha = 0$ by (17.11), this corresponds to the maximum of f.

Turning to the value $q = 1$, we have $S_\delta(1) = 1$, so that $\tau(1) = 0$. Moreover,

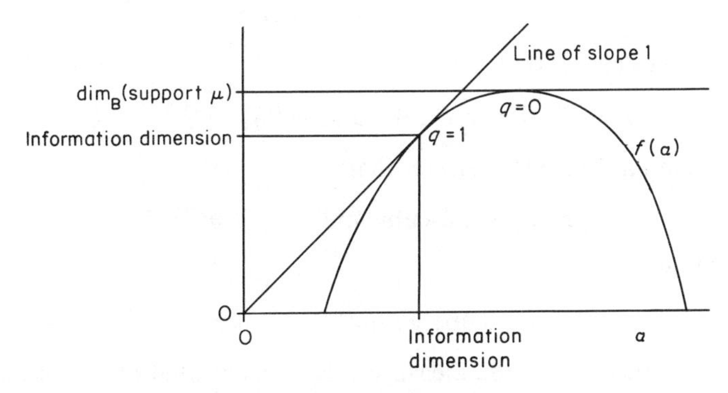

Figure 17.2 Features of the multifractal spectrum—the graph of $f(\alpha)$ against α

$f(\alpha(1)) = \alpha(1)$, with $\mathrm{d}f(\alpha(1))/\mathrm{d}\alpha = 1$, using (17.12) and (17.11). From (17.2)

$$\frac{\mathrm{d}}{\mathrm{d}q}\log S_\delta(q) = \frac{\sum \mu(B_i)^q \log \mu(B_i)}{\sum \mu(B_i)^q}$$

where the sums are over the δ-mesh cubes B_i, so, at $q = 1$,

$$\frac{\mathrm{d}}{\mathrm{d}q}\frac{\log S_\delta(q)}{-\log \delta} = \frac{\sum \mu(B_i)\log \mu(B_i)}{-\log \delta}.$$

Assuming convergence of the derivatives as $\delta \to 0$, (17.7) and (17.13) imply that

$$\alpha(1) = -\frac{\mathrm{d}\tau}{\mathrm{d}q}(1) = \lim_{\delta \to 0}\left(\frac{\sum \mu(B_i)\log \mu(B_i)}{\log \delta}\right). \tag{17.14}$$

The expression $-\sum \mu(B_i)\log \mu(B_i)$, where the sum is over the mesh cubes of side δ, is called the *entropy* of the partition of the measure μ by the δ-mesh cubes. It indicates the average amount of information about the location of a point x, measured by μ, that is provided by knowledge of which cube B_i the point x belongs to. Thus $\alpha(1)$, the rate at which this entropy scales with δ, is often called the *information dimension* of μ.

The number $\alpha(1) = f(\alpha(1))$ has a further important interpretation—it reflects the size of the set on which the measure μ is concentrated. To see this, let $h > 0$ and take $\eta > 0$ such that if $0 \leqslant \alpha < \alpha(1) - h$ then $f(\alpha) - \alpha < -\eta$. Just as in the proof of Proposition 17.2, we may find $\varepsilon < \eta/4$ and $0 \leqslant \alpha_1 < \cdots < \alpha_m \leqslant \alpha(1) - h$ such that $[0, \alpha(1) - h] \subset \bigcup_{k=1}^{m}(\alpha_k - \varepsilon, \alpha_k + \varepsilon)$ and

$$N_\delta(\alpha_k + \varepsilon) - N_\delta(\alpha_k - \varepsilon) \leqslant \delta^{-f(\alpha_k) - \eta/2}$$

for all sufficiently small δ. Then,

$$\begin{aligned}\mu\{\textstyle\bigcup B_i : B_i \text{ is a } \delta\text{-cube and } \mu(B_i) \geqslant \delta^{\alpha(1)-h}\} &\leqslant \sum_{k=1}^{m}(N_\delta(\alpha_k + \varepsilon) - N_\delta(\alpha_k - \varepsilon))\delta^{\alpha_k - \varepsilon}\\ &\leqslant \sum_{k=1}^{m}\delta^{-f(\alpha_k) + \alpha_k - \eta/2 - \varepsilon}\\ &\leqslant m\delta^{\eta/2 - \eta/4}\end{aligned}$$

if δ is small enough, so that

$$\mu\{\textstyle\bigcup B_i : B_i \text{ is a } \delta\text{-cube and } \mu(B_i) \geqslant \delta^{\alpha(1)-h}\} \to 0$$

as $\delta \to 0$. A similar argument shows that

$$\mu\{\textstyle\bigcup B_i : B_i \text{ is a } \delta\text{-cube and } \mu(B_i) \leqslant \delta^{\alpha(1)+h}\} \to 0$$

as $\delta \to 0$. Thus

$$\mu\{\textstyle\bigcup B_i : B_i \text{ is a } \delta\text{-cube with } \delta^{\alpha(1)+h} < \mu(B_i) < \delta^{\alpha(1)-h}\} \to 1$$

as $\delta \to 0$. This means that the measure μ is concentrated on the δ-mesh cubes B_i with $\mu(B_i)$ close to $\delta^{\alpha(1)}$. Thus the 'set of concentration of μ at scale δ'

may be covered by $N_\delta(\alpha(1)+h) - N_\delta(\alpha(1)-h) \simeq \delta^{-f(\alpha(1))} = \delta^{-\alpha(1)}$ mesh cubes of side δ.

We now continue the analysis of our example.

Continued Calculation of Example 17.1. We have already shown that

$$\tau(q) = \log(p^q + (1-p)^q)/\log 3. \tag{17.15}$$

Let us assume that (17.4) holds and the above theory is valid. By (17.13)

$$\alpha(q) = -\frac{(p^q \log p + (1-p)^q \log(1-p))}{(p^q + (1-p)^q)\log 3} \tag{17.16}$$

with (17.12) giving the corresponding value of f in terms of the parameter q

$$f(\alpha(q)) = \frac{\left(\log(p^q + (1-p)^q) - \dfrac{q(p^q \log p + (1-p)^q \log(1-p))}{(p^q + (1-p)^q)}\right)}{\log 3}. \tag{17.17}$$

The graph of the multifractal spectrum $f(\alpha)$ when $p = \frac{1}{3}$ is displayed in figure 17.3.

*[The remainder of this calculation may be omitted.] This approach to finding $f(\alpha)$ avoids the considerable calculation that would be required working from (17.3). Nevertheless, for it to be valid, we still need to show that the limit (17.4) exists. One way of doing this that avoids excessive calculation is to appeal to Chernoff's Theorem for large deviations, from probability theory. Let X be a random variables and let $X_1, X_2, \ldots$ be independent random variables, all with the same distribution as X. A version of Chernoff's Theorem states that, if γ is such that $\mathsf{E}(\log X) < \log \gamma$ and $\mathsf{P}(X > \gamma) > 0$, then

$$\lim_{k\to\infty} \mathsf{P}(X_1 X_2 \cdots X_k \geqslant \gamma^k)^{1/k} = \sup_{0 \leqslant q < \infty} \mathsf{E}(X^q)\gamma^{-q}. \tag{17.18}$$

For each j, let $X_j = p$ with probability $\frac{1}{2}$ and $X_j = 1-p$ with probability $\frac{1}{2}$.

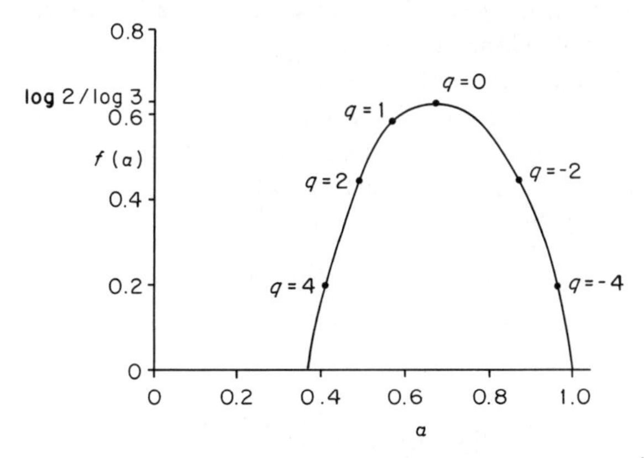

Figure 17.3 The multifractal spectrum of Example 17.1 with $p = \frac{1}{3}$

Since the 2^k intervals of E_k have masses $X_1 X_2 \cdots X_k$ for each of the 2^k possible assignments $X_j = p$ or $1-p$ $(1 \leqslant j \leqslant k)$, equation (17.18) may be interpreted as, taking $\gamma = 3^{-\alpha}$,

$$\lim_{k\to\infty} (2^{-k} N_{3^{-k}}(\alpha))^{1/k} = \sup_{0\leqslant q<\infty} \mathsf{E}(X^q)3^{\alpha q}$$

$$= \sup_{0\leqslant q<\infty} \tfrac{1}{2}(p^q + (1-p)^q)3^{\alpha q}$$

or

$$\lim_{k\to\infty} \frac{\log N_{3^{-k}}(\alpha)}{-\log 3^{-k}} = \sup_{0\leqslant q<\infty} \left(\frac{\log(p^q + (1-p)^q)}{\log 3} + \alpha q \right) = g(\alpha), \tag{17.19}$$

say, provided that $\frac{1}{2}(\log p + \log(1-p)) < -\alpha \log 3 < \max\{\log p, \log(1-p)\}$. By comparing δ-mesh intervals with nearby intervals of lengths 3^{-k}, it is not difficult to see that this implies

$$\lim_{\delta\to 0} \frac{\log N_\delta(\alpha)}{-\log \delta} = g(\alpha).$$

It follows that

$$\frac{\log(N_\delta(\alpha+\varepsilon) - N_\delta(\alpha-\varepsilon))}{-\log\delta} = \frac{\log(N_\delta(\alpha+\varepsilon) + \log(1 - N_\delta(\alpha-\varepsilon)/N_\delta(\alpha+\varepsilon))}{-\log\delta}$$

which tends to $g(\alpha)$ as $\delta \to 0$. A very similar argument using an 'opposite' version of Chernoff's Theorem shows that the limit exists when $\frac{1}{2}(\log p + \log(1-p)) > -\alpha \log 3 > \min\{\log p, \log(1-p)\}$. Thus the limit (17.4) does indeed exist for $\alpha \geqslant 0$. As might be anticipated, the value of q giving the supremum in (17.19) corresponds to the maximizing value of α in (17.6).

Setting $q = 0$ in (17.15) gives $\dim_{\mathrm{B}}(\text{support}\,\mu) = \tau(0) = \log 2/\log 3$, the dimension of the middle third Cantor set F. If $q = 1$, we get the information dimension $\alpha(1) = -(p\log p + (1-p)\log(1-p))/\log 3$, which has a maximum value of $\log 2/\log 3$ when $p = \frac{1}{2}$. For this value of p, a knowledge of the interval of E_k that a point x belongs to locates x to within a set of μ-measure 2^{-k}; for other p rather less information is provided.

It is not hard to see from (17.17) that $f(\alpha(q))$ is positive for $-\infty < q < \infty$ and tends to 0 as $q \to \pm\infty$. Moreover, noting that for $\delta = 3^{-k}$ we have $\min_i \mu(B_i) = (\min\{p, 1-p\})^k$ and $\max_i \mu(B_i) = (\max\{p, 1-p\})^k$, a little calculation using (17.16) shows that

$$\lim_{q\to-\infty} \alpha(q) = \lim_{\delta\to 0} \frac{\log(\min_i \mu(B_i))}{\log\delta} \tag{17.20}$$

and

$$\lim_{q\to\infty} \alpha(q) = \lim_{\delta\to 0} \frac{\log(\max_i \mu(B_i))}{\log\delta}. \tag{17.21}$$

These expressions provide an interpretation of the points at which the $f(\alpha)$ curve meets the axes; by (17.11) the curve has infinite slope at these points. □

Although Example 17.1 seems rather specific, it is typical of a wide class of self-similar measures. With the notation of Section 9.1, let F be the invariant set for the similarity transformations $S_1, \ldots, S_m$ where S_i has ratio $0 < c_i < 1$; for convenience we take the components $S_i(F)$ to be disjoint. Let $p_1, \ldots, p_m$ be positive numbers with $\sum_{i=1}^m p_i = 1$. The mass distribution μ on F defined by setting $\mu(S_{i_1} \circ \cdots \circ S_{i_k}(F)) = p_{i_1} \cdots p_{i_k}$ exhibits multifractal properties; these *self-similar measures* are a measure analogue of self-similar sets. Using ideas in the proof of Theorem 9.3, an adequate estimate for $S_\delta(q)$ is given by $\sum (p_{i_1} \cdots p_{i_k})^q$, where the sum is over all sequences $i_1, \ldots, i_k$ with $1 \leqslant i_j \leqslant m$ and such that $c_{i_1} \cdots c_{i_{k-1}} > \delta \geqslant c_{i_1} \cdots c_{i_k}$. Probabilistic techniques such as Chernoff's Theorem may be used to estimate such sums; in particular it may be shown that the limit (17.4) exists so that the general theory is valid, with formulae such as (17.10) to (17.13) also holding.

There is an alternative approach in this situation, which is computationally simpler, and which takes into account the differing values of c_i. We may form sums over all k-term sequences

$$\begin{aligned} S_k^d(q) &= \sum_{i_1, \ldots, i_k} \mu(S_{i_1} \circ \cdots \circ S_{i_k}(F))^q (c_{i_1} \cdots c_{i_k})^d \\ &= \sum_{i_1, \ldots, i_k} (p_{i_1} \cdots p_{i_k})^q (c_{i_1} \cdots c_{i_k})^d \end{aligned} \tag{17.22}$$

and define $\tau(q)$ by the requirements that $\lim_{k \to \infty} S_k^d(q) = \infty$ if $d < \tau(q)$ and $\lim_{k \to \infty} S_k^d(q) = 0$ if $d > \tau(q)$. This provides a basis for an alternative multifractal theory.

It should be pointed out that rigorous derivation of the results that form the basis of multifractal theory is non-trivial, and here we have done no more than take a few first steps. To treat even self-similar measures, when the full theory can be justified, requires sophisticated techniques from probability theory and the laws of large numbers. Nevertheless, precise use of multifractal mathematics gives a powerful technique in the study of measures.

We end by remarking that multifractal techniques may be employed to advantage in computer and practical experiments. Let $f: D \to D$ be a dynamical system in the plane with μ as the 'residence time' measure

$$\mu(A) = \lim_{m \to \infty} \frac{1}{m} \# \{k : 1 \leqslant k \leqslant m \text{ and } f^k(x) \in A\}.$$

By choosing a suitable range of small δ, and counting the number of times $f^k(x)$ lies in each square B_i of the δ-mesh for a large number of iterates, $\mu(B_i)$ and thus $N_\delta(\alpha)$ may be estimated so that the multifractal spectrum $f(\alpha)$ may be plotted.

A similar procedure may be followed where $f^k(x)$ is replaced by, say, a sequence of experimental observations made at intervals in time. Comparison of the multifractal spectrum obtained experimentally with that resulting from a theoretical model provides a method of assessing the suitability of the model.

We have said little about the physical meaning of the $f(\alpha)$ and $\tau(q)$ curves; indeed, there are considerable problems associated with their interpretation. For example, it is often desirable to extend the multifractal spectrum to allow $f(\alpha)$ to take negative values (which corresponds to regarding the measure as a low dimensional section of a higher dimensional measure). However, this is beyond the scope of this book.

17.2 Notes and references

For other recent accounts of multifractal measures see Tél (1988), Mandelbrot (1988) and Feder (1988). For the theory of large deviations and Chernoff's theorem, see Billingsley (1979).

Exercises

17.1 Assume that μ is a measure on a compact region D of $\mathbb{R}^n$ for which the theory of Section 17.1 is valid. Let $f: D \to \mathbb{R}$ be a continuous function satisfying $0 < c_1 \leqslant f(x) \leqslant c_2$ for constants c_1 and c_2 and such that $\int_D f(x)\,\mathrm{d}\mu(x) = 1$. Define a mass distribution ν on D by $\nu(A) = \int_A f(x)\,\mathrm{d}\mu(x)$ for Borel sets A. Show that the exponents $\tau(q)$ and $f(\alpha)$ corresponding to the measures μ and ν are equal.

17.2 Let μ_1 and μ_2 be measures, each of total mass 1, on a bounded region of $\mathbb{R}^n$ and define μ by $\mu(A) = \frac{1}{2}(\mu_1(A) + \mu_2(A))$. Show that for each $q > 0$ there are positive numbers c_1 and c_2, independent of δ, such that

$$c_1 \max\{S_\delta^1(q), S_\delta^2(q)\} \leqslant S_\delta(q) \leqslant c_2 \max\{S_\delta^1(q), S_\delta^2(q)\}$$

where $S_\delta(q)$, $S_\delta^1(q)$ and $S_\delta^2(q)$ are the sums corresponding to μ, μ_1 and μ_2 respectively. Deduce that $\tau(q) = \max\{\tau_1(q), \tau_2(q)\}$, where the τ are the exponents corresponding to the three measures given by (17.7).

17.3 Let p be a number with $0 < p < 1$ and let μ be the mass distribution on the interval $[0, 1]$ obtained by repeatedly splitting the mass on binary intervals $[m2^{-k}, (m+1)2^{-k})$ in the ratio $p:1-p$ between the binary subintervals $[2m2^{-k-1}, (2m+1)2^{-k-1})$ and $[(2m+1)2^{-k-1}, (2m+2)2^{-k-1})$. Find $f(\alpha)$ and α in terms of the parameter q for this multifractal measure.

17.4 Let F be the set of Exercise 9.7. For $0 < p < 1$ let μ be the mass distribution on F obtained by repeatedly splitting the mass on each interval of E_k in the ratio $p:1-p$ between the two subintervals of E_{k+1}. Find $\tau(q)$ defined by (17.22) in this case.

Chapter 18 Physical applications

Cloud boundaries, mountain skylines, coastlines, forked lightning, . . . ; these, and many other natural objects have a form much better described in fractal terms than by the straight lines and smooth curves of classical geometry. Fractal mathematics ought, therefore, to be well suited to modelling and making predictions about such phenomena.

There are, however, considerable difficulties in applying the mathematics of fractal geometry to real-life examples. We might estimate the box dimension of, say, the coastline of Britain by counting the number N_δ of mesh squares of side δ intersected by the coastline. For a range of δ between 20 m and 200 km the graph of $\log N_\delta$ against $-\log \delta$ is closely matched by a straight line of slope about 1.2. Thus the power law $N_\delta \simeq \text{constant} \times \delta^{-1.2}$ is valid for such δ and it makes sense to say that the coastline has dimension 1.2 over this range of scales. However, as δ gets smaller, this power law first becomes inaccurate and then meaningless. Similarly, with other physical examples, estimates of dimension using boxes of side δ inevitably break down well before a molecular scale is reached.

The theory of fractals studied in Part I of this book depends on taking limits as $\delta \to 0$, which cannot be achieved in reality. There are no true fractals in nature—for that matter, there are no inextensible strings or frictionless pulleys either!

Nevertheless, it should be possible to apply the mathematical theory of 'exact' fractals to the 'approximate' fractals of nature, and this has been achieved convincingly in many situations. This is analogous to the well established use of classical geometry in science—for example, regarding the earth as spherical provides a good enough approximation for many calculations involving its gravitational field.

Perhaps the most convincing example of a physical phenomenon with a fractal model is that of Brownian motion; see Chapter 16. The underlying physical assumption, that a particle subject to random molecular bombardment moves with increments distributed according to a normal distribution, leads to the conclusion that the particle path has dimension 2. This can be checked experimentally using box-counting methods. The motion can also be simulated on a computer, by tracing a path formed by a large number of small random increments. The dimension of such computer realisations can also be estimated by box counting. Brownian motion, which may be observed in reality or on a computer, has a fractal form predicted by a theoretical model. (It should, perhaps, be pointed out that even Brownian paths fail to be described by fractals on a very small scale, since infinite energy would be required for a particle to follow a nowhere-differentiable path of dimension 2.) Linking up experiment, simulation and theory must surely be the aim with other physical manifestations of fractals.

The study of fractals in nature thus proceeds on these three fronts: experiment, simulation and theory. Physical objects are observed and measured, dimensions and, perhaps, multifractal spectra are estimated over an appropriate range of scales, and their dependence on various parameters noted. Theoretical techniques, such as assuming the Projection theorem 6.1 to estimate the dimension of an object from a photograph, are sometimes used here. Of course, for a dimension to have any significance, repeating an experiment must lead to the same value.

Whilst a dimension may have some interest purely as a physical constant, it is much more satisfying if fractal properties can be explained in physical terms. Therefore, the next stage is to devise some sort of mechanism to explain the natural phenomena. Computational simulation then permits evaluation of various models by comparing characteristics, such as dimension, of the simulation and reality. Computational methods are always approximate; this can actually be an advantage when modelling natural rather than exact fractals in that very small-scale effects will be neglected.

It is desirable to have a theoretical model that is mathematically manageable, with basic physical features, such as the apparent dimension, derivable from a mathematical argument. The model should account for the dependence of these features on the various parameters, and, ideally, be predictive as well as descriptive. Fractal phenomena in nature are often rather complicated to describe, and various assumptions and approximations may be required in setting up and analysing a mathematical model. Of course, the ability to do this in a way that preserves the physical content is the mark of a good theoretical scientist! Sometimes differential equations may describe a physical situation, and fractal attractors can often result; see Section 13.5. On the other hand, analysis of differential equations where the boundary conditions are fractal can present problems of an entirely different nature.

There is a vast literature devoted to examining fractal phenomena in these ways; often agreement of dimension between experiment, simulation and theory is surprisingly good. Moreover, analysis of dimension has been used effectively to isolate the dominant features underlying certain physical processes. Nevertheless, there is still a long way to go. Questions such as 'Why do projections of clouds have perimeters of dimension 1.35 over a very wide range of scales?', 'How does the dimension of the surface of a metal affect the physical properties such as radiation of heat or the coefficient of friction?' and 'What are the geological processes that lead to a landscape of dimension 2.2?' should be answered in the framework of fractal modelling.

For most experimental purposes, box-counting dimension has to be used. With N_δ defined by one of the Equivalent Definitions 3.1, the dimension of an object is usually found by estimating the gradient of a graph of $\log N_\delta$ against $-\log \delta$. We often wish to estimate the dimension of a theoretical exact fractal F by box counting on a physical approximation E. To do this, boxes that are large compared with the accuracy of the approximation must be used. More precisely, if $d(E, F) \leqslant \varepsilon$ where d denotes Hausdorff distance, and $N_\delta(E)$ and $N_\delta(F)$ are the number of balls of radius δ required to cover the sets, it is easy to see that

$$N_{\delta+\varepsilon}(E) \leqslant N_\delta(F) \leqslant N_{\delta-\varepsilon}(E)$$

and this may be taken into account when estimating $\dim_H F$ from $\log-\log$ plots of measurements of E. It is also worth remembering, as we have indicated in various instances throughout the book, that there are often theoretical reasons for supposing that exact fractals have equal box and Hausdorff dimensions.

Sometimes other quantities are more convenient to measure than dimension. For example, in the case of a time-dependent signal, the autocorrelation function (see Section 11.2) might be measured, with equation (11.20) providing an indication of the dimension.

We now examine in more detail some specific physical examples where fractal analysis can aid understanding of physical processes.

18.1 Fractal growth

Many natural objects grow in an apparently fractal form, with branches repeatedly splitting and begetting smaller side branches. When viewed at appropriate scales, certain trees, root systems and plants (in particular more primitive ones such as lichens, mosses and seaweeds) appear as fractals. Forked patterns of lightning or other electrical discharges, and the 'viscous fingering' that occurs when water is injected into a viscous liquid such as oil also have a branched fractal form. During electrolysis of copper sulphate solution, the copper deposit at the cathode grows in a fractal pattern.

The biological laws that govern plant growth are far too complex to be used as a basis for a mathematical model. However, other phenomena may be modelled by relatively simple growth laws and we examine some of these.

A simple experiment demonstrates fractal growth by electrolysis of copper sulphate ($CuSO_4$); see figure 18.1. The bottom of a circular dish is covered with a little copper sulphate solution. A copper cathode is suspended in the centre of the dish and a strip of copper is curved around the edge of the dish to form an anode. If a potential of a few volts is applied between the electrodes, then, after a few minutes, a deposit of copper starts to form around the cathode. After half an hour or so the copper deposit will have extended into fractal fingers several inches long.

The mechanism for this electrolysis is as follows. In solution, the copper sulphate splits into copper Cu^{2+} ions and sulphate $SO_4{}^{2-}$ ions which drift around in a random manner. When the voltage is applied, the copper ions that hit the cathode receive two electrons, and are deposited as copper. Copper ions

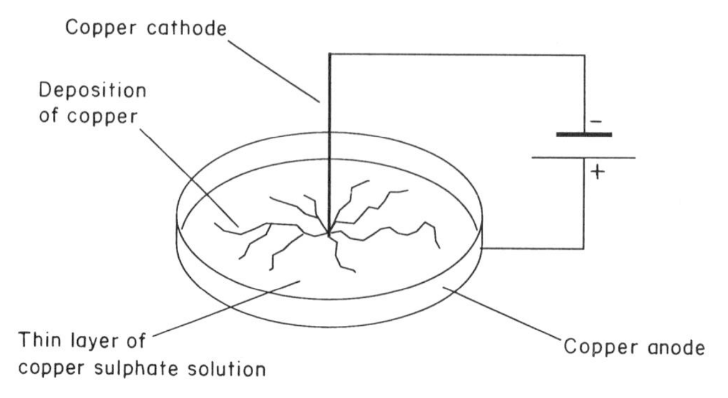

Figure 18.1 Electrolysis of copper sulphate leads to fractal-like deposits of copper growing outwards from the cathode

that hit any copper already formed are also deposited as copper, so the residue grows away from the cathode. Assuming that the copper ions move in a sufficiently random manner, for example, following Brownian paths (see Chapter 16), ions are more likely to hit the exposed finger ends than the earlier parts of the deposition which tend to be 'protected' by subsequent growth. Thus it is at least plausible that growth of the copper deposit will be in thin, branching fingers rather than in a solid 'block' around the cathode.

In the experiment described, the Cu^{2+} ions follow a Brownian path with a drift towards the cathode superimposed as a result of the electric field between cathode and anode. Enriching the sulphate in the solution, for example, by addition of sodium sulphate, screens the copper ions from the electric field. Fractal deposits still occur, but this situation is more convenient for mathematical modelling since the Cu^{2+} ions may be assumed to follow Brownian paths.

The *diffusion-limited aggregation* (DLA) model provides a convincing simulation of the growth. The model is based on a lattice of small squares. An initial square is shaded to represent the cathode, and a large circle is drawn centred on this square. A particle is released from a random point near the perimeter of the circle, and allowed to perform a Brownian motion until it either leaves the circle, or reaches a square neighbouring a shaded one, in which case that square is also shaded. As this process is repeated a large number of times, a connected set of squares grows outward from the initial one. It is computationally more convenient to let the particle follow a random walk (which gives an approximation to a Brownian path), so when the particle is released, it moves, with probability $\frac{1}{4}$ each, left, right, up or down to a neighbouring square, continuing until it leaves the circle or occupies a square next to a shaded one; see figure 18.2. (There are ways of shortening the computation required; for example, if the particle is k squares away from the shaded part the particle might as well move k steps at once.)

Running the model for, say, 10 000 shaded squares gives a highly branched picture (figure 18.3) that resembles the patterns in the electrolysis experiment.

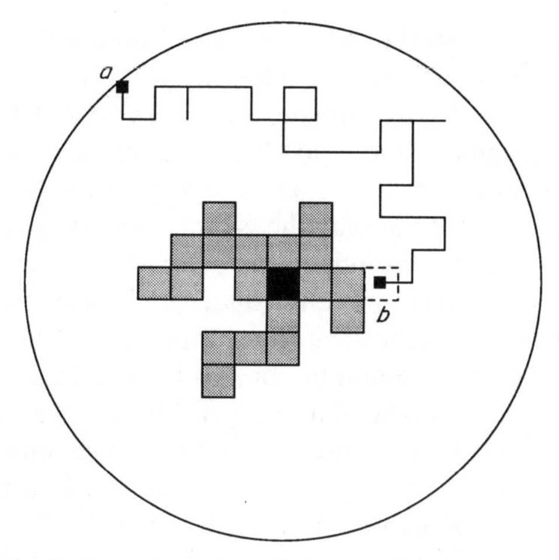

Figure 18.2 The diffusion-limited aggregation (DLA) model. A particle is released from a random point *a* on the circle and performs a random walk until it either leaves the circle or reaches a square *b* next to one that has already been shaded, in which case this square is also shaded

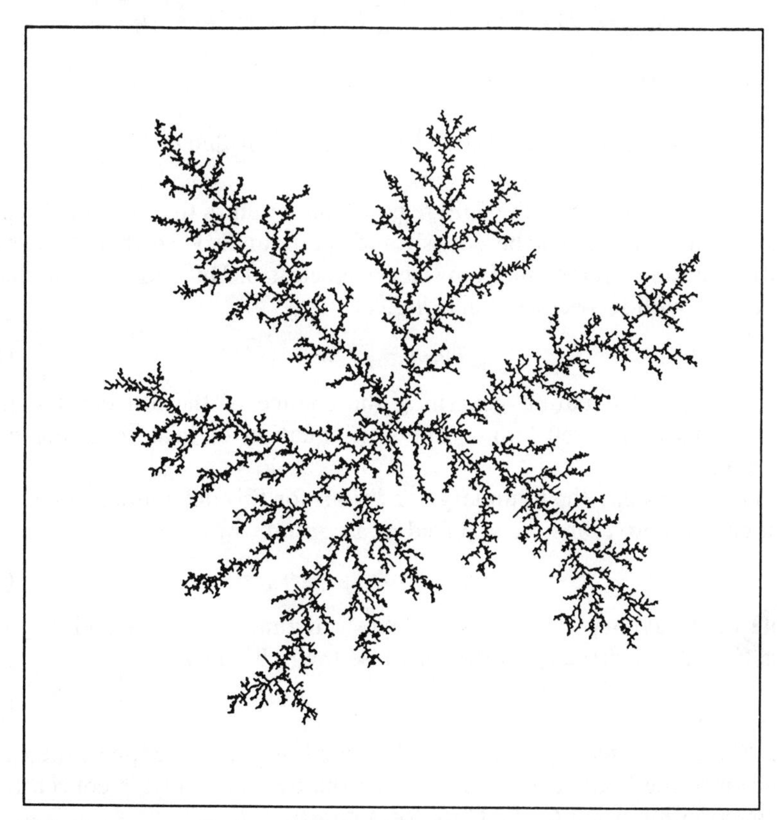

Figure 18.3 A computer realization of diffusion-limited aggregation. The square was divided into a 700×700 mesh from which 16 000 squares were selected using the method described

Main branches radiate from the initial point and bifurcate as they grow, giving rise to subsidiary side branches, all tending to grow outwards. It is natural to use box-counting methods to estimate the dimension of these structures on scales larger than a square side, and there is a remarkably close agreement between the electrolysis experiment and the simulation, with a value for the dimension of about 1.70, or 2.43 for the 3-dimensional analogue.

The DLA model may be thought of as a representation of a succession of ions released from a distance one after another. Whilst this provides a good model for the form of the deposit, it gives little idea of its development with time, which depends on a large number of ions in simultaneous random motion that adhere to the copper deposit on meeting it. Therefore, a 'continuous' version of this 'discrete' model is useful. Suppose that the large number of copper ions in the solution have density $u(x,t)$ at point x and time t, so that the number of ions in a very small disc of area δx and centre x is $u(x,t)\delta x$. Assuming that the ions follow independent Brownian paths, the ions that are in this small disc at time t will have spread out to have density at time $t+h$ given by the 2-dimensional normal distribution

$$\delta u(x',t+h)=(2\pi)^{-1}h^{-1}\exp(-(x-x')^2/2h)u(x,t)\delta x$$

(see (16.3)) and so

$$u(x',t+h)=(2\pi)^{-1}h^{-1}\int\exp(-(x-x')^2/2h)u(x,t)\mathrm{d}x$$

where integration is across the fluid region. This assumes that h is small relative to the distance of x' from the deposit and the boundary, so that the effect of the introduction or removal of ions can be neglected. Differentiating under the integral sign with respect to x' and h gives

$$\partial u/\partial t=\tfrac{1}{2}\nabla^2 u \tag{18.1}$$

as the differential equation governing the change of the ion density in the solution. This is the well known diffusion equation or heat equation in two dimensions.

We need to specify the boundary conditions for this differential equation. At the outer boundary, ions are supplied at a constant rate, so

$$u=u_0 \qquad \text{on} \quad |x|=r_0. \tag{18.2}$$

Denote the boundary of the copper deposit at time t by F_t. Sufficiently close to this boundary, virtually all the ions lose their charge, so

$$u=0 \tag{18.3}$$

on F_t. Since the discharged ions are deposited as metallic copper, the rate of advance v of the boundary F_t must equal the derivative of the concentration in a direction n normal to F_t. Thus, for a constant k,

$$v_n=kn\cdot\nabla u \tag{18.4}$$

on F_t. (We are assuming that F_t is actually smooth on a very small scale.)

Provided that the growth remains a long way from the outer electrode, the diffusion rate is, to a good approximation, time independent, so (18.1) may be replaced by Laplace's equation

$$\nabla^2 u = 0. \tag{18.5}$$

Solving this with boundary conditions (18.2) and (18.3) allows the rate of growth of the deposit to be found, using (18.4).

These equations alone are too idealized to provide an accurate model. First, to prevent the equation being unstable with respect to surface irregularities, a short scale 'cut-off' for the equations is required. This is provided in the square-lattice DLA model—if a particle gets close enough, it sticks to the aggregate. Second, our derivation of the differential equations assumed a continuously varying particle density, rather than a large number of discrete particles. It is the random variation in motion of these individual particles that creates the irregularities that are amplified into the branching fingers. Thus (18.4) needs to be modified to include a random perturbation

$$v_n = kn\cdot\nabla u + p \tag{18.6}$$

where p may be thought of as a 'noise' term. Both of these features are present in the square-lattice DLA model, which is consequently more suitable for simulation of the growth form than direct numerical attempts to solve the differential equations.

One interpretation of the square-lattice DLA model is as providing a spatial solution of equations (18.2)–(18.5) subject to a small random perturbation of the boundary F_t. Surprisingly, the same differential equations and boundary conditions describe several rather different physical phenomena. The DLA model may therefore be expected to apply to some degree in these different cases.

The growth of viscous fingers in a fluid is an example. Suppose two glass plates are fixed a small distance apart (perhaps $\frac{1}{2}$mm) and the region in between is filled with a viscous liquid such as an oil. (This apparatus is called a Hele-Shaw cell.) If a low-viscosity liquid such as water is injected through a small hole in one of the plates, then, under certain conditions, the water spreads out into the oil in thin highly-branched fingers. The patterns resemble closely the deposits of copper in the electrolysis experiment.

Lubrication theory tells us that in this situation the velocity of flow v of the oil is proportional to the pressure gradient.

$$v = -c\nabla p \tag{18.7}$$

where $p(x)$ is the pressure at x. The oil is assumed incompressible, so the velocity has zero divergence $\nabla\cdot v = 0$, giving

$$\nabla^2 p = 0$$

throughout the oil. If the viscosity of the water is negligible compared with that of the oil, then the pressure throughout the water is effectively constant. Thus we have the boundary conditions

$$p(x) = p_0$$

at the fluid interface, and

$$p(x) = 0$$

at a large distance r_0 from the point of injection. Thus the pressure difference $u(x) = p_0 - p(x)$ satisfies the differential equation (18.5) and boundary conditions (18.2) and (18.3) of the electrolysis example. Furthermore, at the fluid interface, the pressure gradient in the oil is normal to the boundary (since the pressure is constant on the boundary), so (18.4) gives the rate of advance of the boundary, $v_n = -kn \cdot \nabla p$, with short-range cut-off provided by surface tension effects. The pressure is analogous to the ion density in the electrolysis example.

It is perhaps, therefore, not surprising that under certain conditions the viscous fingers resemble the patterns produced by the square-lattice DLA model. Whilst the element of randomness inherent in the electrolysis example is lacking, irregularities in the interface are nevertheless amplified to give the fingering effect.

A very similar situation pertains for fluid flow through a porous medium—(18.7) is Darcy's law governing such flow. Fractal fingering can also occur in this situation.

Electrical discharge in a gas provides a further example. The electrostatic potential u satisfies Laplace's equation $\nabla^2 u = 0$ away from the ionized region of discharge. The ionized path conducts well enough to be regarded as being at constant potential, so u satisfies the same boundary conditions as in the viscous fingering example. The (questionable) assumption that the rate of breakdown is proportional to the electric field gives equation (18.4). This is another example with similar differential equations, for which the square-mesh DLA model provides a realistic picture.

Under suitable experimental conditions, the growth patterns in electrolysis, viscous fingering and electrical discharge have a dimension of about 1.7 when estimated over a suitable range of scales. This coincides with the value obtained from computer studies of square-mesh DLA. Although the theoretical explanations of such phenomena are not always entirely satisfactory, the universality of this dimension is very striking.

18.2 Singularities of electrostatic and gravitational potentials

The electrostatic potential due to a charge distribution μ or the gravitational potential due to a mass distribution μ in $\mathbb{R}^3$ is given by

$$\phi(x) = \int \frac{d\mu(y)}{|x - y|}. \tag{18.8}$$

We show that the dimension of the singularity set of the potential, i.e. the set of x for which $\phi(x) = \infty$ cannot be too large.

Proposition 18.1

Let μ be a mass distribution of bounded support on $\mathbb{R}^3$. Suppose that the potential (18.8) *has singularity set $F=\{x:\phi(x)=\infty\}$. Then* $\dim_H F\leqslant 1$.

Proof. Let $x\in\mathbb{R}^3$ and write $m(r)=\mu(B_r(x))$ for $r>0$. Suppose that for $s>1$ there are numbers $a>0$, $c>0$ such that $m(r)\leqslant cr^s$ for all $0<r\leqslant a$. Then

$$\begin{aligned}\phi(x)&=\int_{|x-y|\leqslant a}\frac{\mathrm{d}\mu(x)}{|x-y|}+\int_{|x-y|>a}\frac{\mathrm{d}\mu(x)}{|x-y|}\\&\leqslant\int_{r=0}^{a}\frac{\mathrm{d}m(r)}{r}+\int_{|x-y|>a}\frac{\mathrm{d}\mu(x)}{a}\\&\leqslant[r^{-1}m(r)]_0^a+\int_0^a r^{-2}m(r)\,\mathrm{d}r+a^{-1}\mu(\mathbb{R}^3)\\&\leqslant c(1+(s-1)^{-1})a^{s-1}+a^{-1}\mu(\mathbb{R}^3)<\infty.\end{aligned}$$

Hence, if $x\in F$, we must have that $\overline{\lim}_{r\to\infty}(\mu(B_r(x))/r^s)\geqslant c$ for all $c>0$. It follows from Proposition 4.9(*b*) that $\mathscr{H}^s(F)=0$ for $s>1$, as required. □

Often μ is expressible in terms of a 'density function' f, so that $\mu(A)=\int_A f(x)\,\mathrm{d}x$ for Borel sets A, and (18.8) becomes

$$\phi(x)=\int\frac{f(y)}{|x-y|}\mathrm{d}y. \qquad (18.9)$$

Given certain conditions on f, for example, if $\int|f(x)|^p\,\mathrm{d}x<\infty$ for some $p>1$, similar methods can be used to place further bounds on the dimension of the singularity set.

It is easily verified that, if f is a sufficiently smooth function, then (18.9) is the solution of Poisson's equation

$$\nabla^2\phi=-4\pi f$$

satisfying $\phi(x)\to 0$ as $|x|\to\infty$. For a general integrable function f the potential ϕ need not be differentiable. Nevertheless (18.9) may be regarded as a *weak solution* of Poisson's equation in a sense that can be made precise using the theory of distributions. This technique extends, in a non-trivial way, to give bounds for the dimension of the singularity sets of weak solutions of other partial differential equations.

18.3 Fluid dynamics and turbulence

Despite many years of intense study, turbulence in fluids is still not fully understood. Slowly moving fluids tend to flow in a smooth unbroken manner, which is described accurately by the Navier–Stokes equations—the fundamental differential equations of fluid dynamics. Such smooth flow is termed *laminar*. At higher speeds, the flow often becomes *turbulent*, with the fluid particles following convoluted paths of rapidly varying velocity with eddies and irregularities at all scales. Readers will no doubt be familiar with the change from laminar to turbulent flow as a tap is turned from low to full. Although the exact form of turbulent flow is irregular and unpredictable, its overall features are consistently present.

There is no uniformly accepted definition of turbulent flow—this has the advantage that it can reasonably be identified with any convenient 'singular feature' of a flow description. We consider a model in which turbulence is manifested by a significant local generation of heat due to viscosity, i.e. 'fluid friction', at points of intense activity.

At reasonably small scales, turbulence may be regarded as isotropic, i.e. direction independent. Our intuitive understanding of isotropic turbulence stems largely from the qualitative approach of Kolmogorov rather than from an analysis of differential equations. Kolmogorov's model is based on the idea that kinetic energy is introduced into a fluid on a large scale, such as by stirring. However, kinetic energy can only be dissipated (in the form of heat) on very small scales where the effect of viscosity becomes important. At intermediate scales dissipation can be neglected. If there are circulating eddies on all scales, then energy is transferred by the motion of the fluid through a sequence of eddies of decreasing size, until it reaches the small eddies at which dissipation occurs. If, as Kolmogorov assumed, the fluid region is filled by eddies of all scales, then dissipation of energy as heat should occur uniformly throughout the fluid.

Let $\varepsilon(x)$ be the rate of dissipation per unit volume at the point x, so that the heat generated in a small volume δV around x in time δt is $\varepsilon(x)\delta V\delta t$. Then, on these assumptions

$$\varepsilon(x) = \bar{\varepsilon} \qquad \text{for all } x \text{ in } D$$

where $\bar{\varepsilon}$ is the rate of input of energy into the fluid region D, assumed to have unit volume.

Although such 'homogeneous' turbulence is appealing in its simplicity, it is not supported by experimental observations. Measurements using a hot-wire anemometer show that in a turbulent fluid the rate of dissipation differs greatly in different parts of the fluid. This is the phenomenon of intermittency. Dissipation is high in some regions and very low in others, whereas the Kolmogorov model requires it to be constant. This variation can be quantified using correlation functions. For a small vector h the correlation of dissipation rates between points distance h apart is given by

$$\langle \varepsilon(x)\varepsilon(x+h) \rangle \tag{18.10}$$

where angle brackets denote the average over all x in D. If dissipation were constant we would have $\langle \varepsilon(x)\varepsilon(x+h) \rangle = \bar{\varepsilon}^2$. However, experiment indicates that

$$\langle \varepsilon(x)\varepsilon(x+h) \rangle \simeq \bar{\varepsilon}^2 |h|^{-d} \tag{18.11}$$

for a value of d between 0.4 and 0.5.

The Kolmogorov model can be modified to explain the intermittency by assuming that, instead of the eddies at each scale filling space, the eddies fill a successively smaller proportion of space as their size decreases. Kinetic energy is introduced into the largest eddy and passed through eddies of decreasing size until it is dissipated at the smallest scale. Now, however, the energy and dissipation is concentrated in a small part of the fluid. The cascade of eddies may be visualized as the first k stages E_i of the construction of a self-similar fractal F (see Chapter 9) where k is quite large, with dissipation occuring across the kth stage E_k. For convenience, we assume that each basic set of E_i is replaced by a constant number of sets of equal size to form E_{i+1}.

If A is a subset of D, we define $\mu(A)=\int_A \varepsilon(x)\,\mathrm{d}x$ as the total rate of dissipation of energy in the set A; thus $\mu(D)=\bar{\varepsilon}$, the rate of energy input. Then μ has the properties of a mass distribution on D. Moreover, if we assume that the rate of dissipation in each component of E_i is divided equally between the equal-sized subcomponents in E_{i+1}, we have, as a simple consequence of F being self-similar of Hausdorff or box dimension s, that

$$c_1\bar{\varepsilon}r^s \leqslant \mu(B_r(x)) \leqslant c_2\bar{\varepsilon}r^s$$

if x is in F, where c_1 and c_2 are positive constants (see Exercise 9.8). These inequalities hold for the limit F as the size of the dissipation eddies tends to 0, but also for the physical approximation E_k, provided that r is larger than the dissipation scale.

Then

$$\begin{aligned}\int_{|h|\leqslant r}\langle\varepsilon(x)\varepsilon(x+h)\rangle dh &= \int_{x\in D}\int_{|h|\leqslant r}\varepsilon(x)\varepsilon(x+h)\,\mathrm{d}h\,\mathrm{d}x\\ &= \int_{x\in D}\varepsilon(x)\mu(B_r(x))\,\mathrm{d}x\\ &= \int_{x\in E_k}\varepsilon(x)\mu(B_r(x))\,\mathrm{d}x\end{aligned}$$

since dissipation is concentrated on E_k, so

$$c_1\bar{\varepsilon}^2r^s \leqslant \int_{|h|\leqslant r}\langle\varepsilon(x)\varepsilon(x+h)\rangle\,\mathrm{d}h \leqslant c_2\bar{\varepsilon}^2r^s. \tag{18.12}$$

This is achieved if the correlation satisfies a power law

$$\langle\varepsilon(x)\varepsilon(x+h)\rangle \simeq \bar{\varepsilon}^2|h|^{s-3}$$

for then the integral in (18.12) becomes

$$4\pi\int_{t=0}^{r}\bar{\varepsilon}^2t^{s-3}t^2\,\mathrm{d}t = 4\pi\bar{\varepsilon}^2r^s/s.$$

Comparison with (18.11) suggests that $s=3-d$, so the hypothesis of 'fractally homogeneous turbulence', that dissipation is concentrated on an approximate fractal of dimension between 2.5 and 2.6, is consistent with experimental results.

It is natural to seek theoretical reasons for the turbulent region to have a fractal form. One possible explanation is in terms of the vortex tubes in the fluid. According to Kelvin's circulation theorem, such tubes are preserved throughout the motion, at least in the approximation of inviscid flow. However, the vortex tubes are stretched by the fluid motion, and become long and thin. Repeated folding is necessary to accommodate this length, so the tubes might assume an approximate fractal form not unlike the horseshoe example in Figure 13.5.

The behaviour of a (viscous) fluid should be predicted by the Navier–Stokes equation

$$\frac{\partial u}{\partial t} + (u\cdot\nabla)u - \nu\nabla^2 u + \nabla p = f \tag{18.13}$$

where u is the velocity, p is the pressure, ν is viscosity and f is the applied force density. Deducing the existence of fractal regions of activity from the Navier–Stokes equation is far from easy. Nevertheless, the method indicated in Section 18.2 may be generalized beyond recognition to demonstrate rigorously that, for example, the set on which a solution $u(x,t)$ of (18.13) fails to be bounded for all t has dimension at most $2\frac{1}{2}$. Thus it is possible to show from the equations of fluid flow that certain types of 'intense activity' must be concentrated on sets of small dimension.

18.4 Notes and references

A wide variety of physical applications of fractals are given in the books by Mandelbrot (1982) and Feder (1988) and in the volumes of collected papers edited by Pietronero and Tosatti (1986), Shlesinger, Mandelbrot and Rubin (1984) and Pietronero (1989). For applications to geophysics see Scholz and Mandelbrot (1989) and for applications to chemistry see Avinor (1989). Stanley and Ostrowsky (1986, 1988) and Viesek (1989) contain surveys and papers on fractal growth. Feder (1988) includes a more detailed account of viscous fingering. For an introduction to the ideas of turbulence see Leslie (1973). The homogeneous model of Kolmogorov (1941) was adapted to the fractally homogeneous model by Mandelbrot (1974); see also Frisch, Sulem and Nelkin (1978). Collections of papers relevant to fractal aspects of turbulence include Temam (1976) and Barenblatt, Iooss and Joseph (1983). The book by Temam (1983) discusses the dimension of sets related to solutions of the Navier–Stokes equations.

There are an enormous number of papers on other physical applications. To mention a very few, Nye (1970) applies fractals to glaciology, Berry (1979) considers the effect of fractals on waves, Burrough (1981) discusses the dimensions of landscapes and environmental data, Lovejoy (1982) considers fractal aspects of clouds and Bale and Schmidt (1984) investigate fractal properties of microscopic porosity of surfaces.

Exercises

18.1 Suppose that the DLA square-lattice model is run for a large number of very small squares. Suppose that the set obtained is an approximate fractal of dimension s. What power law would you expect the number of shaded squares within distance r of the initial square to obey? Assuming that during the process squares tend to be added to parts of the set further away from the initial square, how would you expect the 'radius' of the growth after k squares have been added to depend on k?

18.2 Let $m(t)$ be the mass of copper that has been deposited and $r(t)$ be the 'radius' of the copper deposit after time t in the electrolysis experiment described in Section 18.1. It may be shown that the current flowing, and thus, by Faraday's law, the rate of mass deposition, is proportional to $r(t)$. On the assumption that the growth forms an approximate fractal of dimension s, so that $m(t) \sim cr(t)^s$, give an argument to suggest that that $r(t) \sim c_1 t^{1/(s-1)}$.

18.3 Verify that $u(x', t)$ satisfies the partial differential equation (18.1).

18.4 Verify that the potential in (18.9) satisfies Poisson's equation if, say, f is a twice continuously differentiable function with $f(x) = 0$ for all sufficiently large x.

18.5 Show that, if $f(x) = 0$ for all sufficiently large x and $\int |f(x)|^2 \, dx < \infty$, then the singularity set of ϕ, given by (18.9), is empty.

18.6 Show that the argument leading to (18.12) can be adapted to the case when, say, D is the unit cube in $\mathbb{R}^3$ and F is the product of the Cantor dust of figure 0.4 and a unit line segment L. (Dissipation is assumed to occur on the set $E_k \times L$, where E_k is the kth stage in the construction of the Cantor dust for some large k.)

References

Adler R. J. (1981) *The Geometry of Random Fields*, Wiley, New York.

Ahlfors L. V. (1979) *Complex Analysis*, McGraw-Hill, New York.

Anderson R. D. and Klee V. L. (1952) Convex functions and upper semi-continuous functions, *Duke Math. J.*, **19**, 349–357.

Apostol T. M. (1974) *Mathematical Analysis*, 2nd edition, Addison-Wesley, Reading, MA.

Avinor D. (Ed.) (1989) *The Fractal Approach to Heterogeneous Chemistry*, Wiley, New York.

Baker A. and Schmidt W. M. (1970) Diophantine approximation and Hausdorff dimension, *Proc. Lond. Math. Soc.* (3), **21**, 1–11.

Bale H. D. and Schmidt P. W. (1984) Small-angle X-ray-scattering investigation of submicroscopic porosity with fractal properties, *Phys. Rev. Lett.*, **53**, 596–599.

Barenblatt G. I., Iooss G. and Joseph D. D. (Eds) (1983) *Nonlinear Dynamics and Turbulence*, Pitman, London.

Barnsley M. F. (1988) *Fractals Everywhere*, Academic Press, Orlando, FL.

Barnsley M. F. and Demko S. G. (1985) Iterated function schemes and the global construction of fractals, *Proc. R. Soc.*, A **399**, 243–275.

Barnsley M. F. and Demko S. G. (Eds) (1986) *Chaotic Dynamics and Fractals*, Academic Press, New York.

Barnsley M. F. and Sloan A. D. (1988) A better way to compress images, *Byte*, **13**, 215–233.

Beardon A. F. (1965) On the Hausdorff dimension of general Cantor sets, *Proc. Camb. Phil. Soc.*, **61**, 679–94.

Bedford T. J. (1989) The box dimension of self-affine graphs and repellers, *Nonlinearity*, **2**, 53–71.

Bedford T. J. and Swift J. (Eds) (1988) *New Directions in Dynamical Systems*, Cambridge University Press, Cambridge.

Bergé P., Pomeau Y. and Vidal Ch. (1984) *L'Ordre dans le Chaos*, Hermann, Paris.

Berry M. V. (1979) Diffractals, *J. Phys. A: Math. Gen.*, A **12**, 781–797.

Berry M. V. and Lewis Z. V. (1980) On the Weierstrass–Mandelbrot fractal function, *Proc. R. Soc.*, A **370**, 459–484.

Besicovitch A. S. (1928) On the fundamental geometrical properties of linearly measurable plane sets of points, *Math. Annalen*, **98**, 422–464.

Besicovitch A. S. (1934) Sets of fractional dimensions IV: On rational approximation to real numbers, *J. Lond. Math. Soc.*, **9**, 126–131.

Besicovitch A. S. (1938) On the fundamental geometric properties of linearly measurable plane sets of points II, *Math. Annalen*, **115**, 296–329.

Besicovitch A. S. (1939) On the fundamental geometric properties of linearly measurable plane sets of points III, *Math. Annalen*, **116**, 349–357.

Besicovitch A. S. (1952) On existence of subsets of finite measure of sets of infinite measure, *Indag. Math.*, **14**, 339–344.

Besicovitch A. S. (1963) The Kakeya problem, *Am. Math. Monthly*, **70**, 697–706.

Besicovitch A. S. (1964) On fundamental geometric properties of plane line sets, *J. Lond. Math. Soc.*, **39**, 441–448.

Besicovitch A. S. and Moran P. A. P. (1945) The measure of product and cylinder sets, *J. Lond. Math. Soc.*, **20**, 110–120.

Besicovitch A. S. and Ursell H. D. (1937) Sets of fractional dimensions, V: On dimensional numbers of some continuous curves, *J. Lond. Math. Soc.*, **12**, 18–25.

Billingsley P. (1965) *Ergodic Theory and Information*, Wiley, New York.

Billingsley P. (1979) *Probability and Measure*, Wiley, New York.

Blanchard P. (1984) Complex analytic dynamics on the Riemann sphere, *Bull. Am. Math. Soc.*, **11**, 85–141.

Brolin H. (1965) Invariant sets under iteration of rational functions, *Arkiv Math.*, **6**, 103–144.

Bumby R. T. (1985) Hausdorff dimension of sets arising in number theory, in *Number Theory, New York* 1983–84 *Seminar* (*Lecture Notes in Mathematics*, **1135**), pp. 1–8, Springer, New York.

Burrough P. A. (1981) Fractal dimensions of landscapes and other environmental data, *Nature*, **294**, 240–242.

Carathéodory C. (1914) Über das lineare Mass von Punktmengeneine Verallgemeinerung das Längenbegriffs, *Nach. Ges. Wiss. Göttingen*, pp. 406–426.

Carleson A. (1967) *Selected Problems on Exceptional Sets*, Van Nostrand, Princeton, NJ.

Carlin S. and Taylor H. M. (1975) *A First Course in Stochastic Processes*, Academic Press, New York.

Carlin S. and Taylor H. M. (1981) *A Second Course in Stochastic Processes*, Academic Press, New York.

Cassels J. W. S. (1957) *An Introduction to Diophantine Approximation*, Cambridge University Press, Cambridge.

Chayes J. T., Chayes L. and Durrett R. (1988) Connectivity of Mandelbrot's percolation process, *Prob. Theor. Related Fields*, **77**, 307–324.

Cunningham F. (1971) The Kakeya problem for simply connected and for star-shaped sets, *Am. Math. Monthly*, **78**, 114–129.

Cunningham F. (1974) Three Kakeya problems, *Am. Math. Monthly*, **81**, 589–592.

Curry J., Garnett L. and Sullivan D. (1983) On the iteration of rational functions: computer experiments with Newton's method, *Commun. Math. Phys.*, **91**, 267–277.

Cvitanović P. (Ed.) (1984) *Universality in Chaos*, Adam Hilger, Bristol.

Dalla L. and Larman D. G. (1980) Convex bodies with almost all k-dimensional sections polytopes, *Math. Proc. Camb. Phil. Soc.*, **88**, 395–401.

Davies R. O. (1952) On accessibility of plane sets and differentiation of functions of two real variables, *Proc. Camb. Phil. Soc.*, **48**, 215–232.

Dekking F. M. (1982) Recurrent sets, *Adv. Math.*, **44**, 78–104.

Devaney R. L. (1986) *Introduction to Chaotic Dynamic Systems*, Benjamin Cummings, Menlo Park, CA.

Dodson M. M., Rynne B. P. and Vickers J. A. G. (to appear) Diophantine approximation and a lower bound for Hausdorff dimension.

Eggleston H. G. (1952) Sets of fractional dimension which occur in some problems of number theory, *Proc. Lond. Math. Soc.* (2), **54**, 42–93.

Eggleston H. G. (1958) *Convexity*, Cambridge University Press, Cambridge.

Erdös P. and Volkmann B. (1966) Additive Gruppen mit vorgegebener Hausdorffscher Dimension, *J. reine angew. Math.*, **221**, 203–208.

Falconer K. J. (1985a) *The Geometry of Fractal Sets*, Cambridge University Press, Cambridge.

Falconer K. J. (1985b) Classes of set with large intersection, *Mathematika*, **32**, 191–205.

Falconer K. J. (1985c) The Hausdorff dimension of distance sets, *Mathematika*, **32**, 206–212.

Falconer K. J. (1986a) Sets with prescribed projections and Nikodym sets, *Proc. Lond. Math. Soc.* (3), **53**, 48–64.

Falconer K. J. (1986b) Random fractals, *Math. Proc. Camb. Phil. Soc.*, **100**, 559–582.

Falconer K. J. (1987) Cut set sums and tree processes, *Proc. Am. Math. Soc.*, **101**, 337–346.

Falconer K. J. (1988) The Hausdorff dimension of self-affine fractals, *Math. Proc. Camb. Phil. Soc.*, **103**, 339–350.

Falconer, K. J. and Marsh D. T. (1989) Classification of quasi-circles by Hausdorff dimension, *Nonlinearity*, **2**, 489–493.

Farmer J. D., Ott E. and Yorke J. A. (1983) The dimension of chaotic attractors, *Physica*, **7D**, 153–180.

Fatou P. (1919) Sur les équations fonctionelles, *Bull. Soc. Math. France*, **47**, 161–271.

Feder J. (1988) *Fractals*, Plenum Press, New York.

Federer H. (1947) The (φ, k) rectifiable subsets of n-space, *Trans. Am. Math. Soc.*, **62**, 114–192.

Federer H. (1969) *Geometric Measure Theory*, Springer, New York.

Fischer P. and Smith W. R. (Eds.) (1985) *Chaos, Fractals and Dynamics*, Marcel Dekker, New York.

Frederickson P., Kaplan J., Yorke E. and Yorke J. (1983) The Lyapunov dimension of strange attractors, *J. Diff. Eq.*, **49**, 185–207.

Frisch U., Sulem P.-L. and Nelkin M. (1978) A simple dynamical model of fully developed turbulence, *J. Fluid Mech.*, **87**, 719–736.

Frostman O. (1935) Potential d'équilibre et capacité des ensembles avec quelques applications à la théorie des fonctions, *Meddel. Lunds Univ. Math. Sem.*, **3**, 1–118.

Graf S. (1987) Statistically self-similar fractals, *Prob. Theor. Related Fields*, **74**, 357–392.

Graf S., Mauldin R. D. and Williams S. C. (1988) The exact Hausdorff dimension in random recursive constructions, *Mem. Am. Math. Soc.*, **71**, no. 381.

Grimmett G. R. (1989) *Percolation*, Springer, New York.

Grimmett G. R. and Stirzaker D. R. (1982) *Probability and Random Processes*, Clarendon Press, Oxford.

Guckenheimer J. and Holmes P. (1983) *Nonlinear Oscillations, Dynamical Systems and Bifurcations of Vector Fields*, Springer, New York.

Hausdorff F. (1919) Dimension und äusseres Mass, *Math. Annalen*, **79**, 157–179.

Hardy G. H. and Wright E. M. (1960) *An Introduction to the Theory of Numbers*, 4th edition, Cambridge University Press, Cambridge.

Harin V. P., Hruščëv S. V. and Nikol'skii N. K. (1984) '*Linear and Complex Analysis Problem Book*' (*Lecture Notes in Mathematics*, 1043), Springer, New York.

Hayman W. K. and Kennedy P. B. (1976) *Subharmonic Functions*, Volume 1, Academic Press, New York.

Hénon M. and Pomeau Y. (1976) Two strange attractors with a simple structure, in *Turbulence and the Navier–Stokes Equations* (*Lecture Notes in Mathematics*, 565), (Ed. R. Temam), pp. 29–68, Springer, New York.

Holden A. V. (1986) *Chaos*, Manchester University Press, Manchester.

Holden A. V. and Muhamad M. A. (1986) A graphical zoo of strange and peculiar attractors, in *Chaos* (Ed. A. V. Holden), pp. 15–34, Manchester University Press, Manchester.

Hutchinson J. E. (1981) Fractals and self-similarity, *Indiana Univ. Math. J.*, **30**, 713–747.

Jarník V. (1931) Über die simultanen diophantischen Approximationen, *Math. Zeit.*, **33**, 505–543.

Jones P. W. and Murai T. (1988) Positive analytic capacity but zero Buffon needle probability, *Pacific J. Math.*, **133**, 99–114.

Julia G. (1918) Sur l'iteration des fonctions rationnelles, *J. Math. Pure Appl.*, **8**, 47–245.

Kahane J.-P. (1974) Sur le modèle de turbulence de Benoit Mandelbrot, *C.R. Acad. Sci. Paris*, **278A**, 621–623.

Kahane J.-P. (1985) *Some Random Series of Functions*, 2nd edition, Cambridge University Press, Cambridge.

Kahane J.-P. (1986) Sur la dimensions des intersections, in *Aspects of Mathematics and its Applications* (Ed. J. A. Barroso), pp. 419–430, North-Holland, Amsterdam.

Kaufman R. (1968) On the Hausdorff dimension of projections, *Mathematika*, **15**, 153–155.

Kaufman R. (1981) On the theorem of Jarník and Besicovitch *Acta. Arithmetica*, **39**, 265–267.

Kesten H. (1982) *Percolation Theory for Mathematicians*, Birkhauser, Boston, MA.

Kingman J. F. C. and Taylor S. J. (1966). *Introduction to Measure and Probability*, Cambridge University Press, Cambridge.

Kolmogorov A. N. (1941) Local structure of turbulence in an incompressible liquid for very large Reynolds numbers, *C.R. Acad. Sci. USSR*, **30**, 299–303.

Lovejoy S. (1982) Area perimeter relation for rain and cloud areas, *Science*, **216**, 185–187.

Leslie D. C. (1973) *Developments in the Theory of Turbulence*, Clarendon Press, Oxford.

MacKay R. S. and Meiss J. D. (Eds) (1987) *Hamiltonian dynamical Systems*, Adam Hilger, Bristol.

Mandelbrot B. B. (1974) Intermittent turbulence in self-similar cascades: divergence of high moments and dimension of the carrier, *J. Fluid Mech.*, **62**, 331–358.

Mandelbrot B. B. (1980) Fractal aspects of the iteration of $z \to \lambda z(1 - z)$ for complex λ, z, *Ann. N.Y. Acad. Sci.*, **357**, 249–259.

Mandelbrot B. B. (1982) *The Fractal Geometry of Nature*, Freeman, San Francisco.

Mandelbrot B. B. (1986) Self-affine fractal sets, in *Fractals in Physics* (Eds L. Pietronero and E. Tosatti), North-Holland, Amsterdam.

Mandelbrot B. B. (1988) An introduction to multifractal distribution functions, in *Fluctuations and Pattern Formation*, (Eds H. E. Stanley and N. Ostrowsky), Kluwer Academic, Dordrecht.

Mandelbrot B. B. and Van Ness J. W. (1968) Fractional Brownian motions, fractional noises and applications, *SIAM Rev.*, **10**, 422–437.

Marstrand J. M. (1954a) Some fundamental geometrical properties of plane sets of fractional dimensions. *Proc. Lond. Math. Soc.* (3), **4**, 257–302.

Marstrand J. M. (1954b) The dimension of Cartesian product sets, *Proc. Camb. Phil. Soc.*, **50**, 198–202.

Mattila P. (1975) Hausdorff dimension, orthogonal projections and intersections with planes, *Ann. Acad. Sci. Fennicae*, A **1**, 227–244.

Mattila P. (1984) Hausdorff dimension and capacities of intersections of sets in n-space, *Acta Math.*, **152**, 77–105.

Mattila P. (1985) On the Hausdorff dimension and capacities of intersections, *Mathematika*, **32**, 213–217.

Mattila P. (1986) Smooth maps, null-sets for integralgeometric measures and analytic capacity, *Ann. Math.*, **123**, 303–309.

Mauldin R. D. and Williams S. C. (1986a) Random recursive constructions: asymptotic geometric and topological properties, *Trans. Am. Math. Soc.*, **295**, 325–346.

Mauldin R. D. and Williams S. C. (1986b) On the Hausdorff dimension of some graphs, *Trans. Am. Math. Soc.*, **298**, 793–803.

May R. M. (1976) Simple mathematical models with very complicated dynamics, *Nature*, **261**, 459–467.

Mayer Kress G. (Ed.) (1986) *Dimensions and Entropies in Chaotic Dynamical Systems*, Springer-Verlag, Berlin.

McMullen C. (1984) The Hausdorff dimension of general Sierpiński carpets, *Nagoya Math. J.*, **96**, 1–9.

Moran P. A. P. (1946) Additive functions of intervals and Hausdorff measure, *Proc. Camb. Phil. Soc.*, **42**, 15–23.

Nye J. F. (1970) Glacier sliding without cavitation in a linear viscous approximation, *Proc. R. Soc.*, A **315**, 381–403.

Papoulis A. (1962) *The Fourier Integral and its Applications*, McGraw-Hill, New York.

Peitgen H.-O. and Richter P. H. (1986) *The Beauty of Fractals*, Springer, Berlin.

Peitgen H.-O. and Saupe D. (Eds) (1988) *Fractal Images*, Springer, New York.

Peitgen H.-O., Saupe D. and von Haeseler F. (1984) Cayley's problem and Julia sets, *Math. Intelligencer*, **6**, 11–20.

Peyrière J. (1974) Turbulence et dimension de Hausdorff, *C.R. Acad. Sci. Paris*, **278A**, 567–569.

Peyrière J. (1977) Calculs de dimensions de Hausdorff, *Duke Math. J.*, **44**, 591–601.

Pietronero L. (Ed.) (1989) *Fractals*, Plenum Press, New York.

Pietronero L. and Tosatti E. (Eds) (1986) *Fractals in Physics*, North-Holland, Amsterdam.

Preiss D. (1987) Geometry of measures in $\mathbf{R}^n$: distribution, rectifiability and densities, *Ann. Math.*, **125**, 537–641.

Rogers C. A. (1970) *Hausdorff Measures*, Cambridge University Press, Cambridge.

Rogers C. A. (1988) Dimension prints, *Mathematika*, **35**, 1–27.

Ruelle D. (1980) Strange attractors, *Math. Intelligencer*, **2**, 126–137.

Ruelle D. (1982) Repellers for real analytic maps, *Ergod. Theor. Dyn. Syst.*, **2**, 99–108.

Ruelle D. (1983) Bowen's formula for the Hausdorff dimension of self-similar sets, in *Scaling and Self-similarity in Physics*, (*Progress in Physics*, 7) Birkhauser, Boston, MA.

Santaló L. A. (1976) *Integral Geometry and Geometric Probability*, Addison-Wesley, Reading, Massachusetts.

Saupe D. (1987) Efficient computation of Julia sets and their fractal dimension, *Physica*, **28D**, 358–370.

Schmidt W. M. (1980) *Diophantine Approximation* (*Lecture Notes in Mathematics*, 785). Springer, Berlin.

Scholz C. H. and Mandelbrot B. B. (Eds) (1989) *Fractals in Geophysics*, Birkhauser, Boston, MA.

Schuster H. G. (1984) *Deterministic Chaos—An Introduction*, Physik Verlag, Weinheim.

Shlesinger M. F., Mandelbrot B. B. and Rubin R. J. (Eds) (1984) *Proceedings of the Gaithersburg Symposium on Fractals in the Natural Sciences, J. Stat. Phys.*, **36**, (special issue) 519–921.

Smale S. (1967) Differentiable dynamical systems, *Bull. Am. Math. Soc.*, **73**, 747–817.

Sparrow C. (1982) *The Lorenz Equations: Bifurcations, Chaos, and Strange Attractors*, Springer, New York.

Stanley H. E. and Ostrowsky N. (Eds) (1986) *On Growth and Form*, Martinus Nijhoff, Dordrecht.

Stanely H. E. and Ostrowsky N. (Eds) (1988) *Random Fluctuations and Pattern Growth*, Kluwer Academic, Dordrecht.

Taylor S. J. (1961) On the connection between Hausdorff measures and generalized capacities, *Proc. Camb. Phil. Soc.*, **57**, 524–31.

Taylor S. J. (1973) Sample path properties of processes with stationary independent increments, in *Stochastic Analysis* (Eds D. G. Kendall and E. F. Harding), pp. 387–414, Wiley, New York.

Taylor S. J. (1986) The measure theory of random fractals, *Math. Proc. Camb. Phil. Soc.*, **100**, 383–406.

Tél T. (1988) Fractals and multifractals, *Zeit. Naturforsch.*, **43A**, 1154–1174.

Temam R. (Ed.) (1976) *Turbulence and the Navier–Stokes Equations*, (*Lecture Notes in Mathematics* 565), Springer, New York.

Temam R. (1983) *Navier–Stokes Equations and Non-linear Functional Analysis*, Society for Industrial and Applied Mathematics, Philadelphia, PA.

Thompson J. M. T. and Stewart H. B. (1986) *Nonlinear Dynamics and Chaos*, Wiley, Chichester.

Tricot C. (1982) Two definitions of fractional dimension, *Math. Proc. Camb. Phil. Soc.*, **91**, 54–74.

Viesek T. (1989) *Fractal Growth Phenomena*, World Scientific, Singapore.

Voss R. F. (1985) Random fractal forgeries, in *Fundamental Algorithms in Computer Graphics* (Ed. R. A. Earnshaw), pp. 805–835, Springer, Berlin.

Young L.-S. (1982) Dimension, entropies and Liapunov exponents, *Ergod. Theor. Dyn. Syst.*, **2**, 109–124.

Zähle U. (1984) Random fractals generated by random cutouts, *Math. Nachr.*, **116**, 27–52.

Index

Entries in bold type refer to definitions